The Cognitive Paradigm

MARC DE MEY

The Cognitive Paradigm

An Integrated Understanding of Scientific Development

With a new Introduction

THE UNIVERSITY OF CHICAGO PRESS
Chicago and London

First published in 1982 by Reidel in the series
Sociology of the Sciences Monographs

The University of Chicago Press, Chicago 60637
The University of Chicago Press, Ltd., London
© 1982 by D. Reidel Publishing Company
© 1992 by Marc De Mey
All rights reserved. Published 1982
University of Chicago Press edition 1992
Printed in the United States of America

01 00 99 98 97 96 95 94 93 92   6 5 4 3 2 1

ISBN 0-226-14259-0 (paperback)

Library of Congress Cataloging-in-Publication Data

Mey, Marc de, 1940-
    The cognitive paradigm : an integrated understanding of scientific
Development / Marc De Mey.
        p.   cm.
    Reprint. Originally published: Dordrecht ; Boston : Reidel, c1982.
    Includes bibliographical references and index.
    1. Science—Philosophy.   2. Science—Methodology.   3. Cognition.
4. Cognitive science.   I. Title.
Q175.M5447   1992
501—dc20   92-15297
                                                                 CIP

Permission for figures 1 and 2 in the Introduction granted by the University of
Illinois Press. ©1954 by the Board of Trustees of the University of Illinois.
Copyright renewed 1982.

*To the memory of Jean Piaget
and to Leo Apostel*

TABLE OF CONTENTS

Introduction to the 1992 Edition   xi
Preface   xxv

PART ONE: INTRODUCTION TO THE COGNITIVE VIEW

1. The Development of the Cognitive View   3
   Perception, Pattern Recognition and Picture-processing   5
   Communication and Language Processing   11
   The Generic Scheme   16
2. World Views and Models   19
   The Multiplicity of World Views   20
   The Simplest Model of a World View   22
   'Self' and 'I' as Parts of a World View   25
   World Views as Social Entities   27
   Combination and Interaction of World Views   30
   Information Processing and Views on Science   33
   A Prefatory Task   36
3. Positivism as a Monadic View   38
   Positivism and Scientism   38
   Empiristic Units of Knowledge   40
   Helmholtz's Cognitive Model   41
   Scientistic Metaphysics   43
   From Dualism to Neutral Monism   44
   The Status of Concepts   46
   Scientific Concepts: Clusters of Monads   47
4. Logical Positivism: A Structural View   50
   The Structure of Natural Language   51
   The Logical Reconstruction of Language   54
   The Emphasis on Structure   56

The Logical Positivist Model of Scientific Theory   58
Logical Positivist Philosophy of Science   59
5. Contexts of Science: Sciences of Science   63
   Merton's Norms of Science   63
   Intrinsic and Extrinsic Factors   67
   The Personality of the Scientist   71
   The Multiplicity of Arguments in the History of Science   75
   The Multiplicity of Scopes in History of Science   76
   The Science of Science   78
6. The Cognitive View on Science: Paradigms   82
   An Integrated Approach to the Copernican Revolution   82
   The Standard Account of the Kuhnian Model   84
   Paradigms as Cognitive Units   89
   Conceptual Schemes and the Functions of Paradigms   93
   Disciplinary Matrix and Exemplar   96
   The Social Nature of Paradigms   100
   Paradigm-studies   102

PART TWO: THE SOCIAL STRUCTURE OF SCIENCE

7. Bibliometrics and the Structure of Science   111
   Bibliometrics and Research on Science   111
   The Growth of Science   112
   The Detection of Growth Points and Secondary Literature   115
   The Metabolism of Growth and Primary Literature   119
   Citation Networks   124
   Co-citation Clustering   125
   Bibliometrics and Scientometrics   130
8. Informal Groups and the Origin of Networks   132
   Invisible Colleges and Specialties   133
   Characteristics of Invisible Colleges   135
   Invisible Colleges and Small Groups   136
   Communication Patterns and Information Flow   138
   Interdisciplinarity and the Origin of Specialties   140

    Migrations into Psychology: Two Examples   142
    Innovation and Discipleship   145
  9. The Life Cycle of Scientific Specialties   148
    The Stages of the Specialty Life Cycle   149
    Regulative Mechanisms and Growth   154
    Finalization: Cognitive and Social in Sequence?   159
    Escalatory Expansion of Diffusion Studies   162
    Forms of Specialties and Patterns of Life Cycles   165
    Social Studies of Science   168

PART THREE: COGNITIVE STRUCTURE AND DYNAMICS
OF SCIENCE

10. Paradigms and the Psychology of Attention and Perception   173
    Gestalt Perception and Gestalt Switch as Exemplars   175
    Perception and Selective Attention   176
    A Stratified Model of Perception   182
    Interactive and Integrative Processes in Perception   189
    Analysis of a Gestalt Switch in Science: Harvey's Discovery   192
11. Puzzle-solving and Reorganization of World Views   202
    Frames   205
    Defaults and Exemplars   208
    Problem-solving and Debugging   211
    Puzzle-solving and Heterarchical Control   212
    Procedural Aspects of Scientific Knowledge   215
    Self-world Segmentation and Compatibility of World Views   218
    Paradigms and Perspectives   222
12. Conservation and the Dynamics of Conceptual Systems   227
    Scientific Knowledge and Children's Concepts   228
    Action and Adaptation   230
    Table Tennis Ball Expertise   233
    Piaget's Stages and the Finalization-model   237
    Conservation and Closure of Conceptual Systems   238
    Harvey: Conservation of the Blood?   241
    Perspectives on an Object   244

Paradigms and Development   247
Individual Discovery and Social Success   248

Epilogue   252

Notes   260

Bibliography   283

Index   304

INTRODUCTION TO THE 1992 EDITION

Has the Cognitive Revolution Already Begun?

Retrospectively analyzing their accomplishment, authors of innovative papers or treatises often recall some single event as a turning point in the orientation of their work. They remember that very moment when, browsing through books in a random exploration on the shelves of a library, they suddenly hit upon the one that made them see an outcome to the problem they were struggling with. The newly discovered item might trigger an alternative way to state the problem or offer a path to untangle a collection of data and ideas that has grown intricately confused. The experience has a mobilizing effect. It brings new coherence to the endeavor, and one regains confidence and enthusiasm.

Perceptual parallels of such an experience, such as the gestaltswitch with ambiguous figures or seeing the hidden person in a puzzle picture, have a pivotal role in Kuhn's (1962) monograph on *The Structure of Scientific Revolutions.* They are brought up as small-scale examples of processes of reorganization that are supposed to occur in scientific discovery on a much larger scale. With the explosive expansion of the cognitive sciences in the last decade one would expect these phenomena of puzzle solving and gestaltswitch to be well understood. They should be among the most obvious cases for cognitive science or cognitive psychology to handle, and their applicability to scientific discovery should have become a straightforward matter to trace or check. Surprisingly, cognitive science still has little to offer in terms of a generally accepted theory of these modest perceptual examples, let alone of their explanatory value in understanding scientific discovery. Has the cognitive science that one could have in prospect in 1982 not yet matured?

The lightning experience for *The Cognitive Paradigm* came from a 1973 paper by Helga Nowotny, 'On the Feasibility of a Cognitive Approach to the Study of Science'. Here a sociologist of science expressed a keen view on the oncoming cognitive revolution and saw the relevance of both artificial intelligence and Piagetian cognitive psychology for science studies. The conjunction with Kuhn's concepts constituted an appealing challenge and seemed to open an opportunity for developing a genuine cognitive science of science. Why did it not come out that way, up to now?

With some presentiment, we referred in the Preface to William James's bantering distinction between stages in the acceptance of new ideas (James, 1907). One could claim, at least at first sight, that there is some truth to it.

In the first stage, James indicates, opponents will declare the new ideas ridiculous or meaningless. There are indeed powerful voices in the study of scientific knowledge who would consider this to be the case for cognitive analyses of science. A prominent sociologist of science, Bruno Latour, regards them as dangerous and misleading. Some years ago, he proposed a full-scale moratorium intended to ban all cognitive analyses of science for ten years (Latour, 1987). In his case, however, it is not immediately clear what the real purpose is behind his interdiction. Latour himself masters all the rhetorical tricks of the celebrated scientists whom he so cleverly profanes. In the same way that he found out about Pasteur's moves and motives (Latour, 1984), we should try to detect what and whom he is attempting to move with his declarations. The answer might be rather simple and straightforward: eliminate competition. Latour possibly wants to move "cognitivists" out of the way in order to leave social studies of science an uncluttered searching space for exploring fascinating *representation devices,* the very heart of the matter in cognitive science at large! Inscription devices such as maps, instruments, and diagrams are the treasures of science and are also as much constitutive of the cognitive as of the social approach. Latour is not rejecting cognitive science concepts. To the degree that he studies the qualities of representation systems in terms of immutable mobile and combinable symbol structures, he could be seen as an important contributor to cognitive science. Also, his enthusiasm for studies like those of Alpers (1981), Edgerton (1975), and Eisenstein (1980) illustrates involvement with representation issues similar to the ones that cognitive scientists encounter when they compare reading in languages with different writing systems (Bertelson, 1987). They all study different symbol systems whose special expressive qualities and limitations depend upon the material substrate in which the symbols are embodied. Why deny cognitive science access to the rich body of representations and representational concepts and devices that science constitutes? The valuable qualities of immutability, mobility, and combinability for inscriptions invite comparison with the mental operations that yield conceptual invariants at various levels in Piaget's codification of intellectual development. Granted, there are genuine innovations in the consequent constructivism maintained throughout to avoid any epistemological dualism. But on the whole, many core cognitive concepts seem quite compatible with Latour's account of science. With his high esteem for Machiavellian rules, one could safely predict that if he cannot completely control or destroy the cognitive enemy within the ten-year moratorium he declared, he will finally strike up a friendship with it. For the time being, his cry that the cognitive approach to science should be forbidden is camouflaging his respectable and impressive attempt to develop it, as the science of representation, in his own style and categories.

In the second stage of their diffusion, according to James, new ideas are trivialized. Apparently, few authors flatly claim that the cognitive approach contains nothing new. Nevertheless, there are some major works which come close to doing so. In principle, cognitively oriented students of science should feel comforted when a pioneer of cognitive studies like Herbert Simon joins them. One would expect great advances to be made when the impressive conceptual apparatus on problem solving and heuristics derived from computer simulations of human thought is brought to bear upon the structure of scientific discovery, as is apparent from promising volumes such as that of Shrager and Langley (1990). Surprisingly little new stems from this when the master himself embarks upon the task. In an article done collaboratively with Qin, Simon (Qin and Simon, 1991) reports the results of a laboratory reconstruction of Kepler's discovery of the third planetary law. A successful reconstruction is seen in the fact that some first-rate university students in the sciences, after having been given quantitative data comparable to the measures accessible to Kepler, manage to find the mathematical formula for the law in a matter of hours. This performance, remarkably faster than Kepler's but significantly slower than the computer's, exhibits however the same heuristics as encoded in the computer program. Besides several other important points, Hallyn (in press), who studied Kepler from the point of view of semiotics (Hallyn, 1990), seriously criticizes the representativity of the data provided to the students. One doubts whether such a reconstruction in a complete cognitive vacuum, even with numbers identical to the ones available to Kepler, could reveal much about discovery. However, if we accept Simon and Qin's conclusion, we might as well put an end to cognitive studies of science at once, since they produce nothing new. That conclusion is that scientific discovery is just like problem solving in the laboratory or, for that matter, in the classroom. In a sense, there is no need anymore for going through the trouble of analyzing authentic scientific discoveries, since modest substitutes of problem solving in advanced math classes will provide the same information. Such a kind of encouragement coming from one of the founding fathers of cognitive science is like a suffocating embrace to cognitive studies of science.

Obviously, Simon doesn't want to prohibit cognitive studies of science in the way Latour wants to stop them. To the contrary, programs like BACON (a computer program for detecting scientific laws in an amount of scientific data) aim explicitly at rediscovering genuine scientific laws by means of heuristics for data-driven problem solving occasionally inspired by real historical cases. By proving that major scientific discoveries can be achieved through common problem solving procedures accessible to almost everyone, Simon wants to demystify scientific creativity. There is no need for creative genius, which for him, apparently, is almost synonymous with inessential intuition or uncontrollable and unnecessary inspiration. However, to envisage the possibility of a special cognitive status for major discoveries doesn't necessarily entail that one considers them as magic and beyond any scientific explanation. Many cases of famous discoveries analyzed in history of science illustrate the momentous strength of established conceptual systems and the extraordinary effort needed to overcome their grip in order to attain the new. The exploration of such feats is by no means doomed to run aground in inexplicable mystery but constitutes a straightforward scientific challenge to the cognitive study of science. How to explain the gravitational pull of established doctrine on any serious attempt to escape from it? How to explain the discoverer's single achievement in breaking this hold upon him or her? These are fair scientific questions. In thinking of expanded cognitive systems, Simon and Qin consider knowledge bases which would contain collections of several ten thousands of coded facts and rules. Would it be unscientific to allow for the possibility of some more global cognitive features (like some kind of cognitive gravity) to emerge out of such huge collections because of sheer size?

By prematurely declaring scientific achievements squarely within commonplace cognitive performance, Simon reduces cognitive studies of science to an unappealing extension of cognitive science at large. No basically new knowledge is expected to come from detailed cognitive analysis of famous discoveries. This trivialization is unfortunate. One should not exclude the possibility that the rare occasions of major discoveries might reveal characteristics of our thinking that we never witness when observation is restricted to routine laboratory problems. A successful cognitive study of science could constitute an original and genuine subdiscipline of cognitive science providing a window on hitherto unrevealed capacities of our thinking during those very special conditions that constitute the 'context of discovery'.

In the third phase of the diffusion of new ideas, James has them widely acclaimed by proponents who put their previous work under the aegis of the new orientation. Given the explosive increase in the use of the qualification *cognitive* in the last decade, some bandwagon effect might indeed apply. However, in cognitive studies of science, many contributions bearing that label produce a wealth of relevant historical or simulated data to substantiate cognitive models. They have accumulated into an impressively rich collection of pertinent data and doctrine. The psychologist Tweney (1989) has contributed his long-standing methodology and expertise for research on strategies in problem solving to generate a detailed account of Faraday's discovery of electromagnetic induction. Giere (1988), as a philosopher of science, has made the cognitive approach the backbone of a new approach exemplified in an analysis of the discovery of continental drift. Gruber's (1981) earlier work on Darwin as well as Miller's (1984) more recent research on imagery and representation in twentieth-century physics, both bearing the influence of Piaget, equally turn out to be quite relevant. Some panoramic view upon this evolving field, including the computer studies mentioned earlier, is to be found in Nersessian's (1991) report on 'The Cognitive Sciences and the History of Science' for the Conference on Critical Problems and Research Frontiers in History of Science and History of Technology, held at Madison, Wisconsin. Representative samples appear in S. Fuller et al. (1989), which includes a penetrating analysis of discovery issues by Nickles (1989). A crucial collection might well be in the fifteenth volume of the Minnesota Studies in the Philosophy of Science, edited by Giere (in press), which is entirely devoted to "cognitive models of science". Given the amount of such models and their diversity, one can understand Newell's (1990) urge for an all embracing theory. Is the field ready for it? Two kinds of additional considerations indicate a direction: one related to further empirical research and one related to the nature and scope of theory.

Notwithstanding the wealth in data and subjects, some issues still seem banned both as explanatory concept and as research topic. Despite a pronounced interest in discovery, quite typical for the cognitive orientation, there is no focused search for the privileged moment of the discovery itself. A widely spread skepticism affects notions such as 'sudden insight' and 'gestaltswitch'. It might be the same suspicion that Simon expressed for pseudo-explanations based upon some magical process or event. After a long career of research on creativity in science, Gruber (1989) rejects single-event type accounts of discovery for an *evolving systems model,* which goes beyond accommodating very long periods of gestation and equally long periods for follow-through, and encompasses the whole life and personality of the inventor. Still, while recognizing that the flash of insight only occurs to the prepared mind and that the preparation might mean many years of pondering and the working-out might mean many more years, it must be possible to have somewhere along the process a decisive turn where the events qualitatively change. The long-process model or evolving systems model do not exclude a sudden flash of insight that makes one see the direction of an exit out of the crisscrossed paths of the search-labyrinth.

As indicated above, these simple perceptual metaphors of Kuhn have not been seriously developed. No detailed account of the gestaltswitch has been derived, either from perceptual psychology or from full-fledged metaphorical applications in history of science case studies. After having considered it in his analysis of Lavoisier, Holmes (1985, p. 120) reports: "If the Gestalt switch model is highly suggestive, it is misleading, I think, precisely because of the simplicity which allows it to function in the way it does. One can view all at once all of the visual details present in a Gestalt figure, so that one sees it all at once as one or another form. A conceptual structure is so much more

Fig. 1. Modified version of Porter (1954) puzzle figure.

complicated that one cannot perceive its details all at once, but must inspect them over a period of time".

For cognitive psychology, however, it is no easy matter to determine what one can view at once. Analysis of eye movements illustrates the complexity within a few seconds of observing simple objects or line drawings. Take as an example the puzzle picture proposed by Porter (1954).

The detection of the hidden figure in a collection of apparently random shaped black forms undoubtedly qualifies as an authentic aha-Erlebnis. For the naive subject, it might require considerable effort before all of a sudden the features of the face come to the fore (Abercrombie, 1960).

The sudden subjective aha-Erlebnis can be turned into an objective scientific fact by using equipment for eye-movement registration as inscription device in a Latourian sense. If there would be any doubt about the abruptness with which recognition can take place, the dramatic change in eye movement patterns that accompanies the apprehension of the face makes for a very clear-cut case. The diagrams in figures 3 and 4 show the distribution of fixations registered with a semi-naive subject (she knew about the kind of puzzle pictures but did not remember what this particular one represented; it was flipped vertically to reduce chances of familiarity). The subject discovered

Fig. 2. Modified solution after Abercrombie (1960).

the face after one minute and eleven seconds. The numbers indicate the sequence of fixations during two periods of each 17.5 seconds. The first diagram illustrates how during the prerecognition period fixations are scattered over the whole pattern. The second diagram indicates how fixations lock on to parts of the face right after the onset of recognition.

While the prerecognition patterns exhibit erratic search, the postrecognition pattern demonstrates an apparently orderly exploration, presumably evoking and amplifying a conceptual scheme "face-eyes-nose-mouth" in some cyclic fashion. There are few traces of any gradual buildup of features during the prerecognition phase, though more powerful equipment combined with more refined computer analysis might eventually find locations which gradually attract more attention without the subject's being aware of it. Methods of this kind illustrate that sudden discovery and gestaltswitch are perfectly researchable subjects that are no more mysterious than the gradual incremental processes that most authors feel safer with. If extended to other kinds of puzzle pictures, as for example 'impossible figures' varying in degree of impossibility (see front cover), it becomes abundantly clear how complex and diversified perceptual processes can be during the fraction of a second that we experience as a single unitary act in which we see everything at once.

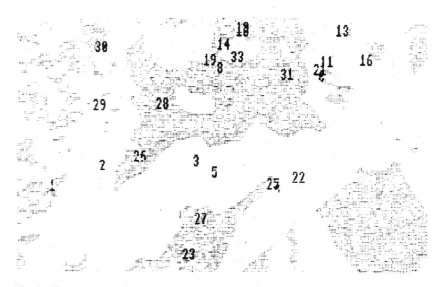

Fig. 3. Fixations in 17.5 secs. before recognition.

Nersessian (1991, p. 94) is certainly right in complaining that "too much of the recent literature in cognitive psychology reports on trivial or narrowly focussed, ecologically unsound experiments; reflecting more the empiricist spirit of the 19th century than the cognitive aspirations of the late 20th". If however the research instruments are oriented toward the inspiring gestalt perceptual examples that Kuhn (1962) refers to, it is not at all obvious that we will end up with "an unfortunate identification of observation and conceptual change with hypothesized aspects of visual perception" (Nersessian, 1991, p. 93). The facts about visual perception might turn out to be more inspiring than these hypotheses. They certainly destroy the delusory immediacy of seeing. New theories of vision might prove more productive. They could come from psychology but they might as well be developed in artificial intelligence. Despite its explosive growth, AI still struggles with basic problems in which a thorough understanding of simple object vision could constitute part of a solution. The affinity between theories of vision, scientific discovery, and core AI problems becomes manifest if we look at AI and cognitive science in a broader theoretical and historical perspective.

Among the by now numerous proponents of a cognitive orientation, Marvin Minsky stands out as both very ambitious and very controversial.

Fig. 4. Fixations in 17.5 secs. after recognition.

While the bulk of current authors in cognitive science keep close to concepts and schemes developed by contemporary colleagues, Minsky aligns himself with such apparently diverging authors as Tinbergen, Freud, and Piaget (Minsky, 1991, p. 393). And where analysts of scientific discovery like Nersessian already confidently report the applicability of common cognitive science findings, Minsky refers to a yet to be developed scientific specialty, "the presently unexplored domain concerned with communication between systems that use different representations" (Minsky, 1991, p. 373). Given the boom in cognitive studies, one would have expected the discipline already well defined and historical affinities direct and unambiguous. Minsky however seems to suggest that the major breakthrough in cognitive science has yet to be made, be it soon. Has cognitive science not started already?

Obviously, artificial intelligence fulfilled a central role but what kind of breakthrough was it really? In the summer of 1980, a memorable meeting at Stanford University established the AAAI, the American Association for Artificial Intelligence. From a social study of science point of view, this was a major event in the development of the discipline. From its very beginnings in the mid 1950s until that year, AI had been an adventurous intellectual undertaking by an elite consisting of mainly university or think tank researchers with access to powerful computers. The intellectual development during that period is indicated in chapter 1. In 1980, very much in line with some concept of 'finalization' of science (Böhme, Van den Daele, and Krohn, 1973), apparently the decision was taken to go 'societal'. The step was justified by Feigenbaum, who acted as the leader, arguing that an American reaction was needed in response to the challenge of the Japanese Fifth Generation Project. Formulated in the style of the Apollo program, this project was intended to provide the Japanese research and industrial communities with funding and guidance for achieving some major goals of AI, such as full MT (mechanical translation), by the early 1990s. To what degree this was a genuine influence in bringing about the shift or only a pretext hiding other driving forces would be an interesting research topic in social study of science. In any case, it did make AI change drastically in character. Membership in AAAS increased significantly, and research proposals became connected with commercial projects, establishing a clear link between AI and industry. AI behaved according to Feigenbaum's dictum that it had become mature enough to turn its findings into profitable products for society. Europe followed, with the EEC establishing the ESPRIT program for similar applications.

Was AI indeed mature enough to take this step? At the 1980 Stanford meeting, the "founding father" of AI, John McCarthy, was present (in the later fifties, he had coined the name of artificial intelligence and developed LISP, the major AI computer language). But while endorsing Feigenbaum's eagerness for enlarging the scope of AI's societal support, he stressed above all the necessity for intellectual innovation. In his own words, McCarthy oracularly said: "We need a .3 Manhattan project and a .7 Einstein, and pray that the Einstein comes before the Manhattan". Probably he now would agree that the .3-Manhattan somehow materialized, but has the .7 Einstein come along as well? The line of thinking suggested by Minsky seems to indicate that Einstein has not yet arrived.

In the original edition of this book, the cognitive paradigm has been introduced as it evolved with the developing field of AI. We have traced it through various approaches in pattern recognition and language processing. In the recent history of AI, one can indeed witness how the search for a powerful general mechanism of intelligence weakened in favor of ad hoc local knowledge. Historically oriented volumes (Smith, 1990) tend to indicate that this shift is not peculiar to AI and rather expresses an oscillatory movement in the study of knowledge and intelligence since antiquity. But the cognitive orientation, while much older than AI, has found its most dramatic expression in the latter. The dominant trend in the second half of the seventies emphasized the pivotal role of substantive amounts of specific knowledge. Backed by promising developments in computer hardware, AI researchers set out to encode it, the prototypical product being the expert system.

While presented as the ultimate AI achievement, the expert system is in a sense the expression of the inability to come to grips with intelligence at large and as such some indication of failure for AI. Being forced to frame skills in terms of their own specific knowledge components without having a capacity to identify the rules of the elusive common sense is indeed a major drawback. Common sense is what keeps it all together and what is supposed to be responsible for coordinating the many experts to be found in natural minds. Descartes claims in the opening sentence of his *Discours de la méthode* that humans feel nothing is as fairly distributed among men as is common sense. Everybody thinks to have just the right amount. But AI computers have none, and McCarthy sees even the need for an Einstein-type of innovation to provide them with some. Is it so complicated?

The winding course of the history of the notion of common sense might seem to increase the concept's complexity, yet it provides a hint. In its origin, the concept is rather technical. It goes back to a coordinating power that Aristotle suggested for combining sensory qualities registered by the different senses. Each sense (vision, hearing, smell, touch, and taste) reports only about its own quality. To perceive objects, sensory qualities need to be combined and brought to bear upon a single entity. For a long time, the *sensus communis* was regarded as the first inner sense, the very specific power that combined the input of different 'external' senses. In medieval anatomy, it was even assigned a specific location, either in the optic chiasma or, as in medieval scholastic psychology, in the anterior ventricles of the brain. Descartes's period indicates a transition. He himself refers both to the technical interpretation as a specific inner sense and to the more modern interpretation of some general power of judgment shared by all human beings. In the so-called Scottish school of philosopher Thomas Reid common sense is attributed a crucial role between perception and science. This attribution is partly preserved in Moore's twentieth-century Cambridge school, where common sense is a source of judgments that constitute a solid start for thinking. The shared meaning throughout the development of the notion of common sense is this aspect of coordination and integration. The earlier technical concept of *sensus communis* keeps quite close to straightforward perception of material objects and involves selection and integration of simple sensory features. The current concerns confront the various opportunities and possibilities for conceptual interpretation and involve the selection and integration of various 'world models' (see chapter 1). While they span very different scopes, these two concepts of common sense might illustrate what it could mean to integrate what Minsky calls "different representations". Notice this common concern in both Minsky's and Latour's preoccupations: to render different (kinds of) representations in such a way that they can be combined. This is the core problem for the cognitive approach. We might think we know at least the apparently simple case of object perception, because we do it thousands of times each day. But we don't. We do not know exactly how the various heterogeneous visual features of an object agglutinate to constitute a solid thing. As Ullman (1989, p. 194) indicates, "Object recognition is one of the most important, yet least understood, aspects of visual perception". As we do not yet know the principles of amalgamation in ordinary perception, nor in the more conceptual realms of common sense, we should not reject the visual metaphor for scientific discovery too readily. As Minsky suggests, the cognitive revolution is yet to come. A breakthrough might originate either in the science of common vision or in the cognitive analysis of the uncommon penetrating visions that

constitute great discoveries in science. We should maintain a synergistic relationship between the science of vision and the study of scientific innovation.

In preparing the Introduction for this edition of *The Cognitive Paradigm*, I
took profit from talks with Steve Fuller and with Peter Slezak, whose symposium on 'Computer Discovery and the Sociology of Scientific Knowledge'
(Slezak, 1989) induced an interesting debate that clarified several issues. I
also greatly appreciate encouragement and help from Margaret Hivnor, Janet
Deckenbach, and Jo Ann Kiser of The University of Chicago Press editorial
department. Special encouragement came from Katrien Buckens and from
Professor Yoshiki Morimoto, who provided me with some feedback on the
Japanese translation that was published with Sangyou Tosho in 1990. The
depicted eye-movement patterns are based upon data registered with the De-
bic 80 system at the Laboratory for Experimental Psychology of the Louvain
University (KUL). I owe thanks to Professor Géry d'Ydewalle for giving me
access to his equipment, and to Johan Van Rensbergen and Caroline Praet for
teaching me how to use it.

References

Abercrombie, M. L. J. (1960). *The Anatomy of Judgement*, London, Hutchinson.
Alpers, S. (1981). *The Art of Describing: Dutch Art in the Seventeenth Century*,
    Chicago, The University of Chicago Press.
Bertelson, P., ed. (1987). *The Onset of Literacy: Cognitive Processes in Reading
    Acquisition*, Cambridge (Mass.), M.I.T. Press.
Böhme, G., Van den Daele, W. and Krohn, W. (1973). 'Die Finalisierung der Wissen-
    schaft', *Zeitschrift für Soziologie 2, 128-44*.
Edgerton, S. (1975). *The Renaissance Rediscovery of Linear Perspective*, New York,
    Harper and Row.
Eisenstein, E. L. (1980). *The Printing Press as an Agent of Change*, Cambridge
    (Eng.), Cambridge University Press.
Fuller, S. et al. (1989). *The Cognitive Turn: Sociological and Psychological Perspec-
    tives on Science*, Dordrecht, Kluwer.
Giere, R. (1988). *Explaining Science: A Cognitive Approach*, Chicago, The Univer-
    sity of Chicago Press.
Giere, R. ed. (in press). *Cognitive Models of Science*, Minnesota Studies in the Phi-
    losophy of Science 15, Minneapolis, University of Minnesota Press.
Gruber, H. (1981). *Darwin on Man*, Chicago, The University of Chicago Press.
Gruber, H. (1989). 'The Evolving Systems Approach to Creative Work', In Wallace,
    D. B. and Gruber, H. E., eds., *Creative People at Work*, New York, Oxford Uni-
    versity Press.

Hallyn, F. (1990). *The Poetic Structure of the World: Copernicus and Kepler,* New York, Zone Books.

Hallyn, F. (in press). 'La troisième loi de Kepler et la "psychologie de la découverte"', *Revue internationale d'histoire des sciences.*

Holmes, F. L. (1985). *Lavoisier and the Chemistry of Life: An Exploration of Scientific Creativity,* Madison, The University of Wisconsin Press.

James, W. (1907). *Pragmatism,* New York, New American Library (1979 edition).

Kuhn, T. S. (1962). *The Structure of Scientific Revolutions,* Chicago, The University of Chicago Press.

Latour, B. (1984). *Les microbes: Guerre et paix,* Paris, Métaillié.

Latour, B. (1987). *Science in Action: How to Follow Scientists and Engineers through Society,* Cambridge (Mass.), Harvard University Press.

Miller, A. I. (1984). *Imagery in Scientific Thought,* Boston, Birkhauser.

Minsky, M. (1991). 'Society of Mind: A Response to Four Reviews', *Artificial Intelligence 48, 371-96.*

Nersessian, N. (1991). 'The Cognitive Sciences and the History of Science', in *Critical Problems and Research Frontiers in History of Science and History of Technology,* Conference precirculation volume, Madison, Wisconsin, 1991, 92-115.

Newell, A. (1990). *Unified Theories of Cognition,* Cambridge (Mass.), Harvard University Press.

Nickles, T. (1989). 'Integrating the Science Studies Disciplines', in Fuller et al., *The Cognitive Turn,* Dordrecht, Kluwer, 225-56.

Nowotny, H. (1973). 'On the Feasibility of a Cognitive Approach to the Study of Science, *Zeitschrift für Soziologie 2, 282-96.*

Porter, P. B. (1954). 'Another Puzzle Picture', *American Journal of Psychology 67, 550-51.*

Qin, Y. and Simon, H. A. (1990). 'Laboratory Replication of Scientific Discovery Processes', *Cognitive Science 14, 281-312.*

Shrager, J. and Langley, P., eds. (1990). *Computational Models of Scientific Discovery and Theory Formation,* San Mateo, Kaufman.

Slezak, P. (1989). 'Scientific Discovery by Computer as Empirical Refutation of the Strong Programme', *Social Studies of Science 19, 563-600.*

Smith, J-C., ed. (1990). *Historical Foundations of Cognitive Science,* Dordrecht, Kluwer.

Tweney, R. D. (1989). 'A Framework for the Cognitive Psychology of Science', in Gholson, B., Shadish, W. R. et al., eds., *Psychology of Science,* Cambridge (Eng.), Cambridge University Press, 342-66.

Ullman, S. (1989). 'Aligning Pictorial Descriptions: An Approach to Object Recognition', *Cognition 32, 193-254.*

PREFACE

> Like many features of a landscape, knowledge
> looks different from different angles. Approach
> it from an unexpected route, glimpse it from
> an unusual vantage point, and at first it may
> not be recognizable. — D. Bloor, 1976, p. 144.

Like movements in art, trends in science are identified and generally recognized only after they have already gone through a substantial development. The inception of this book on cognitivism, currently congenial to branches of computer science as well as to several social sciences, goes back to 1966–1967 when I spent a year at Harvard University and its *Center for Cognitive Studies* directed by Jerome Bruner. The sixties constituted the incubation period of cognitivism and the pioneering and seminal work at the Center expressed the potential of this orientation for psychology. I had been prepared for a confrontation with it through a thorough study of Piaget in a course by Leo Apostel on genetic epistemology which left a strong imprinting effect. At Harvard, it appeared to me that the new conceptual and instrumental means developed by the Center were to be included in an attempt to reach Piaget's ultimate goal: the understanding of the dynamics of scientific thought. Since then, cognitivism has gone through quite an impressive expansion.

There is currently intense interest in views and approaches which, for one reason or another, can be called *cognitive*. Not only in psychology, where there exists a strong cognitive movement opposed to the classic behavioristic viewpoint, but also in other disciplines such as artificial intelligence (Hayes, 1975), linguistics (Lakoff and Thompson, 1975), biology (Goodwin, 1976, 1977, 1978), sociology (Cicourel, 1973; Nowotny, H. and Schmutzer, M., 1974) and anthropology (Colby, 1975), one discovers fashionable trends which are labeled 'cognitive'. These trends are not restricted to philosophical issues with only academic scope. They appear in very practical and crucial areas as well and involve topics as diverse as juvenile justice (Cicourel, 1968) psychotherapy (Beck, 1970; Mahoney, 1977) or economic psychology (Van Velthoven, 1976). They might prove useful in understanding behavior in traffic as well as in democratic decision making. Two books published in 1975 — *Explorations in Cognition*, a collection of papers edited by Norman and Rumelhart (1975) and *Representation and Understanding*, a collection of papers edited by Bobrow and Collins (1975) — have been introduced as the first concerted effort to establish a new field: *cognitive science*. Some of the disciplines mentioned above are indicated in a diagram on the cover of Norman and Rumelhart (1975) which suggests the scope and the multidisciplinary nature of this new field. Among the contributors to these volumes, Schank, Charniak and Collins have taken the initiative in establishing in 1977 a new scientific journal under the same label of *Cognitive Science* which has developed into the official journal of a *Cognitive Science Society* established in 1979.[1] A prestigious new format journal in psychology, *The Behavioral and Brain Sciences*, devoted a special issue of 1980 to the 'Foundations of Cognitive Science' (see Chomsky, 1980; Fodor, 1980; Pylyshyn, 1980) after it already had a discussion of 'cognitivism' (Haugeland, 1978) in its very first volume. Equally on a yearly basis, a cognitive science meeting is organized at the university of Geneva, the center of Piagetian studies.[2] Obviously, 'cognitive' seems promising to several investigators in a diversity of fields and it makes one wonder what the adjective means and from where it derives its power to inspire so many.

Is the 'cognitive movement' only another swing of the familiar pendulum from empiricism toward rationalism, indicating that real understanding of mental processes is still out of reach and that sciences involved with mental phenomena are still doomed to accept an oscillatory pattern of activity around fashionable trends as a substitute for authentic progress? Or is the cognitive view based on new discoveries and insights, made possible through the computer revolution which is, for students of knowledge and cognition, a technological innovation comparable to the discovery of the telescope and the microscope for researchers in the natural sciences? Is it apt to revolutionize the social sciences by becoming the leading paradigm in the original Kuhnian sense?

The cognitive view is not new, but is only now widely discussed, having moved away from heated controversy toward straightforward popularity. Observing this development, one should keep in mind William James' well-known remarks with respect to the acceptance of new ideas. Opponents usually pass through three stages: during the first stage they reject the new ideas as ridiculous or meaningless, in the second stage they look upon them as trivial and self-evident, in the third stage they consider them formulations of the central tenets they themselves have always applied and defended.[3] Popularity does not add much to new ideas, although it might create the receptive atmosphere needed to use and apply them.

The development of new ideas takes place in an unobtrusive manner, ignored by the majority of scientists. For the cognitive paradigm, this development can be traced back several decades to the pioneering work before and during the Second World War by logicians and mathematicians such as A. M. Turing (1936) and N. Wiener (Rosenblueth *et al.*, 1943), theoretical biologists such as W. McCulloch and W. Pitts (1943), and psychologists such as K. W. Craik (1943). They and several others contributed to the elucidation of the cognitive view, which essentially comes down to the principle that any form of *information processing*, whether natural or artificial, requires a device that has in some way or another, an *internal model* or *representation of the environment* in which it operates. Although this principle can be retrieved, almost in its bare form, from Craik's 1943 Ph.D. thesis, *The Nature of Explanation*, it took many years before psychologists dared to exchange their black box for a translucent box in order to take a look at the internal model. It took equally as many years before sociologists and social psychologists took seriously the cognitive or representational nature of concepts like 'self' and 'role' and the importance of a shared model of the world in the process of social interaction. And likewise in anthropology, it took time to realize that rituals and institutions of foreign cultures are not to be considered the real heart of the matter but only the expression of belief systems and world views which have to be regarded as representational systems defining a world and a way of life. It is not the purpose of this book to trace back the history of the cognitive viewpoint in one or more of its currently recognized ramifications. Rather an extension of the cognitive view is attempted to an area in which it has only occasionally been invoked: the study of science.[4]

The study of science is the province of several disciplines whose relation to one another is ambiguous. The main cleavage runs between *philosophy of science*, which seeks to support and to justify science's claim to genuine knowledge by a theoretical analysis of its methods, and a number of empirical *sciences of science* which study a variety of factors related to scientific activity from a diversity of viewpoints. Philosophy of science claims a special position because of the special nature it attributes to scientific knowledge as superior knowledge, this special nature being essential for justifying its existence as a separate discipline. Sciences of science such as the psychology and sociology of science, on the contrary, appear to consist of the application of empirical disciplines not developed with science as their primary object of study, and their products, though sometimes very penetrating and highly interesting, seem unrelated to each other and leave us with a picture of the science of science as a rather fragmented endeavor. But philosophy of science and science of science have not remained without turmoil in recent decades, and one of the main new trends, which is still gathering in momentum, ignores the division between the two approaches, analyzing the normative and justificatory aspects of science in terms of descriptive systems and empirical investigations.[5] It is this trend which is commonly associated with the paradigm theory of Thomas S. Kuhn, which will be presented and developed as a cognitive view on science.[6] This cognitive view does not ignore the distinctive characteristics of science and scientific knowledge, but it does not attribute these to a peculiar methodology but to common cognitive processes in individuals functioning within certain patterns of social organization. Science is a socially organized process of knowledge acquisition. It might be impressive because of the hugeness of the organization it has acquired by now or because of the long list of achievements it has accumulated, but this does not imply that it yields knowledge which is basically different from the knowledge a cook has of spices, or a farmer of the weather, or a salesman of buyers. When the cognitive view is applied to science, it is not because science is a priori considered a superior and very special kind of knowledge, but it is because science is just another kind of knowledge and the cognitive view should hold as well for science as it does for more modest forms of knowledge. As such, the application of the cognitive view to science will require the conceptual apparatus developed in existing branches of *cognitive science* such as Artificial Intelligence (henceforth abbreviated as AI), cognitive psychology, cognitive sociology, etc. The book can thus be considered as an introduction to cognitive science with science as an example for those primarily interested in the cognitive view. But it serves equally as the introduction to a unitary view of science combining different lines of approach and attempting to bridge the gap between philosophy of science and the sciences of science. 'Theorists of science', as D. T. Campbell (1977) prefers to call those receptive to both areas, know how challenging an ambition like this is. For them cognitive science will derive its interest from its contribution to an integrated theory of science.

An integrated approach to the study of knowledge is the hallmark of cognitive science. As such, it affects several major disciplines. If successful, the cognitive view will produce a basic reorganization among them. Dealing with the social sciences, Myrdall (1972, p. 158) apparently feels tempted to suggest that some of their fundamental troubles might well go back from an ill-conceived segmentation into separate disciplines. To have kept economics, psychology and sociology apart from each other might have carried with it an unnatural chopping up of problems into subproblems which turn out to be meaningless when studied in isolation. A similar impression seems to apply to disciplines which touch upon cognition in one form or another. The way in which we consider perception to belong to psychology, language to linguistics and communication to sociology might stem from an equally awkward segmentation forced upon authentic interests in communication and cognition. Cognitive scientists promote the idea that there is no way to understand language or perception apart from communication. In their eagerness to arrive at a unitary view on cognition, they are not afraid of extensive reshuffling of boundaries between traditional disciplines. In that respect, the impact of the cognitive orientation goes far beyond the topic with which we will be concerned. However, though our analysis is prudently restricted to scientific knowledge, we shall not be able to avoid encounters with some turbulent areas in established disciplines such as sociology and, in particular, psychology. Indeed, as Piaget (1979a) indicates: "Psychology occupies a key position in the family of sciences in that it depends upon each of the others, to different degrees, and in turn it illuminates them all in distinct ways".

This book consists of three parts. In Part I, the cognitive orientation is introduced as the most recent stage in a series of stages which characterizes the evolution of theories on perception and communication in AI whose development is used as an exemplar. These stages which apply to our understanding of cognitive processes in everyday life also turn out to be most suitable for the analysis of scientific knowledge and scientific development. After having discussed positivism, logical positivism and sciences of science, including sociology of science, psychology of science and history of science, Kuhn's paradigm-model is introduced as a cognitive approach allowing, in principle, the integration of philosophy of science and sciences of science. Part II and III follow this cognitive program in the light of recent achievements in sociology of science and cognitive psychology.[7]

Part II examines the structure of the scientific literature, communication channels and informal groups in various attempts to locate and to trace communities in science which correspond to a specific 'world view'. Also a life cycle model of scientific specialties is included. Including a substantial part of recent work in sociology of science, these discussions allow us to illustrate the perennial nature of the search for specification of the relationship between social structure and cognitive structure. They also substantiate the need for a powerful conceptual framework to deal with *cognitive structure*.

The elusive nature of cognitive structure is due to the tacit nature of the knowledge involved. In Part III, current cognitive science concepts based upon cognitive psychology (LNR[8]) and AI are brought to bear upon the notion of paradigm as a cognitive structure which shapes expectations and guides search activities. Perception is dealt with as an interactive and integrative process involving schemes contributed by the perceiver. Puzzle-solving is analyzed in terms of recombinations of world view segments involving meta-cognitive processes relating to the self-concept and the segmentation between inner and outer. This way of dealing with perception and problem-solving allows us to overcome the segmentation between 'internal' and 'external' which has been troublesome to both sociology of science and history of science. The risk for fragmentation of knowledge inherent in some current cognitive trends is counteracted by Piaget, the pioneer of cognitive science. This is illustrated in an analysis of his notion of conservation in relation to the dynamics of conceptual systems in science. The epilogue considers how what has been learned from the cognitive paradigm contributes to overcome the dualisms which have impeded progress towards a genuine science of science.

It is obvious that a venture of this scope cannot be undertaken without the help and support of a number of people. Important of course are those whose work is dealt with explicitly and to whom my debt is apparent from the notes and the bibliography. However, citation links only partly reflect intellectual descent and social affinity. Equally important are the friends and colleagues whose work is 'undercited' because of their permanent availability and willingness to share ideas before they are accessible through the more formal channels of scientific communication. In this respect I should single out Leo Apostel, to whom I owe very special thanks for the inspiring teachings and the unlimited support and encouragement for almost all of my endeavors during many years. He not only opened up for me the cosmopolitan community of science and scholarship but also taught me to think for myself through his remarkable combined capacity for unrestricted understanding and constructive criticism. My debt to him is therefore boundless. I am also greatly indebted to my fellow researchers in Communication and Cognition and in particular to Fernand Vandamme with whom I started it as a working group and whose dedication made it flourish, attracting several other members with whom I could cultivate a continuing dialogue. I also benefited greatly from my membership of the Dutch group of 'wetenschapsonderzoekers', periodically brought together by Peter Koefoed at Amsterdam. Preliminary versions of some chapters were presented at various conferences in both Europe and the U.S. and I received important help through reactions and feedback from organizers and participants who offered me their spiritual and material hospitality. Trying to think of them all together makes one sharply aware of the quasi impossibility to list all relevant influences and acquaintances, because they are more numerous and more important than I can express. Nevertheless, I should mention Margaret Boden, Arie Rip, Thomas Place, Mary White, Werner Feibel and Erwin Breusegem, who at various times read various parts of the manuscript and provided me with extensive comments, allowing me to correct mistakes and to find more adequate formulations. This also applies to Ludo Peferoen who made suggestions for improvements while making the final version of the drawings. Similarly helpful were the comments and suggestions of the managing editors Loren Graham, Helga Nowotny, Peter Weingart and Richard Whitley. The book grew out of a course on the theory of science which I taught at the universities of Tilburg (The Netherlands) and Ghent (Belgium). I am thankful to the various groups of students in psychology, sociology and philosophy and the occasional historian who forced me, through their collaboration and contributions, to keep in touch with the variety of disciplines involved in a science of science while at the same time encouraging me, through their enthusiasm, to adopt a more integrated approach. I should also express my gratitude to both universities which provided ample institutional facilities and support. The university of Ghent, as should be expected from the place where George Sarton started *Isis*, has been a rich source in history of science and historical documents. Maria Smeets' library of the new and expanding department of psychology at Tilburg has been invaluable for keeping up with current literature in cognitive psychology. The Belgian N.F.W.O. (Nationaal Fonds voor Wetenschappelijk Onderzoek) supported me through several grants and so did the United States Educational Foundation (now 'Commission for Educational Exchange'), which provided me with a Fulbright scholarship and a Frank Boas award for Harvard. I am greatly indebted to Mr. Frank Boas and his generous family. Loes Muijsenberg, Els Koks and Mady De Sloovere typed parts of the manuscript. The final version from flyleaf to bibliography was handled by Etienne De Vlieger in astonishingly little time. I am very grateful to all of them.

I also thank my wife Hilda and my sons Jon and Tim for bearing the burden of a husband and father absorbed in a book project. At times I discovered myself negotiating with my children for isolation in order to study Piaget, only to discover that they were performing Piagetian scenarios right before my eyes. They have kept me alert for what exists beyond the scope of books. The continuing silence of my son Ben has been my greatest source of wonder.

MARC DE MEY

PART ONE

INTRODUCTION TO THE COGNITIVE VIEW

CHAPTER ONE

THE DEVELOPMENT OF THE COGNITIVE VIEW

> L'étranger nous regarde les yeux grands ouverts,
> mais il ne voit ce qu'il savait de nous (ou pensait
> en connaître) — African proverb quoted by
> Piaget, 1974a, p. 162.

Cognitive science deals with the study of knowledge: i.e. what knowledge is and how it can be represented, how it can be handled by transforming it from one form to another. When applied to science, the problem is to indicate how knowledge is produced and handled in science.

Common descriptions of scientific method tend to promote a picture in which the role of observation is overemphasized. Scientific activity is depicted as a circular process starting from and ending in observation, the classical stages being: the gathering of facts by means of observation, the formation of a hypothesis, the deduction of consequences from the hypothesis and the test of the hypothesis, again by means of observation. Pictures of famous scientists prefer to show them with their favorite observational instruments: Galileo with his telescope, Pasteur with his microscope, rather than with the books they have written. Within the classical approach to science the facts are holy, not the books, and facts are supposed to derive from observation. Yet, when one makes an empirical study on how scientists spend their time, one discovers, as Meadows (1974, p. 91) indicates, that "communication in one form or another usually takes up a significant fraction of a scientist's working life". An important part of this communication consists of consulting the written communications by fellow scientists in journals and books. Most universities have extensive libraries with a special staff to assist the scientist in obtaining the written materials he wants to read. Meadows estimates that a scientist scans roughly 3000 articles a year and studies 10 percent of them more thoroughly (Meadows, 1974, p. 97)! Furthermore, a substantial part of a scientist's time is devoted to exchange of ideas and informal discussion with colleagues. Secretaries at universities handle a lot of mail which in terms of

3

research proposals, preprints and comments on these also serves the purpose of communication in science. When examined more carefully, the communication machinery of modern science is as impressive as the observational capabilities of spectacular devices such as telescopes and electron microscopes and *communication* is as much a constituent of scientific activity as *observation*.

Granted, then, that both observation and communication are key processes in scientific activity, what do we know about their psychology? What have we learned from either psychology or AI[1] about the basic mechanisms of perception and language processing that allows us to understand observation and communication?

At one point in the development of the cognitive view, it was promoted as the *information processing approach*. Although this approach originally in the spirit of Shannon's information theory, has now been rendered somewhat out of date by recent developments,[2] it remains useful to consider both observation and communication as information processing activities. In an act of observation, the observer samples sensory information to find out whether or not a certain event occurs or a certain object can be perceived. In that sense, an analyst looking through a microscope to detect a microorganism in a microscopic slide is processing visual information. In communication, a receiver generally samples visual or auditory information from a message and tries to decode the information in order to arrive at the meaning of the message. In that sense readers process visual information and listeners process auditory information. The central point of the cognitive view is that *any* such *information processing*, whether perceptual (such as perceiving an object) or symbolic (such as understanding a sentence) *is mediated* by a *system of categories or concepts* which for the information processor constitutes a *representation* or a *model* of his *world*. Although this notion has been familiar for quite a while (K. Craik, 1943; Miller *et al.*, 1960), the full implications for our understanding of perception and communication have only been recognized in careful attempts to simulate these processes on computers, and the adoption of the cognitive view is only a recent stage in a series of stages which have been accumulated in thinking about information processing as it has developed in AI. Extending a classification by D. Michie[3] (1974), four stages can be distinguished:

— a *monadic* stage during which information-units were handled separately and independently of each other, as if they were single, self-contained entities;

— a *structural* stage which considered information as a more complex entity consisting of several information-units arranged in some specific way;

— a *contextual* stage where, in addition to an analysis of the structural organization of the information-bearing units, supplementary information on context is required to disambiguate the meaning of the message;

— a *cognitive*, or *epistemic*, stage in which information is seen as supplementary or complementary to a conceptual system that represents the information processor's knowledge or model of his world.

Since this scheme will form the backbone of Part One, we should look in some detail at illustrative examples from both the area of perception and of communication in order to grasp the dynamics of this development.

Perception, Pattern Recognition and Picture-processing

AI aims at understanding cognitive processes in such a manner and to such a level of detail that it can build artificial devices that perform the same cognitive function in a way that, in principle, makes it possible to substitute them for human performers. It is not that AI researchers are necessarily more interested in technical devices than in procedures of the mind. But among their criteria for adequacy for models of mental processes is the requirement that they should be specifiable in terms of programs and run on a computer. Thanks to their stringent standards and tenacity of purpose, they deepen our respect for the human brain by showing us and helping us to appreciate how it is even much more subtle and ingenious than we (and they) thought.[4]

Several AI projects have been oriented toward the construction of what one could call 'artificial scientific observers'. For example, in particle-physics many investigations require the screening of 'bubble chamber photographs'. These are pictures of a heated fluid in which ionized particles passing through trace their path as a string of vapor bubbles. AI devices designed to screen such pictures commonly make use of a camera of some sort to produce a standard representation of the information in the photograph which is then further processed and interpreted by means of a computer program. It is in the development of these and similar basic representations and the related programs that we can retrieve the four stages mentioned above. For an illustration of that development, we shall first look at the simplest subclass of these machines: devices for pattern recognition.

Imagine a machine, something like a photocopy device, into which you would feed a page of your very idiosyncratic handwriting and which, as output, would produce not a copy but a nice type-written version of your text in printed letters. In principle all that is needed is a machine that can recognize handwritten letters. Much effort in pattern recognition has indeed been spent on this apparently simple task producing the following sequence of approaches.

The Monadic Stage: Template Matching

Template-matching is a pattern recognition technique which uses a simple or elaborate set of tricky procedures called *preprocessors* to reduce signals, in this case letters, to some form of standard representation which makes it possible to search for a match with a *template* stored in the system. An image of the signal is projected on a grid and it is described in terms of a digital representation: cells of a two dimensional array being either activated (*1*) or not (*0*), depending on the way they are affected by the signal (Figure 1.1).

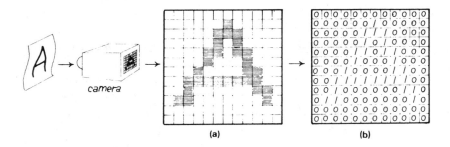

Fig. 1.1. Signal projected on a grid pattern (a) to arrive at a digitized representation (b).

The preprocessing is really to be considered a kind of cleaning up of the 'noise' in the signal, so that it is optimally comparable with the template or canonical form which is specified in the same digital code. Preprocessing might straighten up lines, eliminate blots, close up small empty spaces, rotate the signal into a vertical position, or involve many more tricky procedures which might require quite sophisticated mathematics. However, these procedures are not meant to be analyses of the signal. Preprocessed inputs are compared to the stored templates. Recognition consists of ascertaining a match between the preprocessed signal and a stored template. Template matching is monadic in the sense that it handles patterns entirely as isolated and monolithic units. As a pattern recognition procedure, however, it is not very successful. Its weakness becomes manifest with ambiguous signals such as in Figure 1.2.

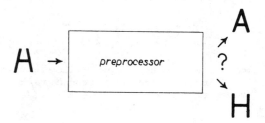

Fig. 1.2. Preprocessors are not equipped to disambiguate ambiguous signals. Should the preprocessor close the space at the top or enlarge it?

The Structural Stage: Feature Analysis

In contrast to template matching, feature analysis does not attempt to conserve and amplify the integrity of the signal. It is not the 'Gestalt' [5] of the signal which has to be congruent with a stored canonical form. The signal is decomposed and recognition involves a structural analysis of certain attributes. A pattern is defined in terms of certain features being present at certain positions. *Features* are to be thought of as elementary patterns, such as *bar, arrow, fork*, etc. . . . Psychologists could think of the kind of units associated with Hubel and Wiesel-type receptive fields. The positions are parts of the figure, *top, middle, bottom*, or regions. A pattern such as *A* can then be characterized by a sharp angle in the top section, a horizontal bar in the middle section and an open space in the bottom section. Feature analysis is superior to pattern matching in that it can make use of redundancies and eliminate some possible interpretations on the basis of structural anomalies (impossible figures). As such, it permits something like a syntax for patterns or figures as Clowes has shown (1969). Nevertheless, an ambiguous signal like Figure 1.2 cannot be 'disambiguated' with feature analysis any more than with template matching.

The Contextual Stage: Contextual Analysis

When a structural analysis leaves open several possible interpretations of a signal, the only way out is an analysis of the context. Making an appeal to context is not just an extension of the domain of features taken into account, but rather refers to the utilization of higher level knowledge. In Selfridge's figure (Figure 1.3) one can call upon his knowledge of consonant-vowel string

THE CAT

Fig. 1.3. Words used as context to disambiguate the character of Figure 1.2. (After Selfridge, O. G., 'Pattern Recognition and Modern Computers', *Proc. West. Joint Computer Conf.*, Los Angeles, 1955, p. 92, Figure 3; © 1955 *IRE* (now *IEEE*), reprinted with permission.)

syntax, or simply his knowledge of words, to disambiguate characters. Thus, contextual analysis makes use of information which is not in the signal (the letter), but which belongs to a larger whole of which the signal is only a part. However, the problem with context is that it is so rich in possibilities, and that it has no well defined boundary. It can always be enlarged and one never knows if a sufficiently large portion of context has been checked. Elaborating the example taken from Selfridge, a phrase such as in Figure 1.4a receives an amplification of its 'cat'-interpretation by having a favorite cat-name in its context. However in a phrase like Figure 1.4b another interpretation is suggested by the context.

(a)   FELIX THE CAT

(b)   CAT – Cylinder Head Temperature

Fig. 1.4. Broader contexts to disambiguate identical ambiguities: (a) additional context which amplifies cat-interpretation, (b) another interpretation suggested by another type of context.

So, although the context at the level of consonant-vowel-combination-rules suggests one interpretation, in a wider context this can apparently be overruled in favor of another interpretation.

The Cognitive Stage: Analysis by Synthesis

The next step, then, is to introduce a well-defined context by having it supplied somewhat arbitrarily or subjectively by the information processing system itself: its *model of the world*. Recognition is guided by what the system (rightly or wrongly) considers relevant features and relevant context on the basis of its 'world knowledge'. It selects only those features which it knows are to be noticed and it analyzes the signal only in so far as it seems necessary in order to check the match between the self-generated expectations and the perceived pattern. Nothing can better illustrate this type of synthesis − experienced as analysis! − than proof-reading by an author who, maybe because of his concentration on the content of his text, reads several misprinted words as correct. 'Analysis by synthesis', strictly speaking, is a model for language perception developed by Stevens, Halle and Liberman (Halle and Stevens, 1962). In its peripheralism − understanding language is listening to our own internal articulations − this view is certainly not representative of the cognitive view. We use the expression here in a broader sense (Neisser, 1967) to refer to a theory of perception which emphasizes the production of expectations on the basis of a world model to such a degree that intake of input can be seen as almost entirely restricted to a few points of control, and to filling in of parameter values. Within such a view *perception* is, to a large extent, the product of *imagination*, the few points of intake of information only safeguarding it from becoming *illusion* by tying imagination to 'reality'. In familiar scenes much of what we see is what we know should be seen but only few things might be perceived directly.[6]

The recognition of letters might seem different from the perception of simple scenes or objects because it is concerned with conventional symbols. Within the cognitive view the 'direct' perception of objects is symbolic as well, in the sense that it is dependent on stored representations. A simple example will illustrate the applicability of the four stages to the processing of pictures. How could one proceed to devise a pattern-recognition device that should recognize eye glasses in a photograph? A template for 'eye glasses' should be thought of along the lines of Figure 1.5a and one could hope that with sophisticated preprocessors actual pictures of eye glasses could be reduced to that form. The more flexible feature analysis procedure would require a decomposition of the figure into a structure of basic constituents.

In this example, a plausible possibility would be a horizontal linear part connecting two circular parts of identical size. However, as with the recognition of letters, ambiguity is the main source of difficulty. If our device is

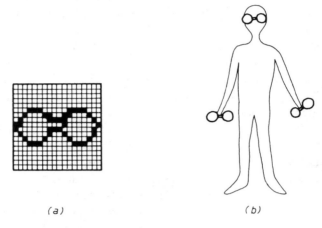

(a)   (b)

Fig. 1.5. A pictorial equivalent of Selfridge's pattern with an ambiguous character (see Figure 1.3). The template indicated in (a) represents dumbbells as well as eye glasses. Its interpretation in (b) depends on its location with respects to parts of the depicted body constituting a local context. Broader contexts which overrule this one are easy to imagine (see text). The eye glasses-dumbbells ambiguity is inspired by Carmichael *et al*. (1932) who provide a whole series of similarly ambiguous stimuli.

to recognize more than one familiar object, how is it going to distinguish eye glasses from something like dumbbells which would have a standard representation very similar to that of eye glasses? Here too, as is apparent from Figure 1.5b, context needs to be invoked to disambiguate the signal. However, no definite context can be indicated which contributes to the assignment of a specific meaning to the signal with absolute certainty. If the picture in Figure 1.5b would illustrate a story of someone who cannot come to a decision with respect to the choice of a new pair of eye glasses, no one would think of seeing dumbbells on the hand locations. A broader context overrules a local one. Again, there seems no other way out than invoking the knowledge of the viewer. In a way, he has to know what the picture is about before he can see it! What he is seeing depends on the world-model used to make sense of the information presented, and the choice of that model is the perceiver's.

Thus, the experiences with pattern recognition and picture processing show that ambiguous figures long known to psychologists, such as Necker's cube, Boring's reversible picture of young lady/mother-in-law, the duck-rabbit figure or the pelican-antelope (see Figure 6.2), are not rare accidents in which the perceptual apparatus is deceived. They are striking illustrations of the fact that ambiguity is the rule and that knowledge derived from a model of the situation is needed to resolve it.

We are now in a position to understand Feldman's synoptic view of the development we have sketched:

> Early efforts in machine perception (and much perceptual psychology) were concerned with visual processes which operated independent of context. We studied edge detectors, pattern classifiers and algorithms for partitioning general straightline drawings. This not only proved difficult but offered no promise of extension to typical real world scenes. There then came a concerted effort to overcome (or circumvent) perception problems by giving programs lots of domain knowledge. This has been carried to the extreme of visual perception without vision, viz. anything black and on a desk is a telephone (Feldman, 1975, p. 92).

A similar development applies to language processing.

Communication and Language Processing

Attempts to program computers to do some information-processing on verbal material have been stimulated — like work in pattern recognition and picture processing — by the idea that some of the work done by scientists with respect to scientific literature might be done better, faster and more cheaply by a computer. The most widely-known example is undoubtedly an ambitious project in mechanical translation (abbreviated as MT) that has received contributions and support not only from AI researchers and linguists but also from nuclear physicists seeking devices that would translate scientific papers from foreign languages into their own.[7] The ambition of more than twenty years ago was straightforward mechanical translation. Now researchers such as Schank (1975) would feel fortunate indeed to have developed programs capable of paraphrasing in a given language, let alone translating from one language to another. Nevertheless, although the goals have not been attained and the community has grown more modest in its ambitions, substantial insights have been acquired which may turn out to be more valuable than the glittering goals of the first years of the MT projects in AI.[8]

The Monadic Stage: Word-to-word Translation

Language translation, at first glance, appears to be a simple problem connected with the use of words as the names or labels that concepts have in diverse languages. To the extent that translation is seen in terms only of an exchange of labels, it is the expression of a monadic view. Units of one kind are exchanged for units of another kind on a one-to-one basis. This is an extremely naïve idea with which one obviously gets into trouble even with simple problems of word-order. Very soon, the AI-language researchers realized that dictionaries are not the core-problem of translation, and attention and effort were concentrated on *the sentence* as the basic unit of analysis.

The Structural Stage: Syntax

The grammatical sentence seems to be the natural unit of speech, although research such as Goldmann Eisler's (1968) shows that only half of the pauses in speech occur at grammatical locations, such as the beginning or the end of a sentence. The isolated grammatical sentence is a highly abstract entity. Within the structural approach, however, it is this grammatical unit, with its organization — more specifically, syntax — which is seen as the level on which languages should be handled and translated. The idea is that there is some set of universal grammatical categories, for the expression of which each language has specific syntactic mechanisms. Something expressed in language *A* by means of word-order can be expressed in language *B* through affixation. Translation, in addition to knowledge of the words, involves procedures for syntactic decoding and encoding. However, the ambiguity of the isolated sentence is the problem that reveals the weakness of the structural approach. Take the well-known example *time flies like an arrow*. At first sight, this may look like a familiar expression of wisdom such as those phrased as proverbs. This may be a natural interpretation of isolated sentences. Proverbs have that self-contained and self-sufficient character which compresses feelings and experiences accumulated over generations into one sentence. However, ordinary sentences in isolation are quite ambiguous. A computer program designed to perform syntactical analysis, but having no conventional wisdom, illustrates this by generating an unexpected number of alternative interpretations. The common interpretation would be represented as in

Figure 1.6a. One can however, as the computer does (Oettinger, 1965), consider *time flies* as a composite noun, and this can yield a perfectly legitimate reading in terms of some kind of insects — time flies — being fond of some kind of object — an arrow. In that case, the 'phrase marker' should be as in Figure 1.6b. Another very productive approach indicated in Figure 1.6c is to consider the sentence as an imperative with *time* used as a verb meaning

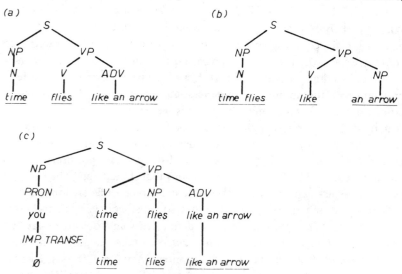

Fig. 1.6. Various interpretations of *time flies like an arrow*, according to different possibilities of syntactical parsing: (a) Common interpretation with *time* as subject; (b) Equally legitimate interpretation with *time flies* as a composite noun; (c) Interpretation with *time* as a verb meaning 'measure the speed of'. Since the specification *like an arrow* can determine either the subject or the verb or the object, several alternative interpretations are possible: *go and time flies as fast as an arrow, time those flies which resemble an arrow* or *time flies as you time arrows*. The example is based on Oettinger (1966). The illustration originally appeared in A. G. Oettinger, 'The Uses of Computers in Science', *Scientific American*, September 1966, p. 169 (reprinted with permission).

'measure the speed of'. How could one decide which syntactic analysis is correct for this sentence? The example illustrates the ambiguity of isolated sentences. On the one hand there is not just one clear and totally unambiguous reading of an isolated grammatical sentence. On the other hand, syntactically anomalous sentences for which no correct structural description can be found might turn out to be very intelligible. And thus, context again appears to be a crucial variable in language processing as well.

Contextual Stage: Indexical Expressions

Isolated sentences do not exist. Sentences relate to a context as figures to a ground.[9] People rarely experience the annoying ambiguity which the syntactic analysis program produces when it generates its many alternatives in a small scale 'combinatorial explosion'. Depending on verbal or extra-verbal context, we almost effortlessly apply, as far as required, the structural analysis connected to that particular context. If we have been talking about tastes and preferences of flies in previous sentences, the sentence *time flies like an arrow* poses no problem since only the analysis we have mentioned as the second structural description is applicable in this context. However, here again, the weak point of context seems to lie in determining its boundary. Efforts have been made to delimit context in terms of lists of presuppositions — require-ments that should be fulfilled, or conditions that should apply, before a given sentence can have any meaning. The list may become endlessly long, especially with isolated sentences. Context grows so large and is apparently so all embracing that it is very difficult to define it in any objective way. In an example of Winograd's, *Sam and Bill wanted to take the girls to the movies but they didn't have any money*, the pronoun *they* is ambiguous. This might be made apparent by proposing a literal translation into French which reads *Sam et Bill voulaient prendre les filles au cinéma mais elles n'avaient pas d'argent*. The translation is contra-intuitive, we did not expect *they* to refer to *the girls* but to *Sam and Bill* and we would translate *ils n'avaient pas d'argent*. As Winograd (1973) indicates, the inference that *they* refers to the boys is not based on syntactic or semantic rules. It is based on our knowledge of social conventions where the rule is that those who invite are supposed to pay. Obviously, one can imagine cultures where, instead of equality between sexes, rules would amplify different sex roles such as: the boys pay whoever might come up with the idea. In a sentence like *Eleonor and Jane wanted to go to the pictures with the boys but they didn't have any money* the inter-pretation for 'they' would be different depending on whether the listener is in a progressive U.S. college context, where equality between sex roles might be widely accepted, or some conservative European 'milieu', where different sex roles would still be fashionable. Are such cultural rules among the presuppositions for understanding sentences of the language and thus part of the context? An example requiring another kind of knowledge to disambiguate pronominal reference is the following sentence. *The box slid off the table because it was not level.*[10] Again, inferring the referent of *it* cannot be based on any grammatical rule. One needs to invoke some common-sense physics in order to understand that *it* has to refer to *table*. Sliding will only occur when the supporting surface is not level and is physically relatively independent from the position of the supported object. Is physics part of the context?

In other types of sentences, which require contextual information of an apparently more modest nature, context might turn out again treacherously simple. This is the case for indexical expressions, sentences such as *today is Monday* or *I am a man*. The truth of such sentences depends on such aspects as *when* they are uttered or by *whom* they are uttered, etc . . . . Bar-Hillel, who tried to cope with such sentences by referring to 'pragmatic context' had also to admit that: "the vagueness in which I left the expression 'pragmatic context' is partly due to the fact that its reference is often intrinsically vague itself" (Bar-Hillel, 1970, p. 81).[11]

The Cognitive Stage: Prerequisite Knowledge

The interpretation of the notion of context tends to expand until it embraces everything we know about a given subject. Again, this is the point where a context-oriented approach changes into a cognitive view in a restricted sense. Context is knowledge brought in by the information processor. In the previous stage it is still considered part of a signal's surroundings, external to the reader or the listener. It has to be perceived together with the signal. In the cognitive view *context* becomes something *supplied by the perceiver*: it is the knowledge he invokes to analyze the signal and to determine its meaning. It is, more precisely, the knowledge that permits him to look for and to detect very selectively those structural characteristics and more specific contextual elements which yield an interpretation of the message congruent with that knowledge. This shift from a contextual to a cognitive stage has been indicated by Winograd (1977, p. 168) in terms of a shift in the meaning of the notion of context: "Context . . . is best formulated in terms of the *cognitive structures* of speaker and hearer, rather than in terms of linguistic text or

facts about the situation in which an utterance is produced" [12] (italics ours).
Bateson (1979) has emphasized a similar point: "it is the recipient of the
message who creates the context" (p. 47). These cognitive structures embody
a perceiver's knowledge by indicating what to discern, what to expect, what
to say at a specific occasion. They constitute *world views* for typical situations
specifying a particular universe in terms of the large amounts of prerequisite
knowledge necessary for understanding even the most simple utterances.

The Generic Scheme

The evolution of both automatic machine perception and mechanical translation exemplifies an underlying development in thinking about information
processing. Table I provides a schematic overview of the major approaches
which AI has accumulated in its short history.

TABLE I

Generic scheme indicating stages accumulated during the development of information
processing in AI

Knowledge and information	AI pattern recognition and perception	AI language processing and communication
Monadic	*Template matching*	*Word-to-word translation*
Structural	*Feature analysis*	*Syntactical analysis*
Contextual	*Context analysis*	*Indexical expressions*
Cognitive	*Analysis by synthesis*	*World models*

It should be clear now how the cognitive view has developed as the expression of two shifts of focus:

First, there is a shift *from the object and the signal to the subject or the
receiver*. Cognition connects a knowing subject to a known object. The crude
empiricist position sees no problem in identifying reality with the object:
reality is out there lying at our feet, unfolding before our eyes. The problem
of cognition is how to get a representation of the object into our heads.
Empiricism favors a rather passive subject, highly receptive to the outside
world. Any cognitive activity is seen as oriented toward a maximal exposure

to 'stimulation'. The cognitive view has gradually developed into a position where it strongly emphasizes the organizing activity of the subject in the cognitive relationship between subject and object. It does not recognize any meaning in an isolated stimulus or message. All stimuli that are perceived are received in terms of signals evoking or confirming a model for sieving experience. Insofar as stimuli are signals evoking representation models, all perception is symbolic. With respect to messages, the meaning of a message is synthesized by the receiver out of his own knowledge. Understanding speech, whether written or spoken, is seen as a constructive activity during which a message only induces a meaning (in the sense of an 'inducer' in embryology), with the bulk of the message content being provided from the knowledge base of the receiver. Neisser's book *Cognitive Psychology* (1967) starts out by referring to the saying "beauty is in the eye of the beholder" as relating to a central problem of cognition.[13] One could characterize the cognitive position by paraphrasing that saying: "meaning is in the world view of the subject".

The second shift is from clearly delineated micro-units handled in *isolation toward handling very complex entities up to the macro-scale of world models*. Larger and larger wholes are involved in understanding even apparently simple messages. Each internal context is rich and dense in knowledge items, variously related to each other into an intricate maze that constitutes a universe in its own. To characterize this trend, one could paraphrase Wittgenstein's famous dictum: "to understand one sentence is to understand a whole language". It should read, "to understand one sentence is to understand a whole world view".

Though the impetus of the cognitive movement in AI might have made us sensitive to a commonly underestimated contribution of the internally provided knowledge of the perceiver in most perceptual processes, the cognitive view could still seem mystifying and even paradoxical. Indeed, roughly stated the cognitive orientation contends that we perceive things and understand states of affairs because we expect them. But if we only perceive what we expect to perceive, what then is the sense of perceiving? As Neisser indicates "There is a dialectical contradiction between those two requirements: we cannot perceive *unless* we anticipate, but we must not see *only* what we anticipate" (Neisser, 1975, p. 43, italics in the original).

An attempt will be made to disentangle this fundamental problem in Part

III (especially Chapters Ten and Eleven). However, before we investigate the
applicability of our scheme to schools of thinking on science, this paradox of
the cognitive view and related questions require a closer look at the notion of
internal models. Such an analysis will correct some possible misunderstandings of the stages of our scheme and suggest a way to solve the paradox.

CHAPTER TWO

WORLD VIEWS AND MODELS

> Die Trennlinie zwischen einer äusseren und
> einer inneren Welt wird . . . eine immer vagere
> und unhaltbarere. – Nowotny, H. and E. A.
> Schmutzer, 1974, p. 15.

One could say that the cognitive view implies an *iceberg-model* of *knowledge*.
To support something like ten percent explicit knowledge, which would
commonly be considered genuine, requires ninety percent prerequisite
supporting knowledge, of a more covered or implicit nature. Obviously, these
percentages only indicate in a metaphoric way that a disproportionate large
amount of implicit knowledge is necessary for receptivity to small portions of
what appears like more explicit knowledge. In the previous chapter we have
illustrated this idea with respect to the recognition of patterns and pictures as
well as the understanding of sentences. The examples showed how difficult it
is to determine the meaning of isolated items of information. However, to
fully appreciate the cognitive view, we should realize the extent to which all
of our behavior is guided by stored knowledge in comparison to external
knowledge acquired on the spot. Let us consider an example from traffic
behavior where the intake and processing of fresh information seems to go on
continuously.

The following situation should be familiar to many. We are used to en-
countering works on the roads. If they are not being renewed or repaired,
they are being improved with road signs and traffic lights. One day on our
usual way home we might notice traffic lights being installed at a crossing
where there were none before. We might take notice of this event because of
the traffic jam created by the works. However, after a few days, the light
posts are there and since they are not yet functioning, we learn to ignore
them. Until . . . one day, the amazed faces of some other drivers and the
screech of tires show us that, apparently, the lights have been put into operation. Despite the well-chosen salience of the red light, it has proven incapable

19

of penetrating into the daily routine in which we have minimized the points of looking at a too familiar environment. Our model of that cross road had the traffic light represented as unworthy of attention because it was inoperative.. Only when the consequences of missing vital signals suddenly emerge in appalling situations do we realize how we are blinded, in a sense, by our own knowledge. In general, world views or world models consist of relatively stable routines for actions, such as driving home according to a daily pattern, with built in selective windows for perception.

The Multiplicity of World Views

The notion of *world view* as used in cognitive science should not be confused with some abstruse Weltanschauung which strives to understand everything from within one all-embracing unifying scheme. As modern cosmology has confronted us with a bewildering physical universe, populated with millions of galaxies, each potentially bearing a comparable multitude of worlds, the cognitive view populates our mind with not one, but a multiplicity of world models. Indeed, from current work in AI and cognitive science emerges the picture of a mind equipped with many 'micro worlds'. Workers in AI usually focus on only one or a few of these and try to realize on a modest level some information processing within that specific world. Schank (1975) has tried to penetrate the world of a 'restaurant diner', Charniak (1975) is trying to make explicit the knowledge needed to shop in a supermarket, Winograd's (1972) and Shirai's (1973) programs have a knowledge base whose scope is limited to a simple 'block world'. This is partly due to the method of working out a well-selected example in detail to investigate the consequences and the results of an approach. However, the general assumption is also that the world of our knowledge is a complex assembly of rather loosely coupled subsystems (Simon, 1969). Ignoring, for the moment, linkages between worlds or models, this corresponds with Boden's (1972, p. 3) formula that "the mind should be thought of as a set of . . . representational models, systematically interlinked in certain ways". We should indeed realize that, as complex information processing systems, most of us change from one representational model to another even more easily and more frequently than we change clothes. Working at our desk, our mind might be dominated by elusive scientific models which we struggle to make explicit. However when we walk out to have lunch, we change to a Schank-type 'restaurant-model' or a Charniak-type 'supermarket-shopper model'. Our actions and perceptions are then guided by what we know about restaurants and supermarkets and having lunch at those places. Another model is selected for going to the swimming-pool which again is another 'world', as are our house, the hotel, the highway, the university, the hospital. All of these environments require, if we are going to behave in them, large amounts of knowledge to direct our information processing and to govern our action. For each of them, that knowledge is predominantly tacit – embodied in practical skills and patterns of attention. The multifaceted reality and autonomy of these worlds is very similar to the various worlds described by James. After having listed several worlds such as the world of sense, the world of science, the world of abstract truths, etc., James' classical description reads as follows

Every object we think of gets at last referred to one world or another, of this or of some similar list. It settles into our belief as a common-sense object, a scientific object, an abstract object, a mythological object, an object of someone's mistaken conception, or a madman's object; and it reaches this state sometimes immediately, but often only after being hustled and bandied about amongst other objects until it finds some which will tolerate its presence and stand in relations to it which nothing contradicts. The molecules and ether-waves of the scientific world, for example, simply kick the object's warmth and color out, they refuse to have any relations with them. But the world of 'idols of the tribe' stands ready to take them in. Just so the world of classic myth takes up the winged horse; the world of individual hallucination, the vision of the candle; the world of abstract truth, the proposition that justice is kingly, though no actual king be just. The various worlds themselves, however, appear (as aforesaid) to most men's minds in no very definitely conceived relation to each other, and our attention, when it turns to one, is apt to drop the others for the time being out of its account. Propositions concerning the different worlds are made from 'different points of view'; and in this more or less chaotic state the consciousness of most thinkers remains to the end. Each world *whilst it is attended to* is real after its own fashion; only the reality lapses with the attention. (James, 1890, Vol. 2, p. 293; italics in the original.)

All this seems to indicate not only that the number of representational models is relatively large, but also that they vary widely in kind and internal structure. And though some of these worlds might temporarily combine and fuse into a complex model, many are so alien to each other that they never meet. Again, as in the physical universe, though they may be internally rich and highly differentiated, world views can be quite isolated and on their own – sometimes separated from each other by staggering oceans of ignorance.

Looked upon this way, the universe of our knowledge evokes the same strange mixture of feelings of richness and emptiness with which modern cosmology inspires us.

Having indicated that there are many world views and that each of them can be quite complicated, in order not to drawn in premature complexity, it is now useful to look at the basic components of representational models in their most elementary forms. We can then add additional features step by step to reintroduce the multitude and variety which we have touched upon.

The Simplest Model of a World View

"Cognitive science is", according to Winograd

based on two assumptions: . . . 1. The human mind can be usefully studied as a *physical symbol system*. . . . 2. It is both possible and revealing to study the properties of physical symbol systems at a level of analysis abstracted from the physical details of how individual symbols and structures are embodied, and the physical mechanisms by which the processes operate on them. (Winograd, 1976, pp. 263–264; italics in the original).

The two assumptions together might seem an odd combination. The first assumption says it is useful to study the *mind as* a *physical* system and the second assumption says that the physical details are not that important. What then is the relevance of looking at the mind physically? The relevance of the approach is that it yields a relatively new concept of knowledge which applies to machines as well as to minds. Let us investigate some simple examples to illustrate how an apparently mentalistic approach such as the cognitive orientation turns out to be quite compatible with objectivistic trends.

To orthodox behaviorists, the cognitive view should still be suspect since under the scientistic cover of computer processing it brings back old-time mentalism. The world model a subject is using in an act of perception is itself not accessible to the psychologist observing the act. How else then could world models be attained than through annoyingly uncontrollable introspection?

In the thirties, S. S. Stevens (1935) enthusiastically introduced *operationism* into psychology as an objective behavioristic method to study mental processes such as perception. How could one know whether a rat observes a red light as being different from a green light? The operationist approach consists in a conditioning procedure where response R_a (e.g., lifting forelegs), is associated with what the experimenter could describe as 'observing red', since it would be coupled with a red light's being on. Similarly, response R_b (e.g., forelegs down), would be associated with 'observing green', since it would be coupled with a green light being on. The responses R_a and R_b are considered operational definitions of the unattainable mental states in a rat. A rat cannot tell us how it sees the world, but it demonstrates that in its world model it makes a distinction between red and green lights whenever, confronted with the appropriate stimuli, it produces the responses R_a and R_b in a consistent and reliable manner. In that way, discriminatory responses can be taken as indicators of mental states, i.e. of components of representational models. One might feel tempted to think that discriminatory behavior will be sufficient to investigate anything that one wants to analyze in terms of a representational model attributed to an organism or a machine. Its discriminatory responses reveal the categories of its world model. This would imply some general principle of attributing knowledge of a certain object or a certain state of affairs to a device whenever that device would discriminate between the presence and absence of that object or that state of affairs. An organism would be considered to know something whenever its behavior indicates that it takes into account the presence or absence of the crucial knowledge-item. It might seem plausible. However, K. Craik (1943, p. 58) noticed an unexpected and troubling hylozoïstic trend in this approach. Attempting to attribute mental qualities such as knowledge on a very objective basis, we can be brought to assign mental states to simple, inanimate things. If one is attributing mental states on the basis of discriminatory responses, why not apply them in the case of water that seems 'to know' how to get down from the mountain and thus discriminates between high and low? Why not attribute some elementary knowledge of geometry to the inclined plane that retains angular objects but not globular bodies and thus can distinguish cubes from spheres? To avoid this unattractive and useless proliferation of mental states, the attribution of representative models should be restricted to systems which process information and not just undergo energetic influences. When does a system process information in a strict sense?

Information processing depends on some form of *interpretation* of physical events. Water streaming down from a mountain does not decide to do so on the basis of an interpretation of the physical forces by which it is influenced. It simply undergoes these forces. Information processing requires that certain forms of energy impinging upon a system are transformed into some other form of energy which then will symbolize an event or carry information about a given state of affairs. In general, *information processing systems* should contain the following components (Amosov, 1967):

(1) one or more *transducers* which transform one form of energy (e.g. light) into another (e.g. electricity);

(2) some elementary form of a *short term memory* which can temporarily hold the transformed energy;

(3) a *long term memory* in which knowledge is built in terms of standards with which the stored inputs in the short term memory can be compared;

(4) an *output system* which can respond differently according to the presence or absence of a match between the received encoded information and the stored standards.

Any information processing system, from the simplest device to complex living organisms, should contain these basic units. One of the simplest examples is a thermostat governing a furnace. The thermostat transforms the heat into motion, usually by means of a bimetallic sensor which, through its inertia, also functions as a short term memory. The registered temperature, symbolized by the position of the bimetallic plate, is compared with a standard fixed by means of another metallic part positioned such that it would be touched by the moving bimetallic plate whenever the desired temperature is reached. The contact between the two plates closes an electric circuit, thus turning off the furnace which has been kept burning as long as the metallic parts were separated. It should be clear that the thermostat does not simply undergo the physical forces to which it is exposed. For one form of energy, thermal energy within strict limitations, it can make an *interpretation* in terms of a built in *world model*. This model is extremely simple, reducing the world to only two possible states: 'warm enough' or 'too cold', corresponding to its two possibilities for action: furnace off or furnace on. Nevertheless, from an information processing point of view, the device contains a *representational model*.[1] The position of the fixed metal plate *represents* the temperature desired, the position of the moving plate *represents* the temperature in the room. There is no intrinsic necessity to take *these* metals as representational devices, and one could think of other means, such as pressure in a closed vessel filled with gas, to register temperature. This illustrates the arbitrary relation between what is represented (in our example, temperature) and the medium by means of which it is done (expansion of a bimetal, variation of pressure in the gas). Such arbitrariness is typical for symbolic representation. The model represents, in arbitrary symbols, a world image or world view, in the sense that it denotes (in a classical term of von Uexküll),[2] the *Umwelt* of the device, i.e. those aspects of its environment to which it can react on the basis of an interpretation. In that sense, one can introduce the notion 'world view of a thermostat'. It is about the simplest form of a representational model imaginable, since it connotes only two states. Does the thermostat know its environment? Strictly speaking, from an information processing point of view, yes it *knows* — not in a metaphorical but *in a literal sense* — whether it is warm enough or too cold! Thus, knowledge is embodied in a relatively simple physical system.

At this point many readers might become suspicious of such a theory of knowledge since it still seems to attribute knowledge too easily and superficially. Many people believe with Kant, and prominent philosophers of today,[3] that 'to know' implies 'to know that one knows', and, obviously, this does not apply to the thermostat. The thermostat does not have a representational model of itself and accordingly cannot make any observation on the relations between itself (the device) and the 'outside' world it knows in terms of temperature.

However, this is not due to a deficiency in cognitive science concepts. Nothing impedes the inclusion of notions as 'self' and 'I' as representational models for certain parts of the world. For a representational system equipped with those categories, reflexive knowledge, i.e. knowing that one knows, is perfectly specifiable. To introduce such categories as parts of representational models of the world (!) might well be contrary to our common habits of thought, but it is neither mysterious nor intrinsically difficult.

'Self' and 'I' as Parts of a World View

Solso (1973, p. x) quotes an apparently simple but also puzzling definition of cognitive psychology by Naomi Weisstein. According to her, cognitive psychology should study "How something out there, gets in here, and what happens to it here, and then how it gets out there". Even when one has no

trouble imagining the gestures which obviously have to accompany the statement, the division between 'inner' and 'outer' is not as handy as it appears. It
opposes a knower to the object of his knowledge, and it suggests that the
knower has no difficulty in knowing himself — the problem being limited to
'how something out there, gets in here'. However, the distinction between
'out there' and 'in here' is itself within a representational model, and should
not be seen as a necessary prerequisite for knowledge in *any* information
processor, but only for a subclass of more complex ones.

Confusion arises here from the Cartesian dualism which still pervades
many basic concepts in psychology, despite the energetic attempts of psychologists to cut loose from such lineage. Concepts such as *I* and *self* are
considered entities to which we — somehow — have direct access by means
of some privileged observation channel called 'introspection'. Notice that
even behaviorists like Watson[4] do not deny the existence of this special form
of observation but only reject its usefulness for the purpose of scientific
investigation! In the cognitive view however, it is hard to see how any basic
distinction between two forms of observation like introspection and extro-
spection could be justified. There is no *direct perception, neither* of parts of
the *world nor* of the *self* in any of its aspects. The cognitive view holds that
all perception is mediated by models.

Our concepts of an 'outside world' and an 'inside self' are conceptual
models used to filter sensations of the same kind. Nothing justifies a special
cognitive status for a notion such as 'self'. The cognitive orientation is in this
respect quite comparable to the pre-behavioristic point of view promoted
by such authors as Wundt, Mach (see Chapter Three) and Avenarius. They
distinguished between psychology and physics, not by invoking different
kinds of experience but by referring to two different schemes for the interpretation of the *same* experiences, i.e. two different filters to apply to the
same raw material. When I reach out to touch a pot of coffee in order to find
out if the drink is still warm, the same experience might be framed in either
the expression *the coffee is hot,* or the expression *I burned my fingertips,* i.e.
two different conceptual representations of the same bodily experience. One
refers to an object in the outside world, the coffee, the other to an object
which belongs to the self, my body. The common mistake is to think of the
cognitive view as applying only to models of the external world. Representation, however, is prior to segmentation in self and world. The distinction

between self and world is a division *within* world models. The domain of the self is a conceptual model giving room to many subdomains of knowledge ranging from body image to one's social images of oneself in different roles.

Recognizing the modelling aspect of the self, one still seems left with the elusive concept of the *I*. However, in an information processing machine with recursive properties, the same segmentation[5] might be repeated in order to arrive at a higher order representational category such as the *I*. The *I* should then be considered to be in a relation to the 'self-outside world compartment' which is similar to the relation between self and outside world (i.e. being part of and at the same time being distinguished from). This brings a third party to the scene which is not unlike the Freudian *I*, confronted with the goals and ambitions embodied in the several selves as well as with opportunities offered by the outside world. This *I* can be conceived of as the executive agent of the multi-purpose business of the self, selecting and combining or changing specific world views.[6] Since the mechanism of its functioning might be relevant for understanding creativity it will be discussed in Part III. Here, it is sufficient to notice that the cognitive approach allows for higher order representational notions such as *self* and *I* and thus, in principle, reflexive knowledge or statements such as 'knowing that one knows'. The cognitive view can accommodate reflexive knowledge but it does not restrict the study of cognition to an analysis of reflexion.[7]

The categories of *self* and *I* have been introduced as rather abstract entities in order to illustrate that the basic problem in understanding knowledge is not 'how something out there gets in here' but how a part of the world develops a model of the whole in which it represents itself as differing from the world. However, to relate these concepts to the multiple world views discussed in previous sections, we need to realize that these concepts reappear in many different world views loaded with the specific prerequisite knowledge that governs these world views. Understanding ourselves as well as others mostly comes down to the identification of the particular models that govern our behavior rather than to some abstract comprehension of notions as self and I.

World Views As Social Entities

Cognitive sociologists and social psychologists have recognized prerequisite knowledge in the many rules of behavior which govern interactions between people. From a cognitive viewpoint, the concept of *role* refers to *models* which represent the knowledge specifying what a subject is expected to do in a certain class of situations. As such, a specific role with the detailed instructions and prescriptions it contains might be a representational model available to the self. When I go out dining in a restaurant the 'I' slips into the role of client, while running the available model for restaurant eating. With respect to perceptual aspects, my world model becomes sensitive to such things as free chairs and available dishes. The 'others' in my restaurant world model are reduced to the stereotypic roles of steward, waitress, cook. The restaurant model specifies the kinds of actions that are appropriate: asking for a seat, asking for some explanation with respect to an unknown dish, giving a stereotype answer to the steward's ritualistic question whether or not we liked the meal ... etc. The model specifies a limited set of possibilities, both with respect to what we should look at and what we should do. The stereotypical, and almost ritual nature of these components is, in principle, applicable to all world views. They all more or less restrict the class of things and events which one should perceive and the class of things one can say or do. The model, however, specifies several roles including indications for what the 'others' are supposed to do. 'Others' can be considered as objects in the outside world to which we assign models of the world comparable to our own. Superimposed on the basic skeleton of self versus *world* populated with *others*, world models thus specify expectations and obligations for all parts involved. The model functions only when all participants know their role and organize their behavior accordingly. World models are social entities and sharing one is a basic requirement for understanding in social interaction. Understanding the behavior of others comes down to identifying the world model that governs their behavior. Behavior for which we know no world model is like an unfamiliar language: a continuous stream of sounds which does not segment into meaningful units. Observing it is like watching unknown rites. We might be able to perceive some unitary fragments, but at the holistic level it appears a senseless combination. What, for a Catholic attending a religious ceremony, is unambiguously to be described as 'a priest celebrating Mass' is described by Sartre as 'a man drinking wine in front of women on their knees'. Here an observer deliberately chooses not to share the world view governing the behavior of the actors he is describing. Confusion and misunderstanding arise when partners invoke different models to understand the same situation. A standard procedure for producing humor is to combine plausible but incompatible world views. An example is the scene from Chaplin's 'Modern Times', in which the film's hero is arrested as the leader of a rebellious mass of people apparently because he is in the first row of a demonstration, holding a red flag in the air. The viewer has been able to learn his true intentions (i.e. the world model he is applying) in the previous scene. Observing a truck losing the flag used as a signalling device for an extra long load, the hero has taken up the flag and sweeps it in the air trying to catch the attention of the truck driver. Meanwhile, the demonstrating mass moves in from behind and, suddenly, he is in the midst of protesting people and the police look upon him as the leader. A single physical act is 'seen' very differently, depending on the point of view of the hero or the point of view of the police.

In psychology, *attribution theory* represents an approach that explains social interaction and understanding of behavior by postulating personal or environmental *dispositions* behind the behavior observed. The term disposition however refers to a purely individual and poorly structured attitudinal entity compared to the highly differentiated world models of cognitive science.[8] The rich and social nature of world views is recognized in sociology by *ethnomethodological* studies which investigate them in terms of the *formal structures* of *everyday activities*. The social character of such entities is obvious from a definition of Garfinkel and Sacks:

... by formal structures we understand everyday activities (a) in that they exhibit upon analysis the properties of uniformity, reproducibility, repetitiveness, standardization, typicality, and so on; (b) in that these properties are independent of particular production cohorts; (c) in that particular-cohort independence is a phenomenon for members' recognition; and (d) in that phenomena (a), (b), and (c) are every particular cohort's practical situated accomplishment (Garfinkel and Sacks, 1970, p. 346).

Conceived of in terms of such formal structures of everyday activities, world views are not laboriously constructed products of the individual mind. They are products cultivated by society and held available to its members. They are characteristic not of an individual but of a group, accessible to the members in the way a specific language is available to a specific language community.

The social nature of world views is not just another feature added to a long list of traits. The introduction of world views in terms of representational models that provide interpretative schemes for certain physical events might have induced the impression that, in some way, perception of the physical world is prior to communication, i.e. that communication is based on pre-established perceptual models. The way we defined 'others' as 'objects in the outside world to which we attribute models of the world comparable to our own' might exemplify such as bias. To restore the balance however, we should emphasize that the other way around is equally plausible, i.e. that communication is prior to perception of the physical world. This would make the notion of 'other' the more basic concept while bringing in the 'outside world' as a subject of communication. Developmental psychology contains sufficient materials to argue both ways.[9] The basic point for our purposes is to realize that perception is shaped by communication as much as communication is shaped by perception. In a phrase which Bohm applies to science: "perception and communication are one *whole*" (Bohm, 1974, p. 375, italics in the original). The locus of that union is the specific world view.

Combination and Interaction of World Views

When we presume that behavior is governed by such a variety of complex cognitive structures as we have suggested above, it is clear that a heavy burden is placed on a mechanism for the selection of world models. Simple observations already give an idea of the kind of complexity we have to face with respect to that problem.

When I enter a restaurant to have dinner, the scenario for restaurant eating constitutes the skeleton for my behavior. However, it does not necessarily determine my entire behavior. I can go out for dinner with colleagues and spend a substantial part of time during the meal at a conversation for which I need to invoke world models which are extraneous to the restaurant scheme. In fact, our behavior seems rarely governed by a single world model. Most of the time one world model seems to function as a host structure which can accommodate other models as guests. We have dealt with those notions in a way too sketchy for tackling this type of question now. We need more detailed descriptions of the structure of world models. Through extensive scanning of a variety of large mental entities, cognitive science should arrive at more tangible descriptions of the kind of cognitive structures that have sufficient relative autonomy to count as a world model. These studies of the scopes and forms of the world models will undoubtedly reveal structural characteristics which are of crucial importance to the question of combinability. Again, in Part III, we will confront this and other questions of combining and assembling world models with some findings and explanatory mechanisms currently available in cognitive science. There is, however, one basic aspect of the combination and interaction of world views that deserves to be mentioned at this stage of our exploration. It brings us back to the stages of information processing concepts dealt with in Chapter One and the paradox of the cognitive viewpoint.

The central claim of the cognitive view is that both perception and communication require the selection of a world model. However, one should *not reduce* the act of *perception* or comprehension to the *selection of* the appropriate *world view*! The world view is nothing more than a selectional device which, by indicating the set of possible events in a given area or situation, keeps information processing below a certain level of complexity and makes search practically feasible. Within the expectations derived from the particular world view, further analysis is required to achieve detailed understanding and to keep the continued applicability of the world view under control. Therefore, it should be emphasized that in the sequence of monadic, structural and contextual approaches, the cognitive stage is to be seen as complementing rather than eliminating less advanced stages of information processing! Successful programs in AI are not solipsistic ones which dream up their own world entirely. Rather, they combine world model knowledge with more small-scope inspection of local context, analysis of structures and identification of specific micro-units. The development of the cognitive view in AI has been a progressive line of discovery in which new *levels of analysis* have been *added* to those already in use.[10] But then, what is the status of these lower levels which we have called contextual, structural and monadic? Is there any one level at which the real outside world enters in a pure and genuine form?

Scrutinizing the different levels of analysis, we discover that each of them presupposes an a priori segmentation of the world. Even the fine-grained template matching procedure requires a grid-like structure in terms of which it is receptive for the signals coming from the world. In fact, the grid structure with the two possible states of each of its cells might well count as a highly diluted world view. Similarly, the forks, arrows, peaks and other basic combinations of line segments which are used in feature analysis might constitute together with their possible positions, another rather elementary world view. If each level involved is to be identified as a separate world view, any single act of perception requires the combination and interaction of several world views. Now, this is not just an extra complication invoked to illustrate once more the all pervasive character of world views. Such a multi-level theory of perception contains both the gist and the zest of the cognitive view. Its challenging character derives from the fact that it surpasses the prevalent belief that there is some elementary level of description for perceptual data, i.e. a type of description independent of higher order theoretical conceptions. Its power derives from its bringing into play the coherence between various levels of description as a criterion for the authenticity of perceptual knowledge. This multi-level approach means again an emphasis on the importance of the mechanisms for selecting and linking world views. Part III will reveal observations and specific explanatory concepts that cognitive science has to offer with respect to those mechanisms. To realize the promising nature of cognitive science for science studies at this point, it should be clear how much the cohesion of multiple levels of description also typifies scientific experience.

Consider a remark by Austin on an act which could exemplify a scientific observation.

Suppose that I look through a telescope and you ask me 'What do you see?'. I may answer (1) 'A bright speck'; (2) 'A star'; (3) 'Sirius'; (4) 'The image in the fourteenth mirror of the telescope'. All these answers may be perfectly correct. Have we then different senses of 'see'? *Four* different senses? Of course not. The image in the fourteenth mirror of the telescope *is* a bright speck, this bright speck *is* a star, and the star *is* Sirius; I can say, quite correctly and with no ambiguity whatever, that I see any of these (Austin, 1962, p. 99, italics in the original).

Austin goes on arguing that the choice of a specific answer depends on what various authors have called 'pragmatic context'. That should already be a familiar story for us. The point is that even within one single world view, several levels of description are retained and constitute *together* an experience reported in a single statement. Consider a comparable description by Duhem:

Enter a laboratory; approach the table crowded with an assortment of apparatus, an electric cell, silk-covered copper wire, small cups of Mercury, spools, a mirror mounted on an iron bar; the experimenter is inserting into small openings the metal ends of ebony-headed pins; the iron bar oscillates, and the mirror attached to it throws a luminous band upon a celluloid scale; the forward-backward motion of this spot enables the physicist to observe the minute oscillations of the iron bar. But ask him what he is doing. Will he answer 'I am studying the oscillations of an iron bar which carries a mirror'? No, he will say that he is measuring the electrical resistance of the spools. If you are astonished, if you ask him what his words mean, what relation they have with the phenomenon he has been observing and which you have noted at the same time as he, he will answer that your question requires a long explanation and that you should take a course in electricity. (Duhem, P., *La théorie physique*, 1914; quoted in Hanson, 1972, p. 4).

While specifying the lower level conceptualizations, Duhem's passage emphasizes the by now quite familiar role of the higher level conceptualizations. In a similar account, basically defending the same view as Duhem, Hanson puts the emphasis on the relative autonomy of the lower level observations:

Granted, in experimental situations involving titrations, or litmus paper reactions, or salivation in response to a bellring, the observer's sensation report *may* be a datum of significance. 'Red now' or 'ding-a-ling' may be observation-signals of primary importance in such contexts. The tastes of acids, the odors of gases, the textures of surfaces, the colors of fluids, the warmth of circuits, etc. – these all require normal observers, with standard sense neuro-circuitry, in order to determine which factual claims are true, and which one false. 'The observer' in these cases is no more than an animated detector; depersonalized, he is just a reticulum of signal receivers integrated with considerable mechanical efficiency and reliability. To this extent and on this account, *any* normal person could make scientifically valuable observations. The color-blind chemist needs help from someone with normal vision to complete his titration work – whether this someone be another chemist, or his six-year old son, does not matter. (Hanson, 1972, pp. 3–4, italics in the original).

Each of these passages stresses the importance of what we could call some 'vertical coherence', some mechanism which relates various levels of description which we learned to recognize successively in terms of the stages through which information processing in AI has evolved. According to the suggestions of authors such as Duhem and Hanson, the key to an explanation for science and scientific observation should be found in the understanding of the mechanisms of this vertical connection which might involve many more levels and world views than the four with which we started. Nevertheless, the figure of four stages is not entirely arbitrary.

Information Processing and Views on Science

Considering the major traditions which have dominated the study of scientific knowledge and cognitive processes in science during the last century, we discover an intriguing correspondence with the four stages of our generic scheme of Chapter One. They fit easily into the boxes of a fourth column that can be added to the scheme (see Table II).

TABLE II
Generic scheme with successive stages of information processing in AI extended to science studies

Knowledge and information	AI pattern recognition and perception	AI language processing and communication	Science studies
Monadic	Template matching	Word-to-word translation	*Positivism*
Structural	Feature analysis	Syntactical analysis	*Logical positivism*
Contextual	Context analysis	Indexical expressions	*Sciences of science*
Cognitive	Analysis by synthesis	World models	*Paradigm theory*

None of these approaches focusses on the subtle interactions which are highlighted by the Duhem and Hanson passages, but as in the development of information processing in AI, each of them contributes an additional level of analysis. Consider the apparent similarities.

Monadic Stage: Classical Positivism

Scientific truth concurs with factual knowledge. Like a collector, the scientist registers the facts, and scientific publications are to be seen as catalogues classifying the only type of self-contained units science consists of: *facts*. Commonly the name of Auguste Comte is associated with classical positivism, but Ernst Mach has been the most eloquent representative of the more critical orientation. His neutral monism reduces all science to *sensations* and concepts are only 'makeshifts', convenient groupings of sense-data which tend to occur together. It is the most fine-grained approach to science. The textbook version is to be found in Karl Pearson's *The Grammar of Science* (1892).

Structural Stage: Logical Positivism

In the first major work of the logical positivist school, Carnap's *Der Logische Aufbau der Welt* (1928), the emphasis on structure is ubiquitous. *"Science deals only with the description of structural properties of objects"* (Carnap, 1928, p. 19, italics in the original).

The similarity between the feature analysis of pattern recognition and the syntactic orientation in language processing on the one hand, and the structural description of science by logical positivism on the other hand is most clearly expressed in the two-level model that became popular during the fifties, and that is now associated with Hempel's name. It makes a distinction between the level of *empirical laws* (in a sense, generalized facts) and the level of *theories* which are to be considered as structured wholes in which the law statements function as units. Very much as the meaning of a sentence in ordinary language is sought by combining the meaning of the separate words with a syntactic description, which indicates the relations between these isolated units, the meaning of a scientific theory is sought by combining the laws in an axiomatic structure, which functions as a structural description for this higher order unit. Although logical positivism has gone through radical changes in its assumptions, in this orientation it sticks to the task Carnap had formulated in the early days of philosophy of science: the *syntactic analysis* of the language of science.

The Contextual Stage: External Factors

The relationship between structural approaches and contextual approaches can be seen in the still lively debate in the history and the sociology of science between *internalists* and *externalists*. Externalists emphasize the necessity of supplementary information about cultural, socio-economic and psychological factors in order to understand scientific development. They study how institutions might induce certain kinds of science, how career patterns influence scientific productivity, how religion affects science, etc. Merton's *Science, Technology and Society in Seventeenth Century England* (1938) exemplifies this type of study and so does the work of Ben-David, Eiduson, Roe and several others. These externalist analyses are presented as the contextual information which makes it possible to understand the 'real' meaning of a development in science. However, despite broad scope investigations and very scholarly research, one is left with the impression that, on the whole, these externalist studies have an anecdotal character. Indeed, one cannot deny that the institutional opportunities, the age of the investigator, his relationship with his parents, the general cultural climate, the geographical location and many other things, might be relevant for the understanding of a particular development in science. The problem is how to define the relevant context in each given case. The diversity of externalist approaches (e.g. psychological versus sociological), and the lively controversy between internalists and externalists indicates the same 'context boundary problem' that has been encountered in the more modest forms of information processing. It is not that context is not relevant: the problem is to delineate it and to locate it.

The Cognitive Stage: Paradigms

The 'context boundary problem' disappears when one no longer tries to locate context in the surroundings of the information, or theory, one wants to understand, but rather tries to have it supplied by the knower, or supplier, of the information as his model of his world. This is the step involved in arriving at the *paradigm*-interpretation of scientific development, promoted by Kuhn in *The Structure of Scientific Revolutions* (1962). Since its first presentation, the concept of paradigm has gone through several reworkings. It becomes increasingly clear that it is not just one key insight, an inspiring metaphor, a striking aphorism or whatever, but a very elaborate cognitive structure. In the same way that world models specify the more mundane prerequisite knowledge necessary for operating in restaurants or super-markets, the paradigm is supposed to specify the prerequisite knowledge necessary for processing the information in a given area of science.

A Prefatory Task

In Chapter One, we have been confronted with the successive discoveries of new levels of analysis in the information processing systems developed in AI. Although each new stage has initially been welcomed as the crucial one that would finally solve the basic problems, the explorations of Chapter Two have made us sensitive for a reappraisal of the lower level stages. Information processing involves a multi-level approach whereby the subtle interactions between levels constitute the heart of the matter rather than the choice of one particular level as fundamental. The bottom up approach of the first chapter is thereby complemented with the top down orientation touched upon in the second chapter. The challenge for cognitive science is to unravel this delicate mechanism of combining bottom up and top down linkages between levels of analysis.

We want to apply the cognitive approach in an attempt to explain scientific thinking. A review of the major traditions in science studies reveals that AI engineers have gone, at an accelerated pace, through a development which in fact is characteristic for the analysis of science of the last century. Positivism, logical positivism, various sciences of science and the paradigm orientation are to be seen respectively as the monadic, structural, contextual and cognitive views on science. As with the stages in the evolution of AI information processing, each of them has been welcomed as the ultimate solution in our understanding of science. However, if the cognitive analysis is going to work in this case as in any other domain of knowledge, we will need all these stages together. To understand them properly and before engaging in the study of interactions, we will introduce them in the next chapters (Three to Six) as self-contained approaches with their own problems and promises.

CHAPTER THREE

POSITIVISM AS A MONADIC VIEW

> The term 'observation' has tended to mislead rather than enlighten. – E. Gombrich, 1959, p. 121.

Now that we have a global view of the cognitive orientation, we can retrace its development more specifically in philosophy of science and science of science. This chapter and the three that follow it aim at a picture of each of the major approaches to science, which, according to the generic scheme of Table II, correspond one to one to the four stages distinguished within the development of the cognitive view. Further on (in Part III), we will argue that genuine progress in the understanding of science will involve the combination of such stages, as separate world views, in some kaleidoscopic model of science. Here, as in Chapter One, the stages are introduced more or less in their historical sequence and on their own. By presenting them as self-contained unitary approaches, we hope to recapture some flavor of the enthusiasm that accompanied the original insights and the convictions they incited. Therefore, we will focus on the germinal ideas rather than aim at historical or systematic completeness.

Positivism and Scientism

Information is, as etymology suggests, something that *fills in forms*. These forms exist prior to the data that complete them and without these forms an information processing device would have no means of registering anything. In classical epistemological literature, these entities, which we call *form* and *data* are known as *reason* and *experience*. Since the work of Kant it is generally recognized that both are important components of knowledge, though there is controversy with respect to their relative importance. *Rationalistic* epistemologies tend to emphasize the role of the forms into which the data are assimilated, while *empiricist* epistemologies tend to take the data as the most basic constituents of knowledge. The cognitive viewpoint, stressing the necessity of elaborate systems of prerequisite knowledge for high level information processing, might appear closer to rationalism than to empiricism. The data seem secondary to the conceptual models that provide meaning for them and that are somehow basic to them. However, there is an empiricist cognitivism possible — one that tries to conserve the primordial nature of the data by reducing forms and structure to residual parts of earlier data. The knowledge that we invoke to account for new data would consist of what we remember from previous encounters with similar data. Though this might seem a very plausible approach, we shall see that it includes risks that are difficult for a strict empiricism to accept. The point is that residues from previous experiences might appear to cluster in relatively stable units that, after a while, start to have a life of their own, creating tension and conflict between new data and the residual forms into which they are to be assimilated. The most radical attempt to avoid this kind of tension is classical positivism that rejects such forms as metaphysical spooks. In this chapter we shall deal with that nineteenth century positivism which unambiguously gives priority to data and which is, in its identification of knowledge with data, to be considered a monadic view on scientific knowledge. Logical positivism, or neopositivism, which is a trend in twentieth century philosophy, has language as its primary concern and sees in the structure of language a formal component that might be considered complementary to the data. Logical positivism constitutes the subject matter of Chapter Four.

Positivism did not start out as a monadic view but as a broad scope descriptive undertaking that would yield a unitary scientific view of the world. Typical representatives of this orientation, which has been called *systematic positivism*, are Auguste Comte and Herbert Spencer. For their systematic endeavors the main problem was to keep the results of vastly expanding sciences integrated into a unitary system.[1] Comte's concern over the growing specialization and division within science has a remarkably modern flavor, as do his arguments for the establishment of a science of science to counteract the tendencies towards fragmentation and divergence within science.[2]

An attempt to integrate and to unify scientific knowledge was well in line with Comte's ambition to replace metaphysics with science. Metaphysics is a branch of classical philosophy that is supposed to deal with the ultimate realities of the universe. In his well-known sequence of *theology, metaphysics* and *positive knowledge*, Comte introduced the third kind of knowledge as a superior stage that would clear the human mind from mysterious forces and magical entities as well as from the manlike gods who were thought to be behind the scenes in more primitive days.[3] Positivism thus became a *scientism*, proclaiming the superiority of scientific knowledge and natural science methods while depreciating the pseudo-knowledge associated with religious myths and speculative philosophy. The substitution of science for metaphysics involved a critical analysis of scientific knowledge motivated by the idea of proving this superiority of scientific knowledge over speculation, intuition or revelation. It is within this *critical positivism* that the monadic view developed, not as an offensive move against religion but as a very sober and ascetic attempt to establish internal purity by eradicating all metaphysical undertones from science.

Empiristic Units of Knowledge

Building blocks of knowledge had already been discussed quite extensively when scientists came to the idea of defining the units of experience for scientific knowledge. The tradition of the British empiricists from Locke on, with Berkeley, Hume and the Mills as famous representatives, had expounded a view on the structure of knowledge in which the basic units were elements of experience held together by means of a rather vague and unspecific bond: mental association. These two entities, the assembly and its constitutive elements, had been distinguished under various names and the relationship between them had been discussed extensively. The units were thought of as elementary sensations such as seeing a patch of color or feeling a drop of cold water or hearing a single sound. It had been noticed, however, that an act of perception could not be reduced to a sum of sensations. When I see a common object, I am not exposed to all the sensations it can yield. If it is an apple, I do not experience any direct sensation of its taste. If I only look at it, I do not feel the pressure of the fruit on my palm and fingers as I would were I to hold it. Even with respect to the visual information, because at any one time I am restricted to a single point of view, I see only half of the apple, the other side being covered by the side at which I am looking. Nevertheless, without any difficulty I see an apple and not half an apple. I see an object and not just a partial set of the sensations it can produce on my perceptual apparatus. Thus, even in the perception of a simple object, I 'see' more than there really is to see on the basis of sensations alone.

The higher order unit of perception was thus considered to be not a sum of the sensations present at any one time but a concept representing all sensations associated with the class of similar objects accumulated over a long time. In an act of perception, authentic sensations call up those concepts in which memories of similar and complementary sensations are stored so that, based on a partial set of sensations, the concept representing the whole is brought into consciousness. When I see an apple, a number of direct sensations calls up a whole in which all my knowledge concerning apples is stored such that, by seeing it, I also know how it tastes when I eat it, what sounds are produced when I bite into it, how it smells, etc. Hume's terminology in this respect distinguishes between *impressions* and *ideas*, and introduces the concept of *projection* as the process of their combination in perception. Impressions are to be considered elementary sensations while ideas refer to the concept in which these are remembered and organized. In an act of perception, the impressions that the object produces on the perceiver activate in him the idea, which he then projects onto the object producing the sensations. The impressions come from the object and the perceiver contributes the idea. However, this is projected onto the object so that it appears as if all attributes are directly perceived in the object. This British empiricism is already a cognitive view: in perceiving an object one does not see solely what one sees, one *sees* partly what one *knows*!

Helmholtz's Cognitive Model

It is this approach of the observation of common things that a nineteenth century positivist such as Helmholtz applied to observation in science. Even more than his British predecessors, Helmholtz promoted, as we now realize, a cognitive view of perception. His view was really cognitive because he emphasizes the *sign-character* of *sensations* (Humean impressions). The classical empiricist did not have at his disposal any detailed scientific theory of nervous functioning. It was thought that the information transmitted from the senses to the brain was more or less similar to the objects producing the sensations. It was supposed that images and colors were transmitted through nerves coming from the eyes, that sounds travelled through nerves from the ear, and that smells were sent from the nose to the brain.[4] Helmholtz, however, was acquainted with the work of his teacher, Johannes Müller, who had discovered that what is transmitted through nerves is 'specific nerve energy' (1838). Nerves signal to the brain that the senses with which they are connected have been stimulated. A blow to the eye will produce a sensation of lights similar to a flashlight, although the stimulus consists of pressure. The nerve energy is neutral with respect to the so-called quality of sensation. Nerve energy *represents* stimulation of the senses just as the movement of a bimetallic plate in a thermostat represents temperature. There is no intrinsic relationship, no affinity between the nerve impulse and the event that it represents. As expressed by Helmholtz: " . . . our sensations are, as regards their quality, only *signs* of external objects, and in no sense *images* of any degree of resemblance" (1885, p. 343). Rather than being like images, sensations are like words of a language: they can denote things without revealing the true nature of what they indicate. Sensations are nothing but labels. As with language, many different label-systems are possible. The same rays of the sun which appear as light to our eyes are experienced as warmth on our skin. Sunlight and warmth are like two words for sun, say *zon* in Dutch and *soleil* in French. Some picture (or, as we have called it, some model of reality) is arrived at however, because the relations between the signs correspond to relations between the objects that produce our sensations. The same signs appear in the same constellations each time we are confronted with the same object. The point is that although our sensations inform us on the presence of a particular object, they do not learn us what the object *really* is. Each time we see the disc of the sun and experience the pleasant warmth, we infer [5] that the object which we call *sun* is again present. But what the sun really is, we do not know. When we bring scientific instruments to bear upon the same object, quite different pictures are arrived at, again, like descriptions in still other languages. Is the real sun the deep red disc which we enjoy on a late summer sunset or is it the cruel hot gaseous balloon 150 million km away from the earth and with 1.3 million times its volume? Or is there, strictly speaking, really no sun at all? Is there anything beyond a huge collection of atoms in various combinations and densities and are familiar notions not reducible to subjective labeling systems for an underlying objective reality which science gradually reveals? Isn't it science that tells us what reality really is?

Scientistic Metaphysics

The sophistication of the Helmholtzian version of what could be called *scientistic realism* resides in its analysis of perception as a complicated chain of intricate events rather than the simple straightforward contact between an object and a knower that naive realism assumes. I know that my perception of the shiny red apple appears deceptively simple and direct. But what I in fact see is sunlight reflected by an object onto my eyes, producing the chemical reaction in the cones of my retinas. This is 'translated' into the electric pulse train that, on the way to my brain must be passed on from neuron to neuron by 'neuro transmitters'. Scientific sophistication drives object and observer apart. In between the known object and the knower are located an increasing number of mediating substances which are supposed to interact with each other in complex and specific ways so that perception appears more and more indirect and inferential. From the point of view of the knower, the Helmholtzian model is one which brings us in contact with the outside world only very indirectly. Where the real world is like a blue sky with a few clouds and birds and one or two airplanes, our perception of it is as if through an aircontroller's office where we, locked up in a half-dark room, *infer* the state of affairs in the air outside from the position of some blips and dots on a flickering radar screen. Our brain is locked up in our body and from events happening to that body it has to reconstruct a picture of the world outside.[6]

For the early positivist, the real world outside was not the world of birds and blue skies. While physiology exposed the complexity of sensory processes, thereby weakening the notion of direct contact with the outside world, physics appeared to reveal the ultimate simplicity of the outside world. It seemed to succeed in reducing all phenomena to matter and motion. Sound was nothing but the vibration of molecules. Temperature turned out to be reducible to the speed of molecular motion. Light could be discussed in terms of moving particles or of undulations produced in a medium composed of such particles. Behind the great diversity of phenomenal worlds revealed by the senses reemerged the old atomist world described by Democritus as the only solid reality:

By convention sweet is sweet, by convention bitter is bitter, by convention hot is hot,

by convention cold is cold, by convention color is color. But in reality there are atoms
and the void. That is, the objects of sense are supposed to be real and it is customary to
regard them as such, but in truth they are not. Only the atoms and the void are real
(Democritus cited by Einstein and Infeld, 1938, p. 52).

Apparently supported by scientific development, this idea became the core of
positivism's materialistic metaphysics. Under the great and confusing variety
of sensory experiences and everyday concepts, the ultimate unity was found
on the level of *particles, forces* and *distances*. Empiricism which had originated in a search for the units of knowledge in terms of the elements of
experience ended up with a doctrine on the basic units of the universe which
were located behind experience. It is no longer true that one partly sees what
one knows, one knows that reality is beyond what one sees.

From Dualism to Neutral Monism

A major problem which naive realism and scientistic metaphysics have in
common is that, either in an explicit or covered form, they are dualisms.
They presuppose two kinds of ultimate realities: on the one hand the outside
world as the collection of entities and events that constitute it, on the other
hand, the mental world as the collection of signs or symbols that are used to
reconstruct a picture of the same world for the knowing subject. Traditional
epistemologies have no trouble with such a division, and define 'true' knowledge as a mental reality supported by a relationship of correspondence with
the 'real' reality. The problem with this approach, however, is that it requires
a third party to judge whether or not this correspondence applies. This can
only be accomplished from the position of the gods and the angels! The
naivety in dualism lies in the assumption that the knowing subject is in a
position to compare 'the world as it is' with 'the world as it is known by
him'. Rather, it is evident that all he knows, is 'the world according to his
knowledge'. There is no way for the aircontroller in our radar scope example,
to go away from his screen and to leave his room through a back door in
order to have a look at the 'real' sky. If there is a real sunny sky filled with
a few scattered clouds and airplanes, he has to infer it from the flickering
patterns on the screen. Never will he be in a position to observe that sky
directly.

This radar operator analogy is an adapted version of an example used by

Karl Pearson in his *The Grammar of Science* (1892). In dealing with the question, "how close can we get to this supposed world outside ourselves" he argues that:

We are like the clerk in the central telephone exchange who cannot get nearer to his customers than his end of the telephone wires. We are indeed worse off than the clerk, for to carry out the analogy properly we must suppose *him never to have been outside the telephone exchange, never to have seen a customer or any one like a customer — in short, never, except through the telephone wire, to have come in contact with the outside universe* (1892, p. 61, italics from the original).

Pearson goes on to argue that the 'reality' at the other end of the line is constructed out of sense-impressions and that the 'outside world' is inferred and *not* observed! All that is observed are sense-impressions that are attributed to that world. What then is the use of such a notion of 'reality'? Can we not manage without such a concept? Critical positivists do away with the notion of 'outside world' as an unnecessary metaphysical construction located behind experience. Are we then locked up in ourselves and doomed to idealism or solipsism?

Many non-philosophically oriented scientists feel rather uncomfortable with philosophers taking doubts about the existence of the outside world seriously. One can maintain the idea for the sake of argument for a while although never believing earnestly that Descartes's fears might turn out to be substantiated. No one considering the idea expects to be forced to accept as a conclusion that he alone exists and that the entire world is his dream. How can positivists with their scientistic orientation be expected to take this type of thinking seriously?

Critical positivists such as Mach and Pearson, reject the notion of 'knowing subject' as some primary reality just as they reject the notion of 'outside world'.[7] Descartes's argument that there is a direct observation of the 'self' or 'I' is rejected on grounds similar to those involved in rejecting the notion of 'outside world', i.e. that observation never reaches anything other than specific sense-impressions. The notion of 'self' is not the product of observation ('self-observation' in this case), but a construction out of sense-impressions. Thus we arrive at a position even worse than Descartes's, doubting the existence of our 'selves' as much as that of the outside world. However, critical positivists have a daring way out through the most radical use of Occam's razor.[8] By doing away with both the central concepts of 'I' and 'world', they leave us with only one ultimate reality: *sensations* or *sense-impressions*. [9] This is the only type of reality accepted by critical positivists. All that exists is somehow constructed by means of sense-impressions and reducible to them. Critical positivism is a *monism*: everything is made out of the same basic substance.

It is important to notice that the monism of positivism is not obtained by unbalancing the common-sense dualism. This monism is neither idealism giving a superior status to mental realities nor materialism giving a superior status to the outside world. Positivistic monism claims to be *neutral* in this respect. Both mind and matter are secondary entities, not directly observable and derived from the same basic stuff. The segmentation between an inside world and an outside world is indeed rather vague (see Chapter Two). Pearson calls it an arbitrary distinction that is "clearly seen to be one merely of everyday practical convenience" (1892, p. 65). Mach also emphasizes, in *Analyse der Empfindungen* (1886), that when we distinguish between self-observation and observation of the world it is "not the subject matter, but the direction of our investigation" that is different in the two domains.

The Status of Concepts

By adhering strongly to the idea that "the world consists only of our sensations" (Mach, 1886, p. 12) critical positivists establish the purest form of empiricism and try to get rid of whatever substance or entity that might be postulated behind these sensations. One should unlearn the habit of asking for the objects that produce the particular sensations. The sensation is the primary thing, and the objects that we suppose to cause these sensations are not 'real'. They are only 'makeshifts', convenient groupings of sensations that tend to occur either together or in a more or less regular sequence. The concept of 'apple' designates nothing more than a group of sense-impressions. It is an abbreviation to indicate a collection of sensations that tend to co-occur: specific forms, colors, smells, tastes . . . etc. According to Mach, we should resist the tendency to think of something that transcends such groups of sensations. Mostly, "it is imagined that it is possible to substract *all* the parts and to have something still remaining. Thus naturally arises the philosophical notion, at first impressive, but subsequently recognized as monstruous, of a 'thing-in-itself' different from its 'appearance', and unknowable"

(1886, p. 6). It is this famous Kantian notion of 'thing-in-itself' that must be abolished. There is nothing to an object or a body beyond the group of sense-impressions that it represents. "All bodies are but thought-symbols for complexes of elements (complexes of sensations)" (Mach, 1886, p. 29). For positivism, the search for knowledge has too long been impeded by the search for elusive *essences* behind the *appearances* of things. The world is as it appears and should not be seen as the expression of forces or powers or whatever other ultimate entities that might be postulated behind it. This phenomenalism is indeed the most radical use of Occam's razor imaginable. Not only should one keep the number of theoretical entities invoked to explain things to a minimum, one should have no theoretical entities at all!

If this doctrine would have been applied only to concepts such as 'mind' and 'matter', it might have been relevant only to a restricted set of philosophical questions, without any deep influence on science. However, Mach and Pearson eagerly scrutinized fundamental scientific concepts as well in order to remove metaphysical undertones from notions such as *space, time, motion, mass, force* . . . etc. It is often quoted how Mach refused to accept the notion of 'atom' in physics since it was not clear to him how it could be claimed to be an economical representation of a specific group of sense-impressions.

Scientific Concepts: Clusters of Monads

According to critical positivists, science is an economic description, not of the outside world but of the only reality which exists: the world of the sense-impressions. Scientific concepts are handy representational units that make it possible to cover the greatest possible number of sense-impressions with the least possible intellectual effort. Mach and Pearson have been criticized for invoking this principle of self-organizing economy in conceptual systems without making any attempt to explain it. According to their own principles, such a criticism is unjustified. Science does not have to explain, it only groups sense-impressions by way of summary. "Science for the past is a description, for the future a belief; it is not, and has never been, an explanation, if by this word is meant that science shows the *necessity* of any sequence of perceptions" (Pearson, 1892, p. 113). Explanations which attempt to conceptualize events in terms of *causes* and *effects* are again a relapse into metaphysics. An explanation of the phenomena is looked for in terms of interaction between supposed entities behind these phenomena. Though entities such as *forces*, *atoms*, *ether*, etc. might seem very scientific to the layman, in so far as they are considered ultimate realities behind the phenomena, they are as obscurantist as black magic rituals. A search for any kind of explanation is doomed to introduce sooner or later metaphysical entities of this kind. Therefore "strong in her power of describing *how* changes take place, science can well afford to neglect the *why*" (Pearson, 1892, p. 303). Restricting science to description and prediction, the critical positivist tries to reduce scientific method to the study of *correlations* between the phenomena. For the student of statistics, this will not be a surprise since he knows that Karl Pearson authored not only *The Grammar of Science*, but also some well known statistical measures that are still widely used. In principle, all theoretical scientific concepts and scientific statements such as laws should be reducible to provisory clusters of phenomena and to provisory associations among such clusters.[10]

The only kind of ultimate realities are the phenomena themselves. It is this basic tenet "that the universe is a sum of phenomena" (Pearson, 1892, p. 173) which makes critical positivism fit neatly into the category of monadic views. Science is nothing more than the most convenient classificatory device discovered up to now, to handle these sensations. It is a file system for the abundant number of minuscule self-contained elements which each on their own constitute a precious little fact and all together add up to our 'outside world', our 'self', everything.

In this high resolution approach to knowledge in which all conceptual models are dissolved, the metaphysical unity of earlier positivism also vanishes. There is no 'intrinsic unity' in a universe where each phenomenon exists on its own, independent of the others. In a chapter on the reception of his views added to the fifth edition of his *The Analysis of Sensations*, Mach indicates "To many readers the universe, as conceived by me, seems to be a chaos, a hopelessly tangled web of elements" (p. 365). Indeed, his views are comparable to Russell's who contends: "Of unity, however vague, however tenuous, I see no evidence in modern science considered as a metaphysic" (Russell, 1931, p. 97) and further "I think that the external world may be an illusion, but if it exists, it consists of events, short, small and haphazard. Order, unity, and continuity are human inventions just as truly as are catalogues and encyclopedias" (p. 98). Though Russell can hardly be considered a pure member of the positivist school, his analyses of neutral monism are highly compatible with Mach's and Pearson's. If critical positivism promotes the idea of a unity of science, it is for the sake of 'economy of thought', not for the sake of scientistic metaphysics.

Assessing the contribution of critical positivism to the advancement of science is difficult. However, considering the doctrines of methodology and the practice of research in various domains of the social sciences, one cannot overlook the substantial influence of methods and attitudes in the spirit of *The Grammar of Science*. The idea that scientific knowledge can be derived from the accurate application of statistical methods on carefully collected massive amounts of data is entirely in line with Pearson's philosophy of science. It is a widespread belief that expresses the high esteem for facts, not for the single conspicuous fact that makes one ponder, but for the common modest facts which count because of their number.

A single conspicuous fact is that Pearson called his major book a *grammar*. Logical positivism developed on the conviction that there is more to grammar than Pearson was willing to consider.

CHAPTER FOUR

LOGICAL POSITIVISM: A STRUCTURAL VIEW

> Philosophy of science. . . is immune to the
> vicissitudes of science — the coming and going
> of particular theories; for those changes have
> to do with the content of science, whereas the
> philosopher is concerned with its structure
> . . . . — D. Shapere, 1965, p. 9.

Some logical positivists have felt uneasy with the name acquired by their movement. They would have preferred the more neutral but less popular label *logical empiricism*. Nineteenth century positivism had been thriving on the enthusiasm of a number of vastly expanding empirical sciences culminating in the triumphant impression around 1900 that the science of physics was almost accomplished. It might have been this over-confidence which later on produced uneasiness with the term 'positivism'. Indeed, logical positivism which developed mainly between 1920 and 1935, originated as a careful attempt to account for and one might even say, to recover from shocking experiences produced by revolutions in physics (Einstein's relativity theory). Reassurance was sought in the use of promising conceptual tools of formal language that had been developed previously in logic and mathematics.

Undoubtedly, a concern with language — formal language especially but also natural language — is the outstanding feature that distinguishes logical positivism from earlier positivism. Language is seen as contributing an autonomous structural component to knowledge. The nineteenth century positivists stuck to the idea that knowledge is nothing more than the collection of data such as those that in the previous section were dealt with under various names as *sensations, sense-impressions, phenomena, data, sense-data* . . . etc. Scientific concepts and laws were explained as psychological realities in terms of groupings of such data at various levels. This psychologism does not appeal to logical positivists who consider it a cheap way to do away with an autonomous formal component of knowledge. Although they recognize, as empiricists, that sense-data should constitute the content of knowledge, they do not accept the notion that these units organize into groups and condense spontaneously into forms that can be readily expressed in language. Language is not simply the reflection of a mysteriously self-organizing bunch of sense-data. It is a separate and basic component of knowledge contributing the forms in which the data are to be collected and handled. Language is a primary aspect of knowledge and is on a par with data.

The relevance of language is all the more important since it is the only way to share knowledge. Although logical positivists are convinced of the idea that all knowledge should be based on sense-data, this does not prevent them from thinking that these sense-data are private entities, the content of which is inaccessible to anyone but the subject that has the experience of them. If we acquire knowledge from others, we receive it in terms of statements. The actual sense-data on which it goes back cannot be transmitted. They constitute personal experiences of which only the description can be made available to others. Therefore, knowledge which we share with others critically depends on what can be conveyed by means of language.

As indicated, this focus on language would not imply a substantial change in approach if it only means that the earlier units of analysis — sense data — are reintroduced in terms of statements. Language is not just a bunch of statements each referring to only one sense-datum. Language is a complex system of statements, and it is in the structure of language that logical positivists seek the principles of grouping and organization of sense-data.

The Structure of Natural Language

The study of language was not initiated by the logical positivists. In the first years of this century, the British philosopher Moore promoted language analysis by pointing out that some difficult problems in philosophy might be difficult because of their phrasing rather than because of intrinsic complexities. He suggested that before taking a philosophical question seriously, one should first determine whether the problem was not due to language itself rather than to entities or events described by language. Two students of Moore, Bertrand Russell and Ludwig Wittgenstein applied this advice to particular questions and also attempted to delineate in an exact way the possibilities as well as the limits of language in general. Since they included with much zeal the language of mathematics and the language of empirical science in their analyses, their work became a jumping off place for the logical positivist endeavor.

One set of typical problems that intrigued Bertrand Russell was *paradoxes*. Where some people think of these as innocent jokes without any importance, Russell took them very seriously and attempted to point out how an apparent correct line of reasoning could lead to annoying and unacceptable results. The most famous example is probably the *liar-paradox* which reads as follows: A Cretan says: 'All Cretans are liars'. Since he is a Cretan, the statement applies to himself as well as to his fellow countrymen. If he is a liar, then what he says is not true since liars do not tell the truth. In that case, the Cretans are not liars, contrary to what our Cretan speaker contends. However, if Cretans are not liars, then what he says should be true since after all, he is a Cretan. But then again, if what he says is the truth, he himself should be lying. . . . Again and again, we find ourselves trapped in a series of seemingly correct deductions which nevertheless lead to a contradictory conclusion.

The first point to be noticed with respect to this and similar paradoxes is that *statements* do indeed *form systems*. Starting with an acceptable sentence, we add sentences that appear to be more or less specifications of the original statement or compatible additions to it. Then, we somehow realize that in spite of apparent compatibility, the sentences do not combine into a coherent system. There are supra-sentential units into which some statements fit and other statements do not fit. In paradoxes misfit seems to be inherent in the way the supra-sentential unit is constructed. This is our second point.

Paradoxes reveal a quality of natural language that, though the label indicates a positive connotation, might be responsible for much confusion. This quality is what has been called the *universality* of natural language. This refers to the possibility of talking about everything — including language itself — in natural language. To use a well-known example from Tarski, in natural language, I can say *snow is white* with the intention of pointing out to the hearer something about an object that we call 'snow'. I can also say *'snow is white' is an English sentence*. Then, I mean to say something about a linguistic object. I am talking about a sentence in the language I am using. I could equally well say *'snow is black' is an English sentence*, without changing the basic content, since the point of the sentence is about the language used. However, when I say *'snow is white' is true* I am making a third kind of statement that informs us about a relationship between a statement of the language and a state of affairs in the world of objects. In natural language, we use these three kinds of statements in a crisscross pattern without being aware of it.

When we take into account the distinctions made, it should become obvious that the liar-paradox is a rather complex form of language use. When saying of someone that he is a liar, we now realize that we indicate the relationship between the statements he is producing and the objects in the world he pretends to describe. Another point is that, as a Cretan, talking about what Cretans say in general, the speaker in our paradox is talking about language in the same language that he is referring to. Russell's diagnosis of paradoxes was that they might occur whenever a statement was referring to a set in which it was itself an element. The specific theoretical solution proposed by Russell need not concern us here. The important point to note is that the paradox could indeed be prevented by keeping different kinds of statements apart. The statement of the Cretan about all statements of Cretans should not be seen as subjected to its own content, the statement is at a different level, and does not belong to the class it is describing. While this separation is made possible by introducing special and separate languages (e.g. *meta-language* to talk about language), the universality of natural language permits confusions, such as paradoxes, to arise.

The example of the paradox illustrates that, although natural language is a system, it is one that can readily be misused in such a way as to produce pseudoproblems: problems due to confused language. This observation has developed into a strong conviction in logical positivism. Besides its universality, natural language has been considered to harbor other ambivalent qualities. It has been blamed for *reification* (assumption that there is a nonverbal reality independent of the language), for *vagueness* (indeterminacy of usage); and for *ambiguity* (different meanings for the same sentence in different contexts) (Mandler and Kessen, 1959). Alone or in combination, these qualities are thought to constitute the mechanism of a number of traps that lure the language user into apparent profundity while in fact he is the victim of the looseness of his language. The idea grows that the language of science and the language of philosophy have to be reconstructed in a clean and purified way, even if the price is artificiality. "Only some poets are happy about being ambiguous" say Mandler and Kessen (1959, p. 17).[1] Where the earlier positivists saw the metaphysical danger in theoretical entities, it now becomes located between the lines of confused language.

The Logical Reconstruction of Language

A form of puzzling propositions that particularly irritated Russell were
statements about non-existent things such as mermaids or dragons. Even
when one says *mermaids do not exist*, merely talking about them provides
these objects with some form of existence. They exist as possible or imaginary
objects, i.e. as mental entities. Should one make an inventory of different
regions of reality each populated with their own kinds of objects? This might
seem a regression into medieval metaphysics. Russell's solution is a proposal
to focus on language and to rephrase such troubling statements into a series
of statements that no longer refer to dubious realities. A proposition such as
there does not exist a fire resistent salamander should be rephrased in such
terms that for the statement *X is fire resistent* no name of any particular
animal can be substituted for *X* and make the statement true. This comes
down to a decomposition of the original statement into simpler ones, *X is fire
resistent* and *X is a particular animal*, whose conjunction is negated. There is
no X such that both *X is fire resistent* and *X is a particular animal* are true.
In the same way, for mermaids it could be said that there is no *X* such that *X
has the body of a woman as upper part* and *X has the body of a fish from
the waist down*. Following the doctrine of *logical atomism* as developed by
Russell and Wittgenstein, all propositions of language can be decomposed
into a logically connected series of similar elementary propositions.[2] This
decomposition thus involves two kinds of basic constituents:

  — elementary propositions: the content of which is as simple as possible
and entirely clear to anyone who understands the language;

  — logical connectives: formal elements that specify connections between
the elementary propositions and provide a linking mechanism for the construction of complex propositions, e.g. *and* (conjunction), *or* (disjunction),
not (negation).

The central idea of logical atomism is that any proposition is, in principle,
a truth function of atomic propositions, that is, the truth of a complex
proposition depends on the truth value of the elementary propositions which
constitute it. The complex propositions *2 + 2 = 4 and Paris is a big city* is
true. This is because the logical connective *and* requires the constituents to be
true in order to form a complex proposition which is true. With the logical
connective *or*, it is sufficient to have only one constituent that is true in order

to have a true composite proposition: *2 + 2 = 5 or Paris is a big city*. However, some complex propositions are true irrespective of the truth value of their elementary constituents. A composite statement such as *Paris is a big city or Paris is not a big city* is true regardless of whether it is the first or the second constituent that is true. Since one constituent is the negation of the other, one constituent will necessarily be true as much as the other will necessarily be false. Thus, the truth of this type of statement depends on the particular combination of logical connectives, and is independent of the truth or content of the elementary propositions that are connected in this way. Such statements, which are true on the basis of their logical form, are called *tautologies* or *analytic statements*. On the same basis, statements might turn out to be necessarily false, e.g. *Paris is a big city and Paris is not a big city*. Statements that are false because of their logical form are called *contradictions*. Statements whose truth depends on an empirical basis, i.e. the truth or falsity of their content, are called *synthetic statements*. To distinguish between analytic and synthetic statements logicians make use of negation. An analytic statement, when negated, automatically becomes a contradiction whereas a synthetic statement can be true in either its affirmative or negative form: one has to check empirically what the actual state of affairs is.

Together with Whitehead, Russell had quite successfully rephrased the basic concepts of mathematics in the *Principia Mathematica* (1910) by means of this decomposition and construction method. In an incisive way, Wittgenstein in his *Tractatus Logico-Philosophicus* (1921) invoked this scheme to characterize empirical science as well as mathematics. All statements in mathematics and logic are to be considered analytic. They do not describe the world, they only constitute acceptable combinations of language units. One could look upon them as empty forms which do not convey any expressible meaning. "All the propositions of logic say the same thing, to wit nothing" (Tractatus 5.43). The use of proof and reasoning in mathematics is "merely a mechanical expedient to facilitate the recognition of tautologies in complicated cases" (Tractatus 6.1262). However, what can be said and what really conveys knowledge are true synthetic propositions. For Wittgenstein "the totality of true propositions is the whole of natural science" (Tractatus 4.11). They do not express 'what' reality is, but 'how' reality is. Should the construction of the true propositions reveal the structure of the world? While Russell and Whitehead had shown that the scheme worked for mathematics, Rudolf Carnap set out to indicate how it might apply to natural science.

The Emphasis on Structure

In *Der Logische Aufbau der Welt* (1928), his first book, Carnap indicated how for the empirical sciences a reconstruction could be worked out which would achieve what the *Principia Mathematica* had provided for the formal sciences. That construction was to be considered a *rational reconstruction* which means that it was set up as an epistemological rather than a psychological or historical endeavor. Epistemology, according to Carnap in his preface, is involved with "the reduction of cognitions to one another". Rational reconstruction is the analysis of the systematic structure of concepts and the central idea is that all concepts can be reduced to what is called the *immediately given*. However, he is convinced not only of the classical empiricist thesis but also of the by now familiar encapsulation in language: "knowledge can be present only when we designate and formulate, when a statement is rendered in words or other signs" (1928, p. 293). How to relate the basic but inexpressible 'immediately given' with the sentences of language?

The key concept in Carnap's attempt is that of *structure*. The content of experience is accessible only to the experiencing subject, hence 'subjective'. What can be objectified and expressed by means of language is the structure of experience. The concept of structure in this context should not be interpreted as the position or interrelations between elements. Carnap (1928) treats the latter as 'relation description' rather than as 'structure'. "In a structure description", according to Carnap, "only the *structure* of the relation is indicated, i.e., the totality of its formal properties" and "by formal properties of a relation, we mean those that can be formulated without reference to the meaning of the relation and the type of objects between which it holds" (1928, p. 21). They are properties such as transitivity, symmetry and reflexivity, studied in relational logic.[3] Carnap is convinced that whatever can be said objectively — whatever science deals with — is concerned with these kinds of structural properties, the content of specific elements and relations being ultimately subjective. It is this concept of structure that should be kept in mind when analyzing Carnap's main thesis that *"science deals only with the description of structural properties of objects*" (Carnap, 1928, p. 19, italics in the original). It refers to that component of knowledge which is, although induced by the object, a product of the mechanisms of logic embodied in the language system. The 'immediately given' enters in the structural framework of science at the bottom. "Since . . . each statement of science is at bottom a statement about relations that hold between elementary experiences, it follows that each substantive (i.e., not purely formal) insight goes back to experience" (Carnap, 1928, p. 296).

It is remarkable that a doctrine which claims to be more empiricist than rationalist needs to introduce the empirical as an inferred entity.[4] The elementary experiences seem evanescent and difficult to specify. Nevertheless, in the popular interpretation of logical positivism, the hallmark of the movement has been taken to be what has been called the *verifiability criterion* of meaning.

One can indeed find in Carnap (1935) such statements as "knowledge consists of positively verified assertions" (p. 453). Carnap also affirms that "what gives theoretical meaning to a statement is not the attendant images and thoughts, but the possibility of deducing from it perceptual statements, in other words, the possibility of verification" (p. 426) This emphasis on the verification criterion implying a subclass of sentences directly linked to units of experience (data-language) should not hinder sight of the structural component that is basic to the programmatic formulation of Carnap (1928) and that makes logical positivism fundamentally different from *operationism*. The latter remains essentially a monadic view on a par with Machian positivism.

Experience provides data, and language (logic) provides structure to assimilate them. The reconstruction of empirical science should be seen as a refinement of its language, i.e. the particular statements it consists of, in order to make manifest its latent structure. This is to be achieved by decomposing the language down to the level of the most elementary statements that correspond to units of experience. The structural component, however, is as essential to knowledge as the experiences reported in the data-language.

Though logical positivism has gone through several substantial revisions and reformulations, this dual composition has remained its most stable characteristic and it still typifies the accepted version of today which has been called the *orthodox* view (Feigl, 1970) or the *received* view (Suppe, 1974).

The Logical Positivist Model of Scientific Theory

In contrast to the positivism of Mach and Pearson, logical positivism considers science not as a loose association of facts but as a well-structured system of sentences specifically linked to facts. Such a system is called a theory.

A theory has two components, the component based on language is known as the *formal system* and the component based on data is labeled the *empirical system*.

The formal system consists of a body of tautologies, expanding and formulating in many alternative forms a number of basic statements called the *axioms* of the system. A classical example is Euclidian geometry in its representation by Hilbert (1897). By means of a set of abstract elements *primitive terms*, such as point, angle, plane, congruence, etc., some of them undefined and some of them introduced by *definitions*, a number of axioms such as the famous parallel postulate can be formulated. When they are well chosen, these permit one to *derive* many other propositions, i.e. theorems, that can be made in geometry. For this derivation, one need in no way invoke empirical data or refer to objects outside the system of abstract elements. A formal system is self-contained and the derivation of propositions is a matter of internal consistency.

The empirical system connects such a formal system with a specific domain of real world objects, e.g. when geometry is used to represent star configurations or social relations between people. The primitive terms are then identified with objects, such as stars or people, and the relations within configurations of such units. The linking of a formal system to an empirical system is achieved by means of *coordinating definitions* or *rules of correspondence*. A primitive term in an empirical system interpretation is called a *construct*. To the axioms and theorems in the formal system correspond *laws* and *hypotheses* in the empirical system. Discussions of whether laws should be seen as analytic or as synthetic statements illustrate that the fit between an empirical system and a formal system is not unambiguous. Usually, the empiricist orientation of logical positivist authors is apparent from their preference for a bottom up application of this scheme. Independently of a formal system, scientific investigators are supposed to discover laws, the basic regularities in their domain as *empirical generalizations*. Only when several laws are firmly established can one look for a formal system to group and to explain the connections between these laws. In some pictorial description – for which Feigl's (1970, p. 6) diagram can count as the picture – Hempel (1952, p. 688) has highlighted this spatial metaphor of theoretical 'elevation' and observational 'soil':

A scientific theory might therefore be likened to a complex spatial network: Its terms are represented by the knots, while the threads connecting the latter correspond, in part, to the definitions and, in part, to the fundamental and derivative hypotheses included in the theory. The whole system floats, as it were, above the plane of observation and is anchored to it by rules of interpretation. These might be viewed as strings which are not part of the network but link certain parts of the latter with specific places in the plane of observation. By virtue of those interpretative connections, the network can function as a scientific theory: from certain observational data, we may ascend, via an interpretative string, to some point in the theoretical network, thence proceed, via definitions and hypotheses, to other points, from which another interpretative string permits a descent to the plane of observation.

Once established in a relative autonomy, the formal system and the empirical system might in their interaction exemplify the *hypothetic-deductive* method. The formal system invoked to investigate the coherence of the previously established laws is used to derive *theorems*. These theorems are amenable to empirical testing when rephrased in the constructs of the empirical system. Probably the nicest example of this procedure in the history of science is the discovery of the planet Neptune by Leverrier in 1846. To account for irregularities in the orbit of Uranus, a hypothetical planet was first proposed on entirely theoretical grounds, only to be discovered empirically afterwards by Galle. This discovery was considered as an almost complete verification of the Newton–Laplace gravitational astronomy.

Logical Positivist Philosophy of Science

Despite their claims that they were *not* interested in and *not* studying actual processes of scientific discovery, the logical positivists' scheme has been accepted and used to develop systematic methods for scientific investigation and theory construction. The logical positivists, however, were well aware that the two components [5] described above are not so easily distinguishable in the actual processes of research and discovery. They saw it as their task to trace them back and to delineate them neatly by an analysis of the results of empirical science. According to Carnap (1928), it often happens in scientific developments that the answers are found before the questions are well phrased. He considered it the task of philosophy of science to rephrase clearly questions as well as answers, even after answers seem already discovered. Let us analyze this task assignment somewhat more systematically.

— A basic distinction is made between what is called the *context of discovery* and the *context of justification.*[6] The philosopher of science is not interested in the actual processes of research and discovery as performed by working scientists. His interest is in the knowledge claims made and in establishing the conditions to be fulfilled for the knowledge to be tenable. He is not involved with the cognitive psychology of the scientist but with scientific knowledge represented in statements expressed by means of language.

— His object of study is the *language* of a science or more specifically, the logical analysis of that language. According to Carnap: "the function of logical analysis is to analyze all knowledge, all assertions of science and of everyday life in order to make clear the sense of each such assertion and *the connection between them*" (1935, p. 425, italics ours).

— The study of the language of a science is considered the *syntactical analysis* of the language of that science. No fundamental difference is seen between logic and grammar (1935, p. 439) and grammar is reduced to syntax: "a formal investigation. . . does not concern the sense. . . or meanings" (p. 436) and "all questions of sense having an actually logical character can be dealt with by the formal method of syntax" (p. 443).

Though it could seem farfetched, it is in fact easy to understand this orientation by seeing its similarity with a more recent trend in the study of language: Chomsky's *generative grammar* as currently studied and defended by 'interpretive semantics'. Using Winograd's (1977) characterization of the latter, one readily discovers the following parallels:

— To the distinction between context of justification and context of discovery corresponds Chomsky's well known *competence-performance* distinction. The Chomsky oriented linguist is interested in a theory of language that is independent of the psychological and social processes that are actually involved in producing or understanding language, just as his colleague in philosophy of science is interested in a theory of science that is independent of the psychological and social processes involved in producing and understanding science.

— In both logical positivist analyses of science and Chomskyan analyses of language, the emphasis is on *grammar*, considered to be the set of principles or rules according to which acceptable strings of language elements can be characterized ('generated' would be the term used by Chomskyans and 'reconstructed' the term preferred by philosophers of science).

— Both orientations tend to reduce the study of grammar to *syntax* and to make meaning subordinate to structural descriptions provided by syntax. In Chomskyan linguistics the semantic component receives its input from the syntactic one which occupies a pivotal position. And remember Carnap's emphasis on philosophy of science as the study of the formal aspects of a scientific language and his explanation of formal as referring to "logical syntax and such considerations or assertions concerning a linguistic expression as are without any reference to sense or meaning" (Carnap, 1935, p. 436).

This parallel is not by accident. It is quite clear that both logical positivism and generative grammar are inspired by the same formalist orientation in logic and mathematics which studies systems of symbols in their possible arrangements independently of what the symbols refer to[7].

In their contributions to the study and development of formal mathematical systems, both have proved suggestive and fruitful. In their application to their respective domains of natural language and empirical science, both have produced only very partial results and considerable confusion. Very few parts of generative grammars for a particular natural language have remained stable over time and very few axiomatizations of empirical sciences have come anywhere near the high standards set by logical positivism. As to the use of their products, both schools have been involved in applications to domains which they claimed, in principle, to be irrelevant and independent of their own purposes. To arrive at legitimate prescriptions for the practice of scientific research, the logical positivist 'reconstructivist' scheme has been used as a description of the actual processes involved in the construction of scientific knowledge in the same way generative grammar has been used by psycholinguists to describe the processes of production and understanding of natural language.

Logical positivism is the most prominent school in philosophy of science. More than the logical positivist program itself, it is probably the confusion between the principles of logical reconstruction and practical rules for the production of science that has made some authors feel embarassed with philosophy of science. In his stimulating *The Art of Scientific Investigation*, Beveridge (1950) does not attempt to hide his reservation: "There is a vast literature dealing with the philosophy of science and the logic of scientific method. Whether one takes up this study depends upon one's personal inclinations, but, generally speaking, it will be of little help in doing research" (p. 7). Undisguised criticism is Radnitzky's (1970) reproach that "logical empiricism has not produced any metascience at all", that "the distance from which they (logical empiricist or positivist philosophers) view science is enormous" and that "they did not study the producers of scientific knowledge, nor the production nor even the results" (p. 188). His complaint that logical empiricists are "like gardeners who for fear of the sot-weed factor do not allow flowers to grow in their garden" (p. 189) applies probably more to zealous users and appliers than to the leaders of the movement. Indeed, a number of researchers in the social sciences have attempted to live up to the high standards of formal rigor only to discover that solid science is not produced automatically through careful hard work along logical positivist lines. But whatever the results, logical positivist philosophy of science has undoubtedly contributed to the high regard in which formal theories and mathematical models are held. Criticism similar to Radnitzky's more fundamental complaint is expressed by the Lachmans in their critical review (1975) of *The Structure of Scientific Theories* edited by Suppe (1974), a book which can be considered highly representative for current philosophy of science. They conclude:

The massive defects we find are not with the book, but with the contemporary philosophy of science it reflects: a preference for debate instead of understanding, the search for unattainable absolutes, and the rejection of the sociology and psychology of science.

Let us investigate whether the inclusion of the latter kind of disciplines could lead to a better view on science.

CHAPTER FIVE

CONTEXTS OF SCIENCE: SCIENCES OF SCIENCE

> Historical research admits of no restriction. –
> Pagel, W., 1967, p. 81.

If scientific achievements were, as Einstein and Infeld (1938) suggest, comparable to mystery stories, positivism would cherish the facts and logical positivism would cultivate the plot. Besides facts and plot however, a classic mystery also contains a motive. Why does one adhere to the scientific method to study a specific problem?

Being preoccupied with the structure in some abstract and a-developmental sense, logical positivist philosophers of science are not interested in the specific conditions of the waxing and waning of specific scientific theories. Neither the reason for being scientific, nor the motive for the choice of a particular problem, are considered relevant. In the words of a former member of the logical positivist school, such a philosophy of science is compatible with a view upon science "as a vast machinery that produces correct views and opinions in a sort of vacuum – completely independent of its setting in a society, of the interests, motives, purposes of those who attend to the machine, oil it, serve it, improve it, spoil it, or neglect it" (Naess, 1972, p. 7). Philosophers of science only pretend to know what the *scientific method* is. Why that method has emerged and for what purposes it is put into use is beyond the scope of their structural approach. Such questions are the concern of empirical sciences of science which study the various *contexts* in which scientific discoveries and developments take place.

We shall start our exploration of some representative sciences of science with an analysis of a major work that seems to support the orthodox view: the contribution with which the sociologist R. K. Merton became the founder of contemporary sociology of science.

Merton's Norms of Science

Merton's first major contribution to the sociology of science refers to specific

63

sociological forces giving shape to science in a well defined period of history
and geographical location: *Science, Technology and Society in Seventeenth
Century England* (1938). In the preface, a proposal for an exact title very
judiciously reads: "some sociologically relevant aspects of certain phases of
the development of science in seventeenth century England". The sociologically relevant aspects dealt with relate *science* to two entities that have a
relative autonomy: *culture* on the one hand and *society* on the other. Culture
should mainly be seen as a mental entity, a set of ideas governing activities. It
includes according to Merton "the scheme of values, of normative principles
and ideals which serve to define the good and bad, the permissible and
forbidden, the beautiful and ugly, the sacred and profane" (1938, p. 209).
Society should be considered in terms of objective conditions and possibilities
and "refers to the pattern of interaction between persons particularly as they
may be thought of as the outcome of natural drives and conditions" (1938,
p. 208). *Civilization*, containing science as a component, is the product of
interaction between culture and society. In this section we restrict our attention to the influence of culture upon science, in the next section we shall
include the societal component in our discussion of Mertonian concepts.

Since Merton's 1938 book is inspired by Max Weber's (1905) investigations
on the sociology of religion − in particular his study on the protestant ethic
and its relation with the development of capitalism − it is useful to recall the
central claim of that influential work. In Chapter Three we studied Mach's
and Pearson's positivism as one way of relating ideas to reality. They looked
for a solution in neutral monism, arguing that both the notions of 'reality'
and 'idea' (or 'concept' for that matter) were secondary to and constructed
out of a more basic and neutral element: sensations. Other solutions with
links to positivism did not remain so neutral: e.g., marxism, which favored a
materialistic scheme. Material reality is seen as the solid real and ideas are
introduced as partial and weak reflections that have no real influence on the
course of history. Weber's (1905) work on the protestant ethic is an attempt
to show, to the contrary, that systems of ideas can be of great significance to
the material activities of people. In particular, Weber tried to illustrate how
capitalism, as a doctrine that governs economic behavior, can be seen as the
expression and extension of ascetic protestantism, a system of ideas that
seems contradictory to it. Merton, in the same spirit, attempts to prove that
the norms and values conducive to science in seventeenth century England,

equally coincide with the norms and values of the very same religion. This protestantism is driven by an ambition to find the religious truth using rational thinking combined with careful investigation to study God in his words (the bible) and even more in his works (nature). Rather than listening to Church authorities and performing prescribed rites, man should use his reason to go out in the world and engage himself in practical activities in order to discover the rules and principles foreordained by God. As basic ideas, or norms, listed by Merton, we should note:

— *rationalism*, in a theological sense rather than in an epistemological sense, i.e. the idea that religious truth can be arrived at by the individual by means of reason instead of revelation;

— *empiricism*, in the sense that careful observation and systematic investigation (i.e. observation combined with reason, Merton uses the term 'empirico-rationalism') of nature can reveal God in his works; there is a divine order to be discovered in nature;

— *ascetism*, engagement in work and worldly affairs is not oriented towards the consumption of its products; success in work indicates that a person is following his true vocation in the predetermined divine plan. His response to success should be taking on more work rather than enjoying the fruits of work. Enjoyment and pleasure are distractions that mask the true vocation.

— *utilitarism*, work and knowledge, although not for the profit of the individual who produces it, should be for the benefit of mankind, concurring with God's plan to alleviate the burdens of life for man in so far as he succeeds in discovering Him.

Max Weber used these notions in order to explain the religious background for the emerging role of the entrepreneur who engaged in work to produce more than he could possibly use or consume for himself and who reinvested his profit to have his business grow and succeed ever more. Merton used them to explain the emerging role of the natural science investigator, whose particular business, in a sense, was the discovery of the laws of nature (as embodiments of the laws of God). Merton did not restrict himself to indicating the compatibility of the basic values of ascetic protestantism with the basic values of scientific research. He made detailed analyses of shifts in occupations and found evidence for a substantial increasing interest in natural sciences (from 1600 to 1650, roughly) and a decreasing interest in clerical roles. By counting the number of ascetic protestants among the newly formed group of scientists (the members of the Royal Society) and by noticing that they constituted a majority, Merton tried to establish empirically a relationship suggested only as a plausible hypothesis on the level of general ideas. Since we are only interested here in the kind of contexts invoked to understand science and scientific development, we need not be concerned with the problem of whether these quantitative methods and data effectively confirm the hypothesis. The main point is that cultural entities considered as conglomerates of ideas, in the case mentioned: the basic values of a religion, are supposed to bear upon the practice of scientific research. More generally, this component can be construed as the *ideology* behind scientific activity if one is willing to use this concept in the broad sense of a body of general ideas governing a particular area of action. In a well-known application to modern science written in 1942, Merton has called a similar body of ideas the *ethos* of science (Merton, 1973, pp. 267–278).

The ethos of science is introduced as

that affectively toned complex of values and norms which is held to be binding on the man of science. The norms are expressed in the form of prescriptions, preferences, and permissions. They are legitimized in terms of institutional values. These imperatives, transmitted by precept and example and reenforced by sanctions, are in varying degrees internalized by the scientist, thus fashioning his scientific conscience or, if one prefers the latter-day phrase, his super-ego (1973, pp. 268–269).

Although science is now clearly detached from religion and constitutes an autonomous part of culture, one can still recognize some kinship between the basic values of ascetic protestantism and the institutional imperatives that Merton ascribes to modern science. They are:

— *universalism*, expresses the idea that the criteria of science are the same everywhere and such that anyone who is interested can apply them; they are *preestablished impersonal criteria*. Science is not bound by any link with race, nationality, religion, class, personality . . . or any form of particularism. "Objectivity precludes particularism", according to Merton (1973, p. 270). The scientific community is one of the most cosmopolitan groups in the world. It is this universalism which supports scientific *objectivity*;

— *communism*, is to be interpreted as the 'common ownership', in this case of the products of science, i.e. scientific knowledge. Scientific knowledge is published, which means that it is in principle available to everyone. Communism in this wider sense and *open communication* go together;

— *disinterestedness*, indicates that scientists do not engage in the scientific enterprise for some extraneous motive such as earning money but are driven by some *intrinsic interest* in research and discovery. Scientists derive their basic satisfactions from recognition and applause by their colleagues as is apparent from the relative ease and frequency with which they become involved in priority disputes (Merton, 1957). In general, an analysis of the reward system in science shows that scientists compete for honor;

— *organized skepticism*, denotes the permanent *critical* attitude toward proposed extensions of knowledge, whether based on old beliefs or on new results. Science scrutinizes everything thoroughly and does not recognize authorities or sacred truths as exempt from critical investigation.

In a number of papers, Merton has analyzed the specific mechanisms by which the scientific community implements and maintains these values, as, e.g., in the refereeing system for scientific publications and other institutionalized mechanisms for evaluation in science.[1] Some of them will be referred to in the chapters dealing with bibliometric aspects of scientific growth and communication in science (Chapters Seven and Eight).

It should now be clear how Merton's analysis of scientific development in seventeenth century England supports the idea of a mentality that functions as a sustaining environment for the 'scientific method'. The scientific method is not self-evidently superior for everyone who is confronted with it. Only members of certain communities with a particular orientation of mind are sensitive to its pecularities.

Intrinsic and Extrinsic Factors

The analysis of the link between the Puritan and the scientific orientation in seventeenth-century England is only one part of the Merton-thesis. The "congeniality of the Puritan and the scientific temper partly explains the increased tempo of scientific activity during the later seventeenth century" says Merton, but "by no means does it account for the particular foci of scientific and technologic investigation" (1938, p. 137). The second part of Merton's book is concerned with those factors that are assumed to constrain the scientific orientation in such a way as to permit only the investigation of a specific and selected set of problems. Granting the existence of this mentality which was conducive to science among Puritans, "which forces

guided the interests of scientists and inventors into particular channels?"
(1938, p. 137).

Studying the minutes of the Royal Society in four selected years in the
later seventeenth century, it seems quite clear to Merton that a non-negligible
part of the investigations pursued by that scientific body was directly or
indirectly related to military problems. About ten percent of eight hundred
or so projects could be classified as military technology. Similarly, the development of mining required better design and performance of pumps and this
attracted the attention of scientists, as did the development of transportation.
This latter, along with the problem of navigation at sea, stimulated those
interested in astronomy to develop a satisfactory method to determine longi-
tude. Detailed investigation in these areas convinces Merton of the importance
of such practical problems and their pull on the mind of scientists. It is not
possible to account for those facts with a model of scientific development
according to which the specific areas of research are the product of "the dis-
interested search for truth coupled with the logical concatenation of scientific
problems". Merton concludes " . . . *some* role must be accorded these factors
external to science, properly so called" (1938, p. 198, italics in the original).

The Mertonian model seems quite simple and straightforward. Given the
scientific method, the development of science depends upon the presence
of a spirituality that is conducive to scientific research (internal factor)
and upon specific technical problems of society which constitute concrete
challenges to the scientific mind (external factor). The spirituality produces
the driving force and eager attitude while the technical troubles provide
focuses to that idle zeal. It looks plausible but again (see Chapter Two), the
distinction between internal and external is not as clear as it seems.

A distinction between *internal* and *external* factors (sometimes called
intrinsic and extrinsic) in the development of science, and in the history
of ideas in general, is quite common. Higham (1962) introduces 'external'
factors in the form of events and behavior influencing thought and intellectual
endeavors, and describes 'internal' factors in terms of ideas influencing other
ideas. Applied to Merton's analysis of science in seventeenth-century England,
the Puritan spirit would indeed count as an internal factor (ideas interacting
with ideas) while economic and military problems would count as external
factors (societal events interacting with ideas). As a general trend with respect
to this distinction, Higham noticed that:

Hardly anyone today would argue the total wrong-headedness of either the internal or the external view of intellectual history. Indeed many scholars seem increasingly concerned with combining the two. The difficulties involved in any real merging of them however, are far more than technical. At bottom each approach expresses a fundamental philosophical commitment. Often accepted implicitly, one commitment or the other directs scholarship more than scholars realise. They may refuse a categorial choice; they may work under the tensions of a divided allegiance. But they can hardly serve two masters with equal loyalty. The issue lies between two ways of conceiving the human mind; and entangled in each is a divergent view of human nature (Higham, 1962, p. 84).

While Merton is equally aware of the ambivalence of the distinction, he sees no problem in 'adding up' both factors. In the 1970 preface to the reedition of his 1938 book, he twittingly deals with commentators who have signalled problems in that respect.

Commentators who concern themselves with segregating theoretical perspectives in historical sociology found them not merely oddly assorted but, on their very face, altogether contradictory. The theme of Puritanism-and-science seemed to exemplify the 'idealistic' interpretation of history in which values and ideologies expressing those values are assigned a significant role in historical development. The theme of the economic-military-scientific interplay seemed to exemplify the 'materialistic' interpretation of history in which the economic substructure determines the superstructure of which science is a part. And, as everyone knows, 'idealistic' and 'materialistic' interpretations are forever alien to one another, condemned to ceaseless contradiction and intellectual warfare.

Thus far Merton has been the devil's advocate. In his opinion, his approach and his findings are not at all troubled by the incompatibility of idealism and materialism, because ... he knows from fact: they are compatible! He counters the incompatibility-argument by calling it wrong in a straightforward manner: "what everyone should know from the history of thought is that what everyone knows often turns out not to be so at all" (1970, p. XIX). For him, his study of science in seventeenth-century England gives ample evidence of the fact that both spiritual and material factors co-determine the development of science. The problem is, however, whether a distinction between internal and external factors, or for that matter between cultural and societal influences, can be based on empirical evidence. Without challenging the validity of the empirical evidence, one can still refuse to accept the subdivision into two groups of influences whereby those stemming from ideas are considered different from those that derive from events and material states of affairs. One can argue that the ambiguities of this distinction are responsible for the kind of confusion which, also in this case, stems from an indiscriminate use of *context* as an explanatory notion. There are several deficiencies to be noted.

First, the distinction between intrinsic and extrinsic factors is *ambiguous*. Introduced in connection with the Mertonian analysis of (external) military and economic factors, one is tempted to consider the other class of influences discussed in the Merton study, i.e., cultural entities such as the Puritan orientation, as internal, or intrinsic, factors in the development of science. However, several historians of science have criticized Merton for his lack of concern with what they call the internal logic of scientific development, i.e. those structural aspects that, according to philosophers of science, are solely relevant for understanding science (see Chapter Four). What should be noticed is that *both* categories discussed by Merton are from that point of view, *external* influences on the development of science. An internal (structural) analysis would try to account for, e.g., Newton's combination of previously unconnected domains of knowledge about falling bodies and planetary motion. The puritan orientation might help to explain why one could be found engaged in the study of such problems in seventeenth-century England, but it could not be considered an explanation of the specific internal dynamics leading to Newton's discovery. Thus, both cultural and societal influences are introduced as contextual factors that are complementary to a structural component provided by the internal logic of science.[2]

Secondly, the distinction between cultural and societal factors is *arbitrary*. Justifying the distinction in terms of the difference between a constellation of ideas influencing another constellation of ideas as opposed to a material event or state of affairs influencing a constellation of ideas (Higham), presents one, in fact, with the intricacies of psycho-physical dualism. Granting that constellations of ideas might interact with each other, how should one conceive of a material event interacting with ideas? In so far as Merton includes in this category economic and military problems to account for the *specific* direction taken in scientific development, they do not restrict the physical conditions of doing science as such. They are influential only in so far as they are *perceived* as relevant problems by some members of the scientific community — either on the basis of their own interest in technology or sensitivity for problems of society, or because their peers or sponsors have pointed these out to them. A perceived technological problem, however, is not so much a plain material fact as it is a conceptual entity, and it is in the form of the latter that it might exert the influence attributed to it by Merton. The difference between cultural factors, such as the puritan attitude, and societal factors, such as the technologies of pumping or navigating, is not a difference between ideational entities and material entities representing two separate realms. Rather, it is a difference between general concepts and very specific concepts as entities from one and the same realm.

Thirdly, Merton's (1938) listing of factors is *non-exhaustive* and one can conceive of other potentially relevant contexts for science as will be illustrated by the other examples presented in this chapter.

However, Merton's own oeuvre also provides ample evidence for the multiplicity of contexts. In fact, in Merton (1977), the term context is recurrent throughout the 122 pages of an intriguing personal account of some major developments in sociology of science. One encounters terms or phrases such as "socio-historical context" (p. 5), "value-context and other cultural contexts of science (such as religious belief-systems, the idea of progress, etc.)" (p. 20), "cognitive contexts" (p. 51), "sociocultural contexts" (p. 75), "social existential contexts" (p. 76), "organizational contexts" (p. 91), "social, political, and economic contexts of science" (p. 113). A general classification of contexts into three categories: cognitive, cultural and social, disguises that rich diversity. Like pictures or sentences in AI, events in science fit into a great variety of backgrounds and there is no general theory of contexts that, in a particular case, indicates which one to choose.

The Personality of the Scientist

Consider now what is usually discussed as 'psychological context'. Unlike the sociology of science, the psychology of science does not have a senior author such as R. K. Merton who is recognized by followers as well as by adversaries as the father of the field. However, it does have available a popular theme: the study of the scientific personality. Among the psychologists who have chosen to apply their methods and concepts to science, many have focussed on the personality of the scientist — investigating what kind of person is attracted to and eventually becomes successful in the career of scientific investigator which is commonly considered a creative profession. According to the popular myth, scientists are depicted as strange, eccentric geniuses with a distracted mind and erratic behavior. Psychologists arrive at more prosaic characterizations. In his book *The Psychology of Science*, Maslow (1966) discusses science as a form of neurosis as well as a form of self-actualization. His analysis of the scientist as an insecure and anxious being "in desperate need for certainty" reminds one of Nietsche's description of the scientist as a despicable weakling. For such personalities, science is "a way of avoiding life", taking refuge in "geometrized little realms" where one is allowed to make things "predictable, controllable and safe" (Maslow, 1966, p. 21). Maslow suggests that this type of immature personality goes together with what Kuhn has called "normal science" while the mature self-actualizing or creative scientist performs "revolutionary science" (see Chapter Six). However, although (according to the Kuhnian description) most scientists could be expected to work in the normal-science-mode, psychologists have preferred to study the *creative* scientist. Using methods of clinical psychology such as projective tests and in-depth interviews on developmental history and social background, attempts have been made to arrive at a list of basic features of the creative scientist.

An example of such a listing can be found in McClelland (1962), who like Merton was greatly inspired by Weber's study of the protestant ethic and noted a frequent background of radical Protestantism in experimental physical scientists even when they were not themselves religious. Eiduson (1962) reports a list comparable to McClelland's on the basis of her own results. Barron (1969), in a confident tone, points at the impressive consistency among the results of the previous authors and of several others and reworks the list, integrating the results of his own studies on creativity (Barron 1963).

In a more recent paper, Eiduson (1973, pp. 11—12) summarizes the work of this tradition by describing scientists as

adventurers, risk-takers, independent and self-sufficient producers who are autonomous in activities, enthusiastic in regard to work, dominant and sensitive . . . , exhibit an interest in things rather than in people or personal relationships, somewhat loose controls in behavior, an acceptance of challenge, unusual drive and commitment to task, high aspirations, confidence and self-esteem.

Fisch (1977, p. 278) has pointed out some methodological problems with respect to these feature-studies. One is the absence of a control-group of non-scientists in some of them. The other is the highly biased sampling that results from focussing on creative, eminent or visible scientists. This makes it difficult to find out whether some of the characteristics found have contributed to the scientist's becoming eminent or are a consequence of his becoming eminent. It could be that autonomy and resistance to group pressures is a characteristic forced upon eminent scientists by their acquired status while group loyalty might characterize the majority of less eminent scientists. A differentiation along these lines could explain the otherwise contradictory finding of both group loyalty and resistance to group pressures. However, even if these methodological precautions were taken care of and if the consistency in findings over studies and techniques were as remarkably high as the authors report, one would still feel confronted with the question how this type of knowledge could contribute to our understanding of science and scientific development.

It should be clear that the characteristics attributed to the scientific personality are very similar in nature to the cultural values which, according to the Mertonian scheme, find an outlet in science. Just as Merton studied how science could develop and grow within a group of people sharing certain values, psychologists study how science can develop within individuals possessing certain psychological traits. Interestingly, the distinction between internal and external is also invoked here to locate psychological factors. Marshall Bush includes in his discussion of "the internal context of discovery" a discussion of the personality profile, motivational base, unconscious factors, etc. (Bush, 1959). For Eiduson and Beckman (1973, p. 631) 'internal' is identical to 'psychological' while 'external' refers to sociocultural or environmental factors. In this way, Merton's values as sociocultural entities turn out to be external, which again illustrates the looseness of this distinction. The internal factors are variously called: interests, attitudes, values, predispositions, capacities, needs, drives . . . etc.

Psychologists complement the clinical instruments they use for the study of these entities with data on birth order, sibling structure and other elements of developmental history that might lead to the detection of precursors of internal factors. Are these external factors? Should we distinguish between external factors that merely reflect internal ones and external factors of another nature which are to be combined with internal ones in order to explain the specific behavior of a scientist? What is the status in this respect of another popular theme in psychology of science: the relationship between age and scientific productivity?[3] Is age an indicator of external restrictions imposed by the physical capacities of the body or is it coupled with a developmental pattern that affects the internal factors? Again, as in sociology of science, we are confronted with a multitude of aspects that are clearly linked, in some way or another, with performance in science. But, somehow, the conceptual scheme for integrating them into a coherent model fails. The distinction between internal and external factors, which seems so obvious and natural at first sight is, in fact, deeply confusing.

Knowledge of the personality profile[4] of the scientist does not provide much explanatory power for our understanding of particular developments in science. One could consider it a version of McClelland's achievement motive which is itself a psychological counterpart of Max Weber's sociological thesis. Maybe it could account for the flourishing of scientific research in countries that are believed to stress achievement. It might thus be supposed to provide a receptive environment for 'scientific' personalities. But how could this kind of general profile be used to gain understanding of an event such as the Copernican Revolution or the discovery of the structure of DNA? In order to explain specific events, we need an understanding of specific personalities. We need models that can account for Copernicus' basic interest in mathematical harmony and structural elegance[5] but also for Kepler's interest in quantitative detail and precision, thus combining personality profiles that are quite opposed on some dimension. As long as the psychology of science is based on an undifferentiated personality profile of the scientist, all it can do is explore the many irrational contexts in which this "monster of rational perfection"[6] can evolve. The negative tone of Fisch's (1977) review of psychology of science seems quite appropriate with respect to studies which have that kind of focus. He points out that: "lacking integration, substantive research in the field has been spasmodic, discontinuous and fragmentary, largely bereft of any cohesive concepts or systematic pursuit of questions and methodologies" (p. 277), ending the review by indicating that "it is lamentably clear that basic concepts are diffuse and contradictory, and rarely become common to several investigations" and that "little scholarly cumulation has resulted" (p. 298). However, Fisch is not totally pessimistic and looks for a solution in terms of a cross-disciplinary integration of psychology of science "with its sociological, historical and philosophical counterparts" (p. 299). Before considering such a multidisciplinary conglomerate, we should investigate what history of science can contribute.

The Multiplicity of Arguments in the History of Science

According to the historian J. K. Fairbank: [7] "by the time you can study it, almost everything is history". When confronted with a representative sample of studies in the history of science one feels compelled to agree with a slightly different but equally general statement, i.e. that everything can be history. The multitude of factors and phenomena invoked to explain steps in discoveries or developments in debates is very great and varies widely from one instance to another. Apparently, scholars, like detectives, have to have an eye for everything, and a special sense for the relevant detail that 'solves' a case. Unlike in general history, there is no privileged area that is considered of primordial relevance for the subject. In general history, national political history is more or less the major framework to which history of education, economy, religion, etc., . . . is hooked. In the history of science, the history of religion is considered relevant at one point; at another time, the history of technology is used; at yet another point, peculiar characteristics of a single person might be invoked and scrutinized. Scholarly research in this area seems peculiarly vulnerable to the problem of 'boundarylessness'. There is always the possibility of finding more potentially relevant material, the possibility of consulting more sources. This is almost typical for the historical method especially when one accepts Cochran's statement that "working from circumstantial inference is a necessary road to new historical ideas, and searching for more evidence to support tentative conclusions is one of the chief sources of historical growth" (1970, p. 182). Consider the following example. In a review of Merton's study of science in the 17th century, Cohen (1973) criticizes his quoting Newton as being partly motivated by potential practical applications of his investigations, although there is a passage in Newton's *Principia* where Newton himself refers to this potential application. Cohen however has found that a Scottish mathematician, John Craig, has made a note in his own copy of Newton's book, mentioning that that specific passage resulted from his questioning Newton on practical applications during a private conversation. Should one consult all copies of all books one studies in order to read what people write in the margin? Cohen claims that "this

example well illustrates the fundamental problem of making precise the
motivation of any scientist, particularly in terms of chance expressions concerning utility" (p. 119). Besides that, the example illustrates even more the
fundamental problem of an endeavor that remains without clear delimitation
of the erudition required to participate in it on an acceptable level. In no
way is this meant to ridicule the scholarship and the ingenuity of the many
analyses that have been done to clarify our understanding of developments in
science. The only purpose is to point out the inherent difficulty in studying
context as it is exemplified in history of science.

The Multiplicity of Scopes in History of Science

The multiplicity of sources [8] and arguments does not necessarily imply that
history of science is a piecemeal undertaking devoid of any wide-scope goal.
History of science research has been practiced for several different and, at
times, contradictory purposes, and it is worthwhile to pay as much attention
to the multiplicity of its contexts of use as to the multiplicity in its methods
and arguments.

Few scientists are deeply interested in the history of their field. Somehow
it appears as if contributing to a field and studying its past are incompatible
activities. Kemble (1966, p. 2) says

... most scientists have little enthusiasm for history. The job of the scientist is to make
discoveries and develop new ideas. When new ideas prove effective they tend to make
old ones obsolete: thus it is not altogether unfair to say that the typical scientist is in the
business of creating the future and destroying the past.

If they employ it at all, most scientists restrict their use of history to a ritual
historical introduction for a mainly a-historical treatise. Obsolete theories are
presented as curiosities which, with their wildly speculative nature, contrast
well with the carefully and empirically elaborated opinions of today. In this
way a neurologist might start an authoritative encyclopedia article on the
brain by referring to the Aristotelian view which attributed to this organ the
role of cooling the blood. Writing on nerves, he might point to Descartes
who considered the nerve-system to consist of hollow tubes through which
travelled 'animal spirits'. Such a ritual is little more than a rhetorical trick
intended to capture the benevolent attention of the audience by telling it

first an amusing story before introducing the serious and solid concepts of science.

Such an attitude combines with a *discontinuity*-theory with regard to the origin of science. It is compatible with a positivist view in which science is seen as the final stage in a long and laborious historical process of combatting and overcoming superstition and prejudice that accompany magic and religion. One should think of Comte's succession of theological, metaphysical and positive (scientific) thinking or magic, religion and science (in terms of the stages proposed by Frazer), to account for the evolution of thinking in different societies. These are views that allow for history and development leading up to science, but not within science itself. They might come up with developmental principles such as the famous idea that a shift from polytheism to monotheism is a step in the direction of scientific thinking. However, science itself remaining basically different because, compared to these unstable systems of pseudo-knowledge, it is characterized by ultimate solidity and stability.

There is a pedagogical use of history of science by teachers of science. Rather than being based on a discontinuity-theory, it stems from the idea that the scientific concepts and theories of our days have been achieved gradually in a continuous development of both prescientific and scientific notions. Obviously, this orientation does not make a sharp distinction between scientific and prescientific knowledge, and it allows for development to occur within science. The teacher presupposes that the now obsolete ideas were quite reasonable intuitions when they were first introduced, and he sees contemporary ideas as corrections and modifications induced by the strenuous and systematic research of later generations of scientists. This pedagogical use of history is quite in line with the *continuity-theory* which has been elaborated especially with respect to the Middle Ages. Contrary to the widely held opinion of nineteenth-century historians of science, this orientation acknowledges substantial medieval contributions to science and does not accept a black and white picture in which this period is seen as a dark age in which any possible development of science has been suppressed by religion.

Continuity versus discontinuity is a major issue in history of science. A related broad scope issue is the autonomy of scientific development. Jean Piaget (whose work is discussed in Chapter Twelve) fits in a French-speaking tradition which tried to understand the dynamics of the laws of thought by studying the history of science. In his inaugural address as a professor in philosophy of science (1925), referring to M. Brunschvicg, E. Meyerson and A. Reymond, he emphasizes that "la vraie méthode philosophique est aujourd'hui la méthode historico-critique" (p. 197). These are all authors who use history of science analyses to discuss epistemological questions and it is interesting to notice that Piaget started out teaching philosophy of science (Neuchâtel, 1925) and history of scientific thought (Genève, 1929) before accepting assignments for teaching psychology.

If Piaget can be considered a representative of a group which looks for insight into the structure and development of cognitive mechanisms by studying history of science, J. Bernal (1957) represents a marxist school of thought which sees science as somehow related to the aims of society at large. His use of history of science is to argue that the scientific achievements of a given society reflect the socio-economic structure of that society so that the development of science is not so much in terms of discovering new ways of thinking as it is in terms of achieving new social goals and ambitions.

The issues around continuity versus discontinuity and autonomy versus social servitude illustrate attempts to arrive at a global view in history of science. However, no broad scope framework has been generally accepted as the backbone of the field. As with sociology of science and psychology of science, we have to recognize again that the multiplicity of scopes and, for history of science in particular, the multiplicity of arguments, only indicate the basic ambiguity of the notion 'historical context'. Has an integration into a multi-disciplinary conglomerate such as 'science of science' proved more promising?

The Science of Science

The idea of a *science of science* has been brought up several times and is revived periodically. In the first pages of Comte's *Cours de philosophie positive* (1830) one can already find arguments and proposals for a science of science. In order to counteract the inevitable fragmentation of knowledge that inheres in the progress of science with its continued creation of specialities, Comte argues that it will be necessary to develop a meta-specialty: a specialty for the study of the specificity of each discipline and for the integration of their results.[9] Comte does not make a distinction between scientific knowledge as a product and the activities and influences involved in the production process. For some, philosophy of science could be considered the discipline that fulfills the coordinating and integrative functions Comte had in mind.[10] However, quite apart from philosophy of science, a century after Comte wrote his famous course, the notion of a science of science came also in discussion as a potential instrument for science policy. In current proposals for science of science, the two components of 'science-knowledge' and science policy can be recognized in the rather large number of subdisciplines that are invoked, as e.g. Walentynowicz (1975). For the purposes of our argument, we can restrict the discussion to the subdisciplines proposed in a well-known paper of Ossowska and Ossowski (1936) instead of Walentynowicz's almost twenty specialties.[11] The Ossowska and Ossowski article, entitled 'The Science of Science', propounds, after a discussion of some alternative subdivisions, the following components: philosophy of science, history of science, sociology of science, psychology of science and organization of science. What they label the 'organization of science' has since then developed into a thriving part of sociology of science, so that the proposal really comes down to the integration of the disciplines we have been discussing in this chapter together with philosophy of science discussed in the preceding chapter. Have attempts at integration of these disciplines opened new perspectives and possibilities for the study of science that none of these disciplines could ever have produced on its own?

There has been no lack of, at times even quite enthusing programmatic proposals for the development of integrated approaches to the study of science. The psychologist S. S. Stevens published, in the same year as Ossowska and Ossowski, an article in *Philosophy of Science* on 'Psychology: The Propaedeutic Science'. Very ambitiously, it argued that since psychology should provide understanding of the human mind, it should also include understanding of the content of the scientist's mind. With another paper, entitled 'Psychology and the Science of Science', it supported the claim that a science of science should be developed out of psychology by means of an integration with philosophy of science. But with respect to these papers, B. F. Singer (1971) has felt obliged to notice "Some 30 years have passed, and we do not as yet have a developed, self-conscious discipline of a science of science". Indeed, others who have taken up the idea and who have made some detailed suggestions, such as R. I. Watson and D. T. Campbell[12] in an editorial preface to a book of selected papers of E. G. Boring (1963), have not been more successful. Neither has a book *The Science of Science*, edited by Goldsmith and Mackay (1964), been able to mobilize the truly integrating forces. D. J. de Solla Price, whose contribution shares the book's title begins: "The science of science, like history of history, is a second-order subject of first-order importance". But science of science has not succeeded in becoming recognized as a first-order endeavor. Characterizations in the style of the title of Fores' (1977) paper 'Science of Science, a Considerable Fraud' indicate an embarassement experienced by several authors, including disappointed science policy officials who want to inspire their decisions on more solid knowledge. The most heroic attempt to remedy this deplorable state is undoubtedly that of Mitroff and Kilmann (1977).[13] Their method is almost one of brute force, using a four by four matrix to consider all possible combinations of the four disciplines dealt with above. A discussion of the integrative systems models based on the results of that analysis is beyond the scope of this chapter. However, it is important to note their remarks with respect to what is troubling the science of science. According to Mitroff and Kilmann: " . . . for the most part, the major traditions of science studies have tried to isolate and to study the system of science in a reductionistic and piecemeal fashion" (p. 125). This is in line with the argument we have been following throughout this chapter. The majority of contributions to history of science, sociology of science and psychology of science is oriented toward the isolation of one or a few factors that are expected to be somehow related to science and scientific progress. Rather than aiming at the establishment of 'universal' statements (law-like), research aims at verifying 'existential' hypotheses showing that there is at least one case in which a certain factor, such as age or religion or sibling structure, early childhood, funding. . . etc, can be proven to be related to developments in science. Untroubled by the vagueness and ambiguity of the notion of 'context', these cases and factors are considered to be illustrations of relevance of all types of contexts: the socio-economic context, the psychological context, the historical context . . . etc. However, 'context' is a term that is easily invoked but rarely explained. And even if it were a clear notion, one would still need a model of the ways in which contexts interact and combine. This is not offered anywhere in the science of science. The distinction between 'internal' contexts and 'external' contexts is only a pseudo-solution to this problem because of the arbitrariness of the distinction between internal and external. Mitroff and Kilmann's paper has as heading a quotation of Ziman that stresses the need for a unifying principle in the study of science. The final lines of the quote read *"The problem has been to discover a unifying principle for Science in all its aspects.* . . . Before one can distinguish separately the philosophical, psychological, or sociological dimensions of Science, one must have succeeded in characterizing it as a whole" (Ziman, 1968; emphasis added by Mitroff and Kilmann). We wholeheartedly agree with Mitroff and Kilmann that "the understanding of science as a total system demands that we understand how the historical, philosophical, psychological, and sociological elements of science all exist as well as act in simultaneous conjunction with one another, not in isolation" (p. 113). The problem is indeed one of specifying the parts in their mutual relations to each other in such a way as to see the whole. Contemporary science of science does not provide that kind of integrative view. It might well be described with the same terms with which Copernicus depicted the astronomical systems of his time " . . . it is as though an artist were to gather the hands, feet, head and other members for his own images from diverse models, each part excellently drawn, but not related to a single body, and since they in no way match each other, the result would be a monster rather than a man".[14]

The view offered by T. S. Kuhn (1962) does not respect the traditional boundaries between the sciences of science. In the introductory chapter, Kuhn almost apologizes for mixing up established areas in an unorthodox way, anticipating the feeling of his reader "can anything more than profound confusion be indicated by this admixture of diverse fields and concerns?" (Kuhn, 1962, p. 9). Is the Kuhnian theory offering an alternative partition that could provide a way to arrive at an integrative picture of science?

CHAPTER SIX

THE COGNITIVE VIEW ON SCIENCE: PARADIGMS

> Le critère de réussite d'une discipline scientifique est la coopération des esprits. — J. Piaget, 1965, p. 44.

Thomas Kuhn's model of scientific development is in more than one way linked with Copernicus. Called by some a Copernican revolution with respect to its own field, it also deals with the structure of scientific revolutions [1] as its major topic (Kuhn, 1962). Furthermore, the most important concepts of Kuhn (1962) are traceable to Kuhn (1957), a book that contains his detailed analysis of Copernicus' work and its impact. Therefore, it seems recommendable to start a discussion of Kuhn with an examination of his approach to that most famous event in the history of science.

An Integrated Approach to the Copernican Revolution

The popular view of the Copernican Revolution is colored by the dramatized account of its rejection and sanctioning by Church authorities. Copernicus is depicted as the keen observer and independent mind who dared to challenge the official doctrine of the Church with a theory that no longer located the earth in the center of the universe. Yet, in fact Copernicus was not bothered by Church authorities during his lifetime. Instead, they were rather favorable toward attempts to reform astronomy in the hope it might result in a better calendar and he was even encouraged by the pope to publish his views. It was only after Copernicus' book had been published and he had been dead for more than seventy years that Galileo had to face condemnation for Copernican views. A black and white picture which presents Copernicus as the heroic scientific mind who breaks away from medieval domination by the Church, thus manifesting for the first time clean 'scientific thinking', is grossly misleading. We see it as breaking away from obscure ideas inspired by religious myths and yet, Kuhn points to Copernicus' sympathy for Neoplatonic Renaissance cults of sun worship (Kuhn, 1957, p. 131). Apparently, occult ideas also played a role in this major scientific achievement.

Kuhn's study of Copernicus is an *integrated* approach: it does not make a strict distinction between the scientific aspects in terms of the conceptual structures of astronomy and the extra-scientific or contextual aspects in terms of the historical circumstances. Copernicus' scientific achievement itself is seen as an integration of various tendencies and traditions which are assimilated and recast in terms of "an apparently petty, highly technical problem" (Kuhn, 1957, p. 5) such as, e.g., the explanation of retrograde motions of planets, in a mathematically appealing way. While the notion of retrograde motion might stem from Ptolemaic astronomy, the emphasis on mathematical harmony stems from a Renaissance involvement with harmony that was alien to the Ptolemaic system. In his formulation of a problem and the requirements set out for a possible solution, the innovating scientist integrates various cultural influences to which he is exposed, some that are considered intrinsically scientific as well as some that are considered external to science. The technicality of the problem in which this integration takes place is, however, a major point in the Kuhnian approach. The Copernican Revolution cannot be reduced to the expression of a new daring idea which subsequently turns our to be true. The idea of heliocentrism had been around before. It is *the combination of various cultural influences into a well-defined technical problem* that constitutes the first stage in the construction of a major scientific revolution while its far-reaching implications for society at large become manifest only after the technical problem and its solution have been accepted, first by the community of specialists directly concerned and secondly by the specialist communities of the other affected disciplines.

In appearance, the Kuhnian approach might seem a paradoxical combination of openness and closeness. By ignoring the difference between internal and external influences or contexts,[2] a heterogeneous mixture of both rational and irrational ideas is allowed to shape the thoughts of the scientific innovator. In Copernicus' case, he is open to the qualities and problems of Ptolemy's system as a mathematical piece of work, he is sensitive to society's need for calendar reform as a matter of practical importance and he is also reported as having, according to our contemporary tastes and standards, a strange sympathy for Neoplatonic sun worship. Yet, by stressing the technicality of his integration of these influences in terms of a specific and highly technical problem, the community of persons who are in a position to judge the merits of the newly proposed system is bound to be very small. Because of its technicality, the new approach is only available and open to very few competent contemporaries. Notice, however, that in this way openness and closeness are sequenced. First, there is openness to various influences in society at large, then closeness follows as a consequence of technicality of the problem formulation.

This sequentialization is Kuhn's solution for the combination of intrinsic and extrinsic factors in scientific development. At times of openness, factors which are considered extrinsic have as much influence upon the minds of scientists as factors that are usually considered intrinsic. Once, however, these influences have been integrated into a highly technical problem representation that is considered fundamental by a small group of highly concerned scientists, a scientific area becomes closed. Its ability to be influenced by extrinsic factors is greatly reduced and its evolution is governed by factors that directly relate to the wider development and applicability of the technique that constitutes the core of the new approach. Those sequences of periods of openness followed by periods of closeness form a major idea in Kuhn (1962), directly related to the standard concepts of his approach: *scientific revolution* and *'normal' science*, which are also labels for these two kinds of episodes.

The Standard Account of the Kuhnian Model

In the previous chapter, we have pointed out that orthodox positivism and education-oriented approaches to history of science tend to differ with respect to the *discontinuity*-thesis. Positivism emphasizes the discontinuity in the development of thinking: science emerges as a new and autarchic way of cognitive functioning. Educational use of history of science is, on the contrary, based on a *continuity*-thesis, presupposing a connection between common sense and science in terms of a gradual development from the former to the latter (Conant, 1951). Positivism allows a continuity-thesis to account for progressive and cumulative development within science once it has been isolated from the contaminating influences of common sense and superstition. Kuhn, who started his work in history of science from the Conantian educational point of view has, in a way, simply reversed these positivistic notions. Starting from the continuity-thesis between common sense and science [3] he felt forced to introduce a discontinuity-theory, not to account for the emergence of science, but to account for developments within science. *Discontinuity* is a distinguishing characteristic of the Kuhnian model for the *internal* evolution of science. According to Kuhn, any mature science follows an oscillational pattern of development, alternating between the two kinds of episodes mentioned above: episodes of paradigmatic growth and episodes of revolutionary turmoil.

During an episode of paradigmatic growth, the activity of researchers is characterized as *normal science*. This kind of research is "research firmly based upon one or more past scientific achievements, achievements that some particular scientific community acknowledges for a time as supplying the foundation for its further practice" (Kuhn, 1962, p. 10). As a suggestion with respect to where to look for such achievements, Kuhn adds to this: "Today such achievements are recounted, though seldom in their original form, by science textbooks, elementary and advanced". Normal science thus defined is the application of textbook concepts and textbook cases with the prospect of extending the domain of applicability of the basic conceptual schemes and of working them out in finer detail to achieve greater precision and more accurate prediction. These particular scientific achievements, which have grown out into classical textbook cases, constitute the most restricted but also most basic meaning of the term *paradigm*. Paradigms in that sense are the standard examples which typify in the most articulate way the conceptual scheme(s) of a specific normal science episode.

In a broader sense, the term paradigm has come to indicate the whole set of shared beliefs and skills which shape the expectations of a community of scientists and establish a common directionality in their activities. It thus indicates a kind of tunnel vision typical for normal science. Normal science is in Kuhn's terms "an attempt to force nature into the preformed and relatively inflexible box that the paradigm supplies" (1962, p. 24). Such 'conceptual boxes' provide directionality, both in the way they help to ignore phenomena — "those (phenomena) that will not fit the box are not seen at all" — and in the way they yield specific expectations: "by focusing attention upon a small range of esoteric problems" (1962, p. 24). Scientific research in such a period is like solving crossword puzzles: the filling in of elements into a pre-arranged pattern according to various clues and restrictions provided by the elements already available. The general framework of the matrix, together with the list of circumlocutions, can be compared with a paradigm in the broader sense while a few anchor terms already filled in can be looked upon as paradigms or exemplars in a restricted sense. Almost literally, this metaphor applies to the search for new elements fitting the periodic table of Mendeleev (Kuhn, 1973, p. 9). Kuhn offers no model of the process of combining more general beliefs or ideas with concrete cases and applications. This process, however, constitutes the heart of the matter since it is the mechanism for the production of a paradigm. A system of boxes to be filled in during a normal science episode is apparently the outcome of a successful welding of general ideas and very specific problems.

Although we are not offered an insight into the mechanism which produces the specific orientations (the boxes) of a normal science period, we should not miss the quality of *closeness* that is a basic element in Kuhn's impressionistic description. The establishment of a paradigm in a given field goes together with the introduction of a well-defined 'search space' and the specification of permissible moves to explore that space. This is an AI way of saying that a paradigm seems to induce a *game*-like character in a given area of science. Suppose that, since war is such a horrible thing, we are interested in a scientific understanding of battles. Maybe surprisingly, but not improperly, we might introduce the game of chess as a paradigm for research in that area. Chess represents in a highly selective way some basic features of ancient warfare. However, chess is today also a closed system of possible positions and moves. Though, at some time, it might have been developed as a fairly faithful representation open to assimilate features of real armed conflicts, it is now closed and immune to developments in the 'external' world. The replacement of horses by all kinds of technical equipment and a controversy over the neutron bomb do not lead to the introduction of new pieces and newly defined moves in the game of chess. In the sense that it fixates basic units of representation (pieces) and possible operations (moves) a paradigm establishes a similar kind of closeness.

If the set of basic entities and possible operations is not indicated in a suggestive way or if there exist alternative ways to represent these basic entities, then the corresponding scientific discipline is in the state of confusion or competition typical for the *extraordinary science* which characterizes revolutionary periods. Though a paradigm for some time provides orientation and direction to a community of researchers, it cannot prevent them from accidentally hitting upon findings which do not fit the 'conceptual boxes' provided by the paradigm. As long as the paradigm-compatible findings keep accumulating, it is rather easy to ignore or suppress discordant data as irrelevant or just noisy elements. Once the majority of conceptual boxes has been filled in, the 'anomalies' — as these discordant findings are called by Kuhn — stand out in a more irritating way and may be looked upon as indications of major flaws in the paradigm itself, rather than as noise superimposed on paradigm compatible data. Once their major task turns out to involve the facing of these anomalies[4] and the problem of how to assimilate them into the paradigm, scientific communities tend to diverge and as the divergence within the community increases, the paradigm disintegrates and ultimately even for the individual researcher, the sense of common directionality disappears.

Revolutionary science applies to episodes during which such fundamental changes are made to basic concepts that they constitute a radical breakaway from the governing paradigm. Because the forging and shaping of a new paradigm goes together with a kind of conceptual turmoil and anarchy in the scientific community, the term revolution is most appropriate. According to Kuhn, paradigms do not change gradually, adapting to new facts or findings in such a way that over a longer interval of time they end up substantially modified. Paradigms are not basically modified but they are simply replaced and the replacement of one paradigm by another is indicated by a revolutionary episode which clearly separates two paradigmatic episodes of normal science.

This biphasic model of evolution in science constitutes the backbone of Kuhn's popular theory. Mature sciences are characterized by this alternation between periods dominated by rigid tradition and periods typified by a clash of traditions and revolutionary turmoil. Sciences that are immature according to the Kuhnian view have never been through a normal science episode. Their state is in some respect comparable to the revolutionary periods of mature sciences in so far as they are characterized by a similar lack of directionality among the majority of participants. Although the activity in those areas might be intense, it is of a spasmodic character. The basic openness of this period is explicitly recognized by Kuhn in the reference he makes to the possible 'external' origin of paradigms. In describing the latter as "some implicit body of intertwined theoretical and methodological belief that permits selection, evaluation and criticism", he emphasizes that "if that body of belief is not already implicit in the collection of facts — in which case more than 'mere facts' are at hand — it *must be externally supplied*, perhaps by a current metaphysic, by another science, or by personal and historical accident" (1962, pp. 16—17, italics ours). (See Figure 6.1.)

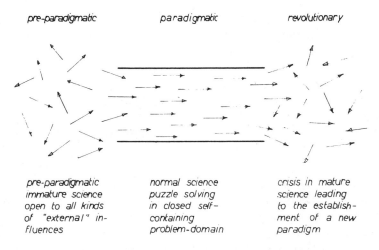

pre-paradigmatic   paradigmatic   revolutionary

| pre-paradigmatic immature science open to all kinds of "external" influences | normal science puzzle solving in closed self-containing problem-domain | crisis in mature science leading to the establishment of a new paradigm |

Fig. 6.1. Representing individual scientists as vectors, paradigmatic or normal science periods can be represented as episodes during which all participants in a given area work in the same direction while pre-paradigmatic and revolutionary periods alike can be considered episodes during which a common directionality and coordination are lacking. Some additional terms are added in the scheme to indicate further differentiations either implicitly in Kuhn (1962) or suggested by some commentators. The basic skeleton, however, remains a biphasic model in which paradigmatic and revolutionary episodes alternate. Only in non-paradigmatic periods are sciences vulnerable to 'external' influences because in fact the distinction between internal and external dissolves during such episodes. The establishment of a paradigm comes down to the establishment of a clear boundary between the real (internal) problems of the area and the irrelevant (external) questions stemming from unrelated areas which become ignored.

It should be clear that these revolutionary episodes are essential to Kuhn's discontinuity-model of scientific development since they constitute the boundaries of paradigmatic episodes. Without revolutions, paradigms would develop in a continuous way from one into another. An important part of the controversial nature of Kuhn's model stems from a coupling of this discontinuity-model to a rejection of the notion of 'scientific progress' and cumulative growth of scientific knowledge. According to Kuhn, cumulative growth and progress make sense only with respect to a given paradigm. These notions cannot be used to describe the acquisition of knowledge over a series of paradigms. Each paradigm stands on its own and a subsequent paradigm is so different from a preceding one that it cannot be judged either superior or inferior to its predecessor. Paradigms within a discipline differ from each other in so many respects that they are *incommensurable*.

For many readers of Kuhn, this relativism has been so manifestly in contradiction with the genuine sense of progress they have with respect to scientific development, that it becomes sufficient ground to reject the Kuhnian model entirely. This is unfortunate since in that way attention is also withdrawn from the very suggestive ideas which the model provides on the cognitive and social organization of science. These ideas and suggestions should be worked out and followed through in great detail before any attempt at a global evaluation of the paradigm-model should be made. We shall now focus on these ideas.

Paradigms As Cognitive Units

The cognitive orientation, as exemplified in AI, emphasizes that in almost all areas of human endeavor, perception and communication are possible only on the basis of 'conceptual models' which account for selectivity in information processing and directionality in action. The Kuhnian description of normal science indicates that scientific research does not constitute an exception. What Kuhn calls 'conceptual boxes' (when saying that normal science is "an attempt to force nature into the preformed and relatively inflexible box that the paradigm supplies", 1962, p. 24) corresponds to constituents of the internal models or world models which AI researchers have located in the brain of restaurant-diners and supermarket shoppers. Paradigms are internal models, cognitive structures which give shape to the specific expectations that guide the research of 'normal' scientists. In order to illustrate this cognitive orientation of Kuhn as well as to explore his further contribution to a more detailed understanding of cognitive mechanisms, we should analyze his suggestions somewhat more systematically.

When psychologists talk about cognitive processes, they think of the use of capacities such as *perception, memory, language, attention* and *thinking*. Though Kuhn does not deal with these separately in his 1962 monograph, there is sufficient material available to reconstruct his opinion on most of them.

Perception is a major category in the discussion of paradigm shift. A recurring theme is the notion of the *gestalt switch* which is worked out especially in Chapter 10 of Kuhn (1962) entitled: 'Revolutions as Changes of World View'. After a change of paradigms "scientists see new and different things when looking with familiar instruments in places they have looked before" (1962, p. 110) and "though the world does not change with a change of paradigm, the scientist afterward works in a different world" (1962, p. 120). Kuhn argues that "the scientist who embraces a new paradigm is like the man wearing inverted lenses. Confronting the same constellation of objects as before and knowing that he does so, he nevertheless finds them transformed through and through in many of their details" (1962, p. 121). Prototypical examples to describe such a change in perception are available in terms of ambiguous figures studied by Gestalt psychology. Kuhn alludes to phenomena such as the well-known reversible cube of Necker and very often to the duck-rabbit reversible figure that is also known from Wittgenstein's *Philosophical Investigations*. The most convenient example is probably the pelican-antelope figure which Kuhn (p. 85) quotes from Hanson (1958) (see Figure 6.2). The general idea is simple and straightforward: perception in science is like seeing the pelican in one paradigmatic episode and seeing the

Fig. 6.2. The pelican-antelope figure. One example of an assortment of ambiguous figures used by psychologists to illustrate how the same data can evoke quite distinct interpretations. (After Hanson, N. R., *Patterns of Discovery*, Cambridge University Press, Cambridge, 1958, p. 13, Figure 4; reprinted with permission.)

antelope in another paradigmatic episode. The gestalt switch shows that the world can look very different even when the data remain the same.

Language is equally important in Kuhn's treatment of paradigmatic changes and the general idea is similar to the one prevailing in the area of perception. Though the languages of different paradigms may share a common vocabulary, even while the terms are identical, the meanings are different. The term *space* is a basic term in both the Newtonian and Einsteinian paradigm. Its meaning in the latter is, however, fundamentally different from what it was in the preceding paradigm: "What had previously been meant by space was necessarily flat, homogeneous, isotropic and unaffected by the presence of matter. . . . To make the transition to Einstein's universe, the *whole conceptual web* whose strands are space, time, matter, force, and so on, *had to be shifted* and laid down again on nature whole" (1962, p. 148, italics ours). A similar argument is made with respect to the meaning of the term 'planet' before and after the Copernican Revolution (1962, p. 127).

Cognitive processes such as *attention* and *thinking* are equally affected by paradigms. Normal science, which is characterized as puzzle-solving, is a highly directed activity. "The man who is striving to solve a problem defined by existing knowledge and technique is *not . . . just looking around*" (1962, p. 95, italics ours). Nature is in Kuhn's terms "too complex and varied to be explored at random" (p. 108). Random search is typical for pre-paradigmatic research: "In the absence of a paradigm . . . all of the facts that could possibly pertain to the development of a given science are likely to seem equally relevant. As a result, early fact-gathering is a far more nearly random activity than the one that subsequent scientific development makes familiar" (p. 15). This does not mean that paradigmatic research is to be considered as a highly systematic or exhaustive search through all possibilities (p. 125). It involves "focusing attention upon a small range of relatively esoteric problems" (p. 24). Thinking in normal science episodes has much in common with the *conservative focussing* strategy discussed by Bruner *et al.* in *A Study of Thinking* (1956). A domain of possibilities is explored by means of a highly specific hypothesis about the expected solution. What is selected for inspection and what is analyzed in thought are only those features and those possibilities that are directly relevant to the temporarily sustained hypothesis.

Since they affect almost all cognitive processes, there can be no doubt about the cognitive nature of paradigms. But obviously, the qualities which we have mentioned above with respect to those cognitive processes do not prove the existence of paradigms nor do they yield an understanding of their structure and functioning. None of the aspects we have dealt with stems originally from Kuhn. The 'theory-laden' character of perception had been vividly described by Hanson (1958) whose concept of *retroduction* could also have accounted for selectivity in attention and directionality in thinking. Feyerabend had already applied a *network theory of meaning* [5] to scientific change. Kuhn's contribution lies not so much in the fact that he supports these claims and observations made by others before him. His innovation is the introduction of the term 'paradigm' to indicate a hypothetical unit accounting for all those phenomena together. This entity is not invoked as a purely hypothetical factor to explain a variety of facts about cognitive functioning. A paradigm is not just an entity or unit. It also provides unity, it has a cohesive force not only in the social sense of bringing scientists together in one group but also in the psychological sense of linking and coordinating various cognitive capacities into a coherent whole within the individual scientist and providing an integrated character to his behavior.

The holistic nature of paradigms is indicated by the emphasis on their 'global' nature (p. 43) and the characterization of normal science as "a single monolithic and unified enterprise" (p. 49). A paradigm is not just the juxta-position of "law, theory, application and instrumentation" (p. 10). It cannot be reduced to a set of rules (p. 45). It is "theory, methods, and standards together, usually in an inextricable mixture" (p. 108).

This unitary notion of paradigm, and the supposed cohesion and coordina-tion it indicates between various cognitive processes involved in scientific research, might well have been responsible for a substantial part of the popularity of Kuhn's 1962 monograph. But it has also been a major point of serious criticism. Though one might very much appreciate the global grasp expressed in Kuhn's picture of normal science, one feels disappointed when left so often without an answer as far as implementation and detailed mechanisms are concerned. In his review of Kuhn's book, Shapere (1964) called paradigm 'a blanket term' accusing Kuhn of "inflating the definition of 'paradigm' until that term becomes so vague and ambiguous that it cannot easily be withheld, so general that it cannot help explain, and so misleading that it is a positive hindrance to the understanding of some central aspects of science" (p. 393). For those looking for a serious understanding of paradigms, these impressions might well seem appropriate if they restrict their analysis to Kuhn's 1962 publication. However, some additional insights can be gained from scrutinizing his 1957 book and the 1969 postscript to his famous monograph.

Conceptual Schemes and the Functions of Paradigms

Though the term paradigm had been used to indicate "standards of rationality and intelligibility" in science by Toulmin (1961), its application to integrative schemes which provide specific expectations for research and permit selective assimilation of results is a major feature of Kuhn (1962). This does not mean that the concept of paradigm is entirely absent in Kuhn's previous work. In the 1957 book a similar notion is repeatedly invoked and discussed under the name of *conceptual scheme*. As such, it is clearly adopted from Conant who, in his *Science and Common Sense* (1951) elaborated the theme that "science is an interconnected series of concepts and conceptual schemes" (p. 25).[6] Conant's use of the term seems inspired by William James, who introduced conceptual schemes as "a sort of sieve in which we try to gather up the world's contents". James thought such a sieve to function as follows:

Most facts and relations fall through its meshes, being either too subtle or insignificant to be fixed in any conception. But whenever a physical reality is caught and identified as the same with something already conceived, it remains on the sieve, and all the predicates and relations of the conception with which it is identified become its predicates and relations too-(1890, Vol. 2, p. 482).

Realizing that Kuhn deals with conceptual schemes in terms of something which "discloses a pattern among otherwise unrelated observations" (Kuhn 1957, p. 37) and "frameworks for the organization of knowledge" (p. 41) which "can guide a scientist into the unknown, telling him where to look and what he may expect to find" (p. 40), we discover that the precursor of Kuhnian paradigms is both historically and conceptually linked to the old and familiar themes of conceptualism. In the analysis of conceptual schemes presented in Kuhn (1957) however, there is an important distinction that does not reappear in the 1962 book. A conceptual scheme is supposed to serve two functions: an *economical* function and a *psychological* function.

The conceptual scheme is considered as economical whenever it makes it possible to represent a vast amount of known facts by means of a relatively simple scheme. It is then "a convenient device for summarizing what is already known" (p. 38). Though Kuhn calls this also a *logical* function (p. 39), he considers it a problem of memory, i.e. how to store or conserve a huge collection of facts. We would now label this aspect of conceptual schemes the *representation* function.

The psychological aspect of the conceptual scheme is related to its *guiding* function with respect to the exploration of the unknown. Why would this be a more 'psychological' aspect than representation or memory for what is already known? Apparently, it is more psychological because it involves such purely subjective entities as *belief* and *cosmological satisfaction*. In order for a conceptual scheme *to guide* the research of a scientist, he has *to believe* that the scheme represents the real world: "a scientist must believe in his system before he will trust it as a guide to fruitful investigations of the *unknown*" (p. 75, italics in the original).

Apparently, the distinction between economical (or logical) and psychological functions of conceptual schemes coincides with the distinction between useful fictions and metaphysical beliefs.[7] When a ski-instructor suggests to the learner to bend knees and back as if he were going to lift two pails filled with water, the idea of lifting pails is recognized by everyone as a useful fiction. Nobody believes that water or buckets are really involved. It is only that the idea of lifting them induces the right tension in the muscles and the right posture of the body. A clearly false idea functions as an economical designation for an otherwise cumbersomely describable set of entities or states. A metaphysical belief, on the contrary, cannot bear the knowledge that it is possibly false. Its functioning critically depends on its being taken as indubitably correct and dealing with the ultimate constituents of reality. A hardheaded psychologist might entertain the metaphysical belief that all behavior goes back to firing patterns of neurons. For him, the ultimate knowledge of his field will be stated in terms of descriptions of these basic units which underly all the phenomena in his scientific universe. A metaphysical belief is coupled to an ontology: a doctrine on the kind of processes, objects or beings, that constitute the real world behind the various worlds and appearances of things in daily life.

Clearly, it is the psychological function which transforms the conceptual scheme into a paradigm. The representation function is independent of belief and does not disappear when the paradigm looses its hold on a scientific community. It might be kept in use as a useful fiction without troubling those who entertain it though they surely know that it is false. In that way, the ancient two-sphere universe, though completely outdated from an astronomical point of view, is still highly practical for the task of navigators. Such old conceptual schemes "are still a useful aid to memory, but they are ceasing to provide a trustworthy guide to the unknown" (Kuhn, 1957, p. 265).

If it is belief that transforms a conceptual scheme into a paradigm, what is it that belief adds to such schemes? Is it sufficient to consider it a motivating force in the way indicated by Uhr's description of science?

> Science does the following: we provisionally entertain a set of working assumptions, and see where they lead us. They may well lead to their own downfall. That is, we do not need to *believe* our assumptions. What we do need is enough motivation to pursue them, and belief is a good, strong motivator (Uhr, 1973, p. 11).

We should recognize that although belief can induce a person to trust a conceptual scheme, that will not make the scheme anymore specific. If a conceptual scheme is to have a *guiding* function, it should not only produce a diffuse 'cosmological satisfaction' but also provide specific expectations and suggestions. If it turns out that beliefs cannot add detail to a conceptual scheme, they might well provide some priority status to it so that if a person is supposed to have several at his disposal, the believed conceptual scheme might affect his behavior in quite a different way than the ones that have the status of 'useful fictions'. It might acquire some pervasiveness, being applied everywhere, 'overgeneralized' in psychologists' terms, with more tenacity and in more detail than the others. Obviously, if belief is a crucial attribute of paradigmatic conceptual schemes, the psychology of paradigmatic thinking has to reveal the cognitive functioning of belief.

Again, though Kuhn clearly holds belief responsible for the mobilizing and integrative force of paradigms, his 1957 book does not yield any more specific insight into the cognitive functioning of beliefs than the 1962 monograph, however much his impressionistic descriptions trigger our curiosity about it.[8] Nevertheless, the notion of conceptual scheme and the distinction between the representation function (called 'economical' by Kuhn) and the mobilizing function (called 'psychological') of such schemes provides a useful background for understanding Kuhn's 1969 attempt to arrive at a more detailed description of paradigms.

Disciplinary Matrix and Exemplar

'Anatomy' is a metaphor which Kuhn uses occasionally to indicate one of his major preoccupations: "the anatomy of scientific belief" (see Kuhn, 1957, pp. 73–77). Critics, however, have complained about the lack of a precise anatomy of paradigms in Kuhn (1962). What the monograph seems to provide is more a global physiology than a detailed anatomy. The *1969 Postscript* (Lakatos and Musgrave, 1970, pp. 174–210) is partly intended to remedy that deficiency. A dissection of the paradigm now yields four different components:
— symbolic generalizations,
— metaphysical beliefs,
— values, and
— exemplars.
We shall deal with them in a slightly different order.

Symbolic generalizations is the label for either formulas or verbal statements which express in a condensed way fundamental laws or principles of the area governed by the paradigm. Kuhn's examples range from formulas such as $f = ma$ to statements such as *action equals reaction* or *all cells come from cells* (see Kuhn, 1974, p. 464). Though he discusses subtle shifts with regard to their status as either analytic or synthetic expressions as basic to their nature, it should be sufficient for our purposes to consider them as the statements in some language, either natural or formal, which scientists would use to list the general principles as well as the general findings of their area in an epitomized form. The list might be long or short. According to Kuhn "the power of a science increases with the number of symbolic generalizations its practitioners have at their disposal" but some sciences might have very few or none at all (Kuhn, 1974, p. 464).

Metaphysical beliefs or *models* provide *preferred analogies* combined with an *ontology*. Preferred analogies serve the heuristic function of producing specific suggestions and expectations in the study of a given object. A gas might be fruitfully regarded as 'a collection of billiard balls in random motion'. The ontology expresses the *metaphysical commitment* to certain models; in Kuhn's example: "the heat of a body *is* the kinetic energy of its constituent particles" (Kuhn, 1974, p. 463, italics in the original).

Keeping in mind our discussion of the *conceptual scheme* in the preceding section, one might wonder whether symbolic generalizations on the one side and metaphysical beliefs on the other could not be considered reincarnations of the previously discerned economic and psychologic parts or aspects of paradigms. Obviously, there are important differences. As an example, Kuhn (1957) seems concerned with various forms of economic representation of a stable set of more or less neutral facts whereas Kuhn (1962) emphasizes how the paradigm partly determines its 'own' facts. But in general, one cannot deny some homology between those functions of conceptual schemes and those of the newly discussed components of the disciplinary matrix.

The *exemplar*, though also clearly retraceable to Kuhn's previous work, is a component whose prominent role is more emphatically indicated in his recent work. An exemplar is a paradigm in the restricted etymological sense of the term, i.e. a prototypical example such as the ones used in grammar to show all inflectional forms of a word. In science, similarly concrete problems such as the pendulum, serve a crucial role. They constitute the concrete embodiment of the two other components in a highly successful and specific application. This does not mean that the exemplar results from the use of those principles and models. The exemplar is both prior to and more fundamental than the statements and metaphors it embodies. A novice in science acquires first a thorough acquaintance with a very specific case and then learns to discover its wider applicability by developing a capacity to identify family resemblances [9] among an apparently heterogeneous set of problems and phenomena. The pendulum, initially studied and understood as a solitary case is seen as akin to free fall, satellite orbiting and other phenomena which turn out to embody the same laws of motion. Rather than these laws of motion ('symbolic generalizations' in the new terminology), the exemplar is the main carrier of knowledge. According to Kuhn: "scientists solve puzzles by modeling them on previous puzzle-solutions, often with only minimal recourse to symbolic generalizations" (1970, pp. 189–190).

The model of a paradigm which we have now arrived at has some resemblance to a cell structure: the exemplar in the position of the vital nucleus with floating around in the cytoplasm other components such as symbolic generalizations and metaphysical models. Notice the heterogeneity of this

conglomerate. A paradigm cannot be reduced to one or more general ideas
(symbolic generalizations), neither can it be reduced to one or more inspiring
metaphors (metaphysical beliefs) nor to a set of specific skills for handling
particular problems (exemplars). To arrive at a paradigm, one needs all these
various components together and in a successful combination.

In more recent publications and interviews, Kuhn has deplored the fact
that his 1962 monograph has induced an inflation of the term paradigm so
that it has become a label for the entire set of beliefs governing the behavior
of a scientific community (Kuhn, 1978). He now considers the exemplar
so pivotal that in his current opinion only the restricted sense of the term
paradigm is really justified. In its broader sense, it only designates a loose
conglomerate of entities which are "no longer to be discussed as though they
were all of a piece" (Kuhn, 1970, p. 182). But the composite nature of a
paradigm in the broader sense is not necessarily in contradiction with the
unifying function attributed to it. Remember how Kuhn's own work on
the Copernican Revolution revealed how much the new paradigm was the
outcome of an amalgamation of widely varying interests and influences. The
fascinating aspect of paradigms is that such amalgamations seem able to
develop into stable cognitive structures that bring coherence and directionality in the work of scientists. The most intriguing question is what it is
in those heterogeneous collections of elements that brings them together,
producing units that present a certain closeness to the mind. A rejection of
the broader notion of paradigm at this stage could be like rejecting the notion
of face after discovering that what we had first tentatively recognized as a
face now turns out to consist of a nose, two eyes and lips in an asymmetric
smile.

Values constitute the fourth kind of element in the disciplinary matrix.
One should think of them in terms of the familiar methodological requirements of *simplicity, consistency, accuracy, testability* and other criteria with
which scientists try to comply. They are less directly involved in the disciplinary matrix since on the one side "they are more widely shared among
different communities" (Kuhn, 1970, p. 184), and on the other side, with
respect to their interpretation "vary greatly from individual to individual",
even within the same paradigmatic community (p. 185).

Though Kuhn apparently thinks of these values as providing "a sense
of community to natural scientists as a whole" (p. 184) one should not

underrate the contribution of a specific value profile in the obtainment of paradigmatic identity. Priorities can be quite different in related areas. Compare the attitude toward parsimony in animal psychology and ethology. In terms derived from Lorenz, "the strategy of psychologists is to assume that the processes of learning in two animals are the same until proved to be different" while "the strategy of ethologists is to assume that the processes are different until they are proved to be the same" (quoted in Bitterman, 1975, p. 708). A similar difference is invoked by Minsky (1975) to distinguish approaches to intelligence developed in psychology from those developed in AI. All those groups claim to be scientific and, in general, adhere to the principles of simplicity, consistency, testability, precision etc. In practice, however, they have their own 'methodological' profile, expressing their particular interpretation of the norms and values which constitute the 'scientific attitude'. However, one could argue that a specific methodology is secondary to the more substantial assumptions and presuppositions provided by the other elements of the disciplinary matrix. Kuhn (1962) seems to suggest that methodological principles and rules are rather governed by the paradigm instead of the paradigm being selected and governed by such rules. Explicit discussion and concern with methodological rules is an indication of the disintegration of a paradigm rather than a regular aspect of its installment. Methodology detached from the cognitive content of the paradigm makes no sense and therefore it should be noticed that the values in the disciplinary matrix can have only a relative autonomy.

While discussing Kuhnian paradigms as world models governing the behavior of scientists, we have emphasized their status as the cognitive entities that populate the minds of scientists, allowing them to approach the world loaded with specific expectations and special techniques. But we should remember that selective perception or tunnel vision is only one aspect of paradigmatic functioning. In his appreciation of the trend represented by Kuhn, Bohm stressed the point that "the very act of perception is shaped and formed by the intention to communicate, as well as by a general awareness of what has been communicated in the past by oneself and by others" (Bohm, 1974, p. 374). Each scientist does not develop his paradigm as a solitary enterprise. Paradigms result from interactions with others and attempts to understand them. Understanding comes down to perceiving the world in terms of the world model of the other. The locus of a paradigm is not the brain of the individual scientist but the community of those scientists who can understand each other because they see their world in a highly similar way and who see their world similarly because they have evolved as one group through intensive interaction. The *paradigm* is *both cognitive* and *social* in nature and as an explanatory unit entirely in line with Bohm's position that "perception and communication are one *whole*" (p. 374). To complete our discussion of the basic Kuhnian concepts, an analysis of Kuhn's ideas on social interactions in science is indicated.

The Social Nature of Paradigms

The construction of a new paradigm seems to be a more or less solitary enterprise. It is done, according to Kuhn (1962) by one or a few individuals "so young or so new to the crisis-ridden field that practice has committed them less deeply than most of their contemporaries to the world view and rules determined by the old paradigm" (p. 143). The penetrating hold of a paradigm upon the mind of scientists seems to involve a mechanism of social control and social amplification which the innovator partly escapes. In Kuhn (1962), this mechanism is identified as education. Scientific education is both the admission into a social group and an initiation in its basic rites which secures that the basic concepts of the paradigm are drummed into the minds of new members. Education provides so much drill in the use of the paradigm that those who endure the long and elaborate process end up with a deep commitment to the underlying beliefs.

The nature of education differs markedly, depending on whether we investigate paradigmatic or non-paradigmatic disciplines. The difference can be made operational by scrutinizing the following indicators: the use of textbooks and exercises, the use of historical materials and the use of current literature.

Textbooks constitute a major vehicle for the transmission of paradigms. They provide a streamlined introduction into the paradigmatic concepts presented as the modern up to date views in the field. Paradigmatic disciplines have a few well-reputed textbooks that are regarded as suitable by virtually all members of the field. In non-paradigmatic disciplines, the textbook has not the same pivotal role as in the paradigmatic ones and when it is used in non-paradigmatic fields, the teacher is troubled with more diversity in opinion, both among textbooks and among users.[10]

Exercises accompany each chapter of a paradigmatic textbook. They involve the repeated use of the same conceptual scheme in different kinds of problems so that the basic concepts are more firmly established and their applicability in specific cases is recognized and experienced. In non-paradigmatic areas such textbook-exercises are rare or non-existent.

History of the field is shunned in paradigmatic disciplines. It is of no use and even distracting or misleading to confront novices with inadequate concepts. Even the famous historical achievements are indirectly learned from the textbook through eponymous labels rather than through thorough analysis of original papers or books. Non-paradigmatic disciplines to the contrary teach the history of the field as a regular component in their education and consult the original writings of out of date eminent authors.

Current literature, such as the papers and books which have just become available to the scientific community, is similarly kept away from the students in paradigmatic fields. It might undermine the paradigm, reporting and discussing anomalous findings. Accordingly, only at an advanced stage in their education are students who have already developed a deep commitment to the paradigm allowed to confront such literature. In non-paradigmatic fields again, the opposite is true. Recent papers are drawn into introductory courses as well as historical texts so that the very new and the very old are mixed up in one educational program and seem of equal relevance to the field.

Many distinctions made by Kuhn have been criticized for being too radical, the distinction between normal science and revolutionary science as the most prominent one being the most assailable. The sharp distinction between the revolutionary scientist and his normal science colleagues has offended several critics who, apparently, located themselves in between those categories. Whereas indeed the revolutionary scientist is depicted as a relatively free and creative individual, the normal scientist appears to them as the product of indoctrination and victim of authoritarianism to whom no room is left for creative and personal contributions[11]. However, the sharpness of Kuhn's distinction might result more from the one-sidedness of his description than from its incorrectness. Education may possibly account for the acquisition of new members in the paradigmatic community but it cannot be solely responsible for the dynamics within the group of mature scientists. In combination

with the massive effect of education, the behavior of an accepted member of
a scientific community is governed by subtle mechanisms of interaction with
other members of the group. These social relations among mature scientists
are not dealt with on equal terms with education in Kuhn (1962). However,
the *1969 Postscript* restores the balance by assigning an important role to
the small group social psychology of scientific communities. It means the
recognition by Kuhn of a line of research which developed partly independently of his work and with which we will deal in some detail in Part II.
It constitutes an important part of the studies incited or intensified by the
paradigm-theory.

Paradigm-studies

Has Kuhn's 1962 monograph revolutionized the science of science and is it
providing the integrating view which has been looked for so often? Scanning
the *popularity profile* of the book, one cannot deny that it is apparently
suggestive to a large and still increasing number of authors. It is an intriguing
question to find out what they see in it and how they have been able to work
out and apply the Kuhnian notions.

A balanced view would require that we consider enthusiast followers as
well as unconvinced critics. In addition, there might exist earlier versions of
Kuhnian concepts which, though less popular or even virtually unknown,
offer more elaborate views on one or more aspects of the model.

We have already referred several times to critics who have pointed out
many weak points in the theory of paradigms. Masterman's (1970) analysis is
often quoted because it is supposed to reveal twenty-one *different* meanings
of paradigm. Though it does provide twenty-one *contexts of use* (Margaret
Masterman is in computational language processing since the pioneer-days of
MT), it is on the whole a positive appraisal, quite sensitive to both the social
and cognitive aspects of the notion. She concludes: "It may be difficult both
to ascertain Kuhn's thought and to develop it; but if we do not make the
effort to do this, then it seems to me that we are left in a very disturbing
position indeed. . . . However much we may cavil at Kuhn's conclusions in
detail, we are not going to be able to go back to where we were before Kuhn
. . . " (1970, p. 87). Philosophers and sociologists in particular have taken
great pains in analyzing both its merits and defaults. Sociologists have been

more sensitive to the historical relevance of Kuhn's work in revitalizing
the dormant field of sociology of knowledge (King, 1971; Martins, 1972),
whereas philosophers have been less willing to accept it as a landmark in
epistemology. Scheffler's (1967) analysis clearly recognizes the cognitive
import of Kuhn's concepts but fails to accept them as a turning-point. The
recurrent theme in many criticisms concerns the vagueness and impressionistic
character of Kuhn's theory. But lack of specificity is not necessarily definitive
and irremediable. Keeping in mind that "the early versions of most new
paradigms are crude" (Kuhn, 1962, p. 155), we should allow the theory to be
judged in its own terms and have our evaluation depend more on what those
who use it do with it than on what those who denounce it say about it.[12]

Another statement of Kuhn that is equally applicable to his own theory
concerns rejecting or ignoring predecessors. According to Kuhn, "the historian can often recognize that in declaring an older paradigm out of date or in
rejecting the approach of someone of the pre-paradigm schools, a scientific
community has rejected the embryo of an important scientific perception
to which it would later be forced to return" (1963, pp. 358–359). Models
further developed than the embryonic stage have anticipated Kuhn's theory.
Contemporaries advocating similar views have already been mentioned:
Hanson, Toulmin, Feyerabend. So have members of the *Harvard General
Education Program*, initiated by James Conant and carried on today by
Gerald Holton and Everett Mendelsohn. The most remarkable anticipator of
Kuhn's 1962 book however is a previously almost unknown publication of
Ludwik Fleck (1935), entitled *Entstehung und Entwicklung einer Wissenschaftlichen Tatsache; Einführung in die Lehre vom Denkstil und Denkkollektiv*. This book, which Kuhn acknowledges in his 1962 preface to have
been very influential in the development of his thinking, combines a history
of the concept of syphilis in medicine with a most original analysis of the
structure and dynamics of scientific communities. The central idea that a new
school of thought derives from a more or less ad hoc combination of various
lines of thought forming a compound that starts to have a life on its own, is
illustrated with respect to various independent influences that lead up to the
concept of syphilis. In his combination of moral ideas about sexual behavior
with primitive medical notions of blood defects and infectual diseases, Fleck
overcomes the distinction between external and internal aspects in a way very
similar if not identical to Kuhn's. At times, in a style as vivid as Kuhn's, he

introduces subtle differentiations which are lacking in Kuhn (1962) but which post-Kuhnian sociology of science has rediscovered as quite important, such as the distinction between esoteric and exoteric circles in relation to the concept of invisible colleges. Reacting against Carnap's philosophy of science, Fleck even makes the remark that one could only wish that he (Carnap) might discover the social conditions of thinking (Fleck, 1935, p. 99, note). One wonders why the scientific community had to follow through a Carnapian line of thought for so long before it became sensitive to the Fleck–Kuhn line of thinking. It is also intriguing to discover how different the fate can be of books which are so similar in spirit and orientation.[13]

Leaving critics and precursors aside to focus on those followers who have seen in Kuhn's 1962 monograph an anchoring point for their own endeavors, we should distinguish between three groups. Each of them relates in some sense to the concept of paradigm and they could properly be characterized by labeling their activity as *paradigm-hunting, paradigm-detection* and *paradigm dissection* respectively.

Kuhn's popularity derives mainly from his success in the social sciences which are usually considered more soft and less scientific than the hard natural sciences. *Paradigm-hunters* [14] indicates the majority of enthusiastic Kuhn followers who hope to remedy a deplorable state in their field by providing a paradigm or by promoting the search for it. Though trouble is associated with the lack of a paradigm, a precise model of what a paradigm is, is mostly lacking as well, so that the application of the remedy is anything but clear. Often paradigms are reduced to slogan-like apodictic statements which have to express a new key insight or an inspiring metaphor considered capable of attaining or recovering the status of uncontested science for the field. How strongly desired paradigmaticity might be for some nostalgic researchers is obvious from such titles as Morin's (1973) book *Le paradigme perdu* and Boneau's (1974) paper on *Paradigm Regained*? . . .[15] Instructive as the numerous searches for paradigms might be for the understanding of both the areas to which the notion is applied and the difficulties encountered with a direct application of Kuhn's theory, they offer little that contributes to an implementation of the notion of paradigm.

Paradigm-detection studies concern a much smaller group of researchers who have combined the use of quantitative techniques, developed independently from Kuhn, with the search for and delineation of paradigmatic groups.

By means of bibliometric or sociometric methods, groups of interacting scientists are tracked down. Their interaction is taken as an indicator for their sharing a particular world view which is the crucial attribute for constituting a paradigmatic community. In the preface of his 1962 monograph, Kuhn had suggested some of these techniques, pointing out that "a shift in the distribution of technical literature cited in the footnotes to research reports ought to be studied as a possible index to the occurrence of revolutions" (p. xi). Substantial developments in these and similar types of analysis, mainly through the work of Derek de Solla Price, led Kuhn in the 1969 Postscript to the position that the study of paradigms should be preceded by the delineation of paradigmatic communities with the use of such detection procedures. This type of approach has joined with other traditions in sociology of science and developed into an area that greatly contributes to our understanding of the social side of the paradigm concept, i.e. the structure and dynamics of paradigmatic communities in science. We shall therefore discuss it in some detail and make an attempt to come to an overall assessment in Part II. Its relevance for the cognitive paradigm derives from the possibility of considering it a particular model for the description of paradigmatic communities in general.

Paradigm-dissection studies indicate the work of an equally small number of researchers who consider Kuhn's notion of paradigm to be a major contribution to the study of cognition and who see it as their task to analyze its structure in such detail as to understand the mechanisms of its functioning. Already in 1964, the great historian of psychology E. G. Boring clearly recognized the cognitive scope of paradigms, discussing Kuhn's monograph in terms of cognitive dissonance (Boring, 1964). As a psychologist who devoted a long career to the study of cognitive processes, Edna Heidbreder stressed even more Kuhn's relevance for cognitive psychology:

Without unreservedly accepting Kuhn's concept of paradigm and his discussion of it, I believe that in presenting his case for it he has made an important contribution to the psychology of cognition. In my opinion, he has placed in a revealing context the role and vicissitudes of complex cognitive structures, not only as they occur in science, but also, at least by implication, as they occur in prescientific and extrascientific commonsense knowledge (1973, p. 284, Note 12).

Some cognitively-oriented studies of paradigms have opted for a differential approach in the perspective of revealing the nature of these cognitive

orientation devices by studying several cognitive styles. A major attempt
in this direction has even been baptized as *paradigmatology* by its author
Maruyama (1974). This approach equally stresses the position that paradigms
play a major role in all cognitive processes and not just in sciences alone. But
again, this study and similar ones such as Royce's (1976) do not seem to be
based on the most suitable methods for revealing the internal mechanism of
a paradigm, interesting as they might be for the study of the structure and
development of paradigmatic groups.

The most straightforward attempt to crack the notion of paradigm is
undoubtedly the attempt of AI investigators to specify and to reconstruct the
complex implicit cognitive structures required for success in even the simplest
cognitive tasks. Driven by an adventurous ambition to synthesize human
cognitive processes, they have no way to circumvent the need for a very
fine-grained analysis of prerequisite knowledge which comes down to some
kind of dissection of paradigms. A major indication of that preoccupation is
Minsky's *frame*-paper (1975). This is not to say that AI researchers are the
sole contributors to the cognitive study of paradigms. They are indebted to
both earlier and contemporary psychologists who have tried to tackle with
human cognition. Outstanding among those is Jean Piaget who, in a career
of more than sixty years, devoted all his time and energy to the study of
cognitive processes. However, the AI contribution seems more or less to
function like a catalyzer, bringing together various lines of thought and
offering a possibility for an integration which has not been attained before.
Even though Piaget's work might appear more specifically connected to the
study of science and scientific development, AI work seems highly suitable to
dramatize its relevance for our understanding of paradigms. A similar point
can be made with respect to Gestalt psychology and some current psychological models of perception which, in view of Kuhn's description, should also be
highly relevant for understanding paradigms.

In general, one can argue that Kuhn's model of paradigms has much in
common with another theory that has intrigued many researchers: the Sapir
—Whorf hypothesis of linguistic relativity.[16] A fate similar to Kuhn's has been
conferred upon Whorf who has been praised for the highly inspiring character of his ideas by some and blamed for the vagueness and emptiness of the
same ideas by others. A fair evaluation of Kuhn's ideas should consider the
remark of Brown who after a highly critical study of a Whorfian hypothesis

concludes: "I still find his ideas powerfully stimulating and would bet that many are true. But they have never been tested. Not because they are untestable, but because no psychologist has yet been sufficiently ingenious" (Brown, 1976). Despite the diffuse nature and the many deficiencies of the concept of paradigm, some groups of researchers have been willing to accept it as a challenge to their ingenuity and see it as their task to make it sharper and more substantial. They are to be located in both the groups of paradigm detectors and paradigm dissectors. In Part II and III we will therefore focus on their approaches in order to grasp how their work contributes to the implementation of a cognitive orientation.

PART TWO

THE SOCIAL STRUCTURE OF SCIENCE

CHAPTER SEVEN

BIBLIOMETRICS AND THE STRUCTURE OF SCIENCE

> Es steckt ein gut Teil Wissenschaftsgeschichte
> und Wissenschaftspsychologie in der Geschichte
> der Terminologie. Die Worte sind gleichsam
> Leitfossilien, aus denen heraus sich ganze
> Schichten früheren, primitiveren Denkens re-
> konstruieren lassen. – Muller-Freienfels, R.,
> 1934, p. 157.

Kuhn's definition of the concept of paradigm has been criticized several times for being circular. Paradigms are introduced as that which is shared by the members of a scientific group, but, when asking for characteristics of such groups, one is referred back to the presence of a paradigm as the criterial attribute. In the *Postscript 1969*, however, this circularity is removed by the unambiguous statement that the detection and delineation of groups is a prerequisite for the analysis of paradigms: "Scientific communities can and should be isolated without prior recourse to paradigms; the latter can then be discovered by scrutinizing the behavior of a given community's members" (1970, 176).[1] Kuhn thereby alludes to developments in sociology of science and history of science which have taken as their research subject the community structure of science, mentioning authors who published pioneering works in the early or mid-sixties: Price (1961, 1963) and Hagstrom (1965) or settled a new tradition with their dissertation: Crane (1964) and Mullins (1966). Since then, the social study of science has grown rapidly and has developed into an autonomous field with its own organizations, regular conferences and meetings.[2] In this second part (Chapters Seven, Eight and Nine) we shall deal with what that field has contributed to the analysis and elucidation of the concept paradigm.

Bibliometrics and Research on Science

Under the label of *social epistemology*, the American library scientist Jesse

111

Shera has promoted the idea of a study of knowledge as some form of 'econo-metrics' of products of intellect. His 'social epistemology' is defined as "the analysis of the production, distribution and utilization of intellectual products in much the same fashion as that in which the production, distribution and utilization of material products have long been investigated" (Brookes, 1974). *Bibliometrics* could be considered a part of this daring endeavor using statistical methods to study quantifiable aspects of publications. For the orthodox epistemologist, this orientation might appear atrocious since it leads to judgements upon texts derived not from reading them but from measuring them in such superficial attributes as their number, length, citation links and popularity. There is, however, one argument in favor of bibliometrics based on an experience familiar to almost anyone in science: the information explosion. There is far more potentially relevant literature in any field than any single individual can ever read. The bibliometric approach to the study of science starts out with a quantification of this – for many scientists – discouraging experience, expressing the growth of scientific knowledge in terms of the number of publications produced within each time unit. This apparently meaningless counting reveals a more structured picture of science than one would expect.

The Growth of Science

More than anyone else, the historian of science Derek de Solla Price (further referred to as 'Price') has promoted the idea that, while we might not be able to read all the scientific publications that appear, we still might learn something about science by merely counting them. Like Kuhn, a physicist involved with history of science, he has suggested studying scientific develop-ment with quantitative methods using demographic and bibliometric data. Within the classical approach dominated by an 'eternistic' image of science (science describes the world as it is), science is timeless and detached from the history of its discoveries. Within Kuhn's scheme, depicted in terms of a sequence of paradigms separated from each other by revolutions, science becomes an endeavor which changes over time. A new dimension of this change is revealed however, when one studies, instead of the conceptual systems expressing science, the number of people involved in producing them and the number of publications registering them.[3] Through the widely discussed reports of the Club of Rome (Meadows, 1972) mathematical concepts of growth have become widely known. Many people have been alarmed by the exponential growth curves which seem to apply to the use

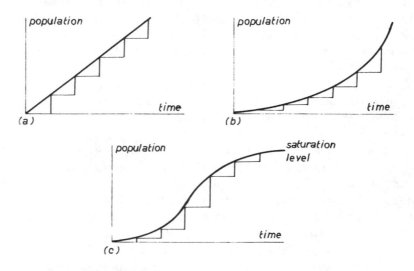

Fig. 7.1 Basic concepts of growth: (a) *Linear growth*: an increase with a fixed quantity of units for each unit of time, e.g. a library acquiring the same number of books every year; (b) *Exponential growth*: an increase with a fixed proportion of the total population for each unit of time, expressed either in *percentages*, e.g. 7% growth per year, or *doubling times*, the time required for a population to reach twice its original number, e.g. a doubling time of 10 years corresponding to 7% growth; (c) *Logistic growth* or *S-shaped growth*: a combination of exponential growth and asymptotic growth whereby, after an initial exponential phase, growth gradually slows down as the population approaches saturation level.

of such vital things as fossil fuels and several rare ores. In two influential monographs, Price (1961, 1963) has illustrated how this same concept of exponential growth applies to science, in terms of the number of people employed in science as well as in terms of the number of units of scientific productivity: scientific papers. The *exponential growth* curve appears to be applicable to the last *three hundred years*, with a rather impressive growth rate of five to seven percent a year (or a *doubling* between *ten* and *fifteen years*). However, all growth depends on resources and if the resources do

not increase as fast as the products, any growth, whatever its nature, will
slow down.

Since the population of scientists is growing faster than the total population on earth, at some time in the future the whole population will be
involved in science if the present growth rates continue, and, according to
the calculations of Price, this would involve very few generations. As such,
science like any other growth phenomenon, seems subject to a description
in terms of *logistic growth* rather than pure exponential growth. There
are saturation limits, and the exponential growth which has been going
on for three hundred years cannot go on endlessly. At the time of writing
his *Science Since Babylon* (1961), Price had the impression that science was
then at the onset of slowed down growth. His opinion might have seemed
in contradiction to the enormous expansion of the sixties but now, after
the seventies, it appears to have been prophetic. However, despite important
practical considerations and suggestions implied by this growth model of
science, its importance for our understanding of paradigms lies not so much
in the characteristics of the growth of science on a global scale as in the
analysis of the detailed mechanism of this growth.

While the growth of science might seem rapid and impressive, it does not
occur like an explosion, expanding at the same rate in all directions. When
the totality of scientific literature doubles every fifteen years, it does not
necessarily mean that this growth rate applies to every single discipline and
subdiscipline. From an analysis of growth data it appears, on the contrary,
that scientific disciplines and subdisciplines may vary widely with respect
to growth rate. A comparison between the subdisciplines of optics and
acoustics on the one hand and atomic-, molecular-, particle- and solid state
physics on the other shows that the growth rate of the latter is much higher
than the growth rate of the former (Menard, 1971, p. 51). Even within
subdisciplines, areas of fast growth can be distinguished from areas of slow
growth, and, when scrutinizing the growth of science in its details, two points
become rather clear: (a) the *rapid growth of science* should be attributed to
the *extremely fast growth of some subdisciplines and specialties,* levelled out
by slower growing areas, and (b) *fast growing areas* in science tend to be
relatively *young areas*. The logistic curve which, according to Price, applies
to the growth of science as a whole seems equally applicable to the smaller
units of growth on the micro-level. After a while, research areas or specialties,

with a doubling time of a few years or even less, slow down gradually —
when the amount of accumulated literature becomes large. We consider the
mechanism of this slow-down below when we study the 'life cycle' of scientific specialties (Chapter Nine). At this point it is important to realize that the
continuous exponential growth of science requires the *continuous creation of
more and more growth points* which, each time again, by their fast initial
growth, counteract the general tendency toward slow-down introduced by
many established subunits. If one should think of these growth areas as
reflecting new paradigms, this description would seem to contradict Kuhn's
original theory of science as proceeding within the established boundaries
of disciplines, rocked only by a revolution from time to time. Indeed, the
growth pattern of science as it is reflected in the growth of its literature is like
the growth pattern of a tree, branching out in ever more new twigs. If the
concept of paradigm is to be coupled to the notion of growth area, opinions
on the origin of paradigms must be enlarged and at the same time the scope
of paradigms has to become more restricted. One paradigm might be followed
by several others instead of just one, and this unit of analysis should not be
localized at the level of disciplines, but at the level of much smaller entities:
scientific specialties and *research areas*. One crucial attribute of Kuhnian
paradigms however, viz. the notion of a consensus permitting productive
communication and coordination among a group of researchers, seems to be
highly compatible with the notion of growth area. At any rate, that is what
is apparent from a closer look at the metabolisms of fast and slow growth.
Before analyzing those metabolisms, we should look at bibliometric methods
for detecting points and areas of fast growth in scientific literature.

The Detection of Growth Points and Secondary Literature

Growth points in science are to be seen, at this point of our analysis, as
flourishing research areas and specialties, where the number of publications
accumulates much faster than in other areas of science. Are there any bibliometric means for detecting and locating points of fast growth? It is very
difficult to detect growth points at an early stage in the development of a
new specialism. While the number of publications might increase very quickly,
the absolute number of items is generally too small to be visible for people
not conversant with the area. So, the first developments occur within the

shadow of established areas, whose more slowly growing bulk of articles masks the small but highly significant developments. Once a certain visibility threshold has been passed, however, the growth of fast growing areas is reflected with some delay [4] in the *secondary scientific literature*.

Secondary literature is literature about literature. It consists of *abstracts, indexes, reviews* and *bibliographies* written or compiled to make *primary literature* (consisting mainly of articles and reports in scientific journals) more accessible and surveyable to the potential user. Secondary literature yields some kind of description of primary literature and is in a more or less systematic way connected to it. In his calculations on the growth of scientific literature Price (1961) did not attempt to count or to estimate the number of publications directly. Rather, he counted the growth of the number of bibliographies — on the assumption that the number of bibliographies is a fixed ratio of the total number of items in an area, and that new ones are only compiled when a substantial literature has accumulated. Abstract journals which are designed to cover domains as broad as disciplines are more likely to trace fast growing areas earlier in their development. Menard (1971) who has studied the development of oceanography as a branch of geology that has become a fast growing autonomous area, indicates how this topic was at first hidden in the buffer category of the classification system for abstracts, i.e. under *varia*. Later on there was a period in which it was classified as a separate topic without a proper name, indicated as 'ocean-'. Afterwards, during the period in which the expansion was most obvious, the area received a proper name, being labeled 'oceanography'. Similar developments can be traced for biochemistry with respect to chemistry in general, and for computer mathematics as a subdiscipline of mathematics — both becoming visible at first as an unusual unidentified swelling of the residual category of 'other'. It takes rather long before a fast growing area is recognized as being sufficiently important to deserve a proper entry in some classification of abstracts or in an index. Frequently the fast growth has already ceased at the time of recognition. When one knows where to look and how to combine categories, fast growing areas might be traced through manifest changes in the number of abstracts appearing each year within several stable categories. If one counts the number of abstracts in *Psychological Abstracts* covered by the categories of *Language and Communication*, together with its subcategories, and adds to this the number of abstracts listed under *Verbal Learning*, which is a subcategory of *Experimental Psychology*, a growth pattern can be retrieved that clearly illustrates the fast growth of psycholinguistics after 1964 (see Figure 7.2). This kind of analysis requires extra-bibliometric knowledge of the areas one is studying in order to be able to select the categories one intends to trace.

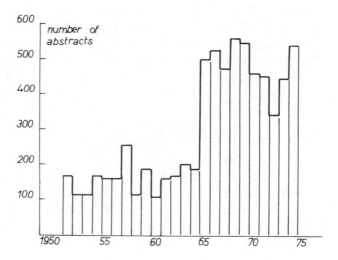

Fig. 7.2. Number of abstracts produced by *Psychological Abstracts* each year in the category of *Language and Communication*, including its subcategories, to which are also added the abstracts in *Verbal Learning*, a subcategory of *Experimental Psychology*. The combination of these categories allows to observe the uprise of psycholinguistics.

Some secondary literature publications make the detection of growth points easier by reviewing at regular intervals (something in the order of five years), a substantial part of the primary literature in a number of subfields. *Annual Reviews* cover many subfields in several disciplines. Since the reviews are written by authors acquainted with the field, chances are fairly good that they will spot new growth points rather early in their development. Some monographs with a title beginning *New Horizons in* ... or *Contemporary Scientific* ... , followed by the name of a specific field, usually contain introductions to a bundle of new areas of fast growth. This applies equally to popular general scientific journals such as *Scientific American* or *Recherche*. These publications frequently amplify the *visibility* of newly developed growth points. However, since their selection does not reflect a regular review of the primary literature on an exhaustive scale, they present a picture which is highly influenced by the idiosyncrasies of the editors.

Bibliographies are pre-eminent secondary literature items. Once a bibliography on a certain subject is compiled, it probably has become a self-contained area — after a period of fast growth. Bibliographies derive some extra interest from the fact that they permit additional bibliometric analyses (see below). However, they pose a problem which reflects a basic difficulty of the concept of growth area or specialism, i.e. the boundary-problem. Key publications for a given area are easy to locate, but it is more difficult to decide whether or not fringe publications belong to the area in question. In this respect, growth areas or specialisms are 'fuzzy sets'. Different bibliographies for the same area yield different sets of literature items. Their intersection will clearly indicate publications central to the area, but the amount of material outside the intersection will illustrate the 'fuzziness' of the same area's boundary. Compare the growth curve based on the 1962—1964 bibliographies of the literature on 'diffusion of innovations' with that from the 1971 version (see Figure 7.3). Although the bibliographies have E. M. Rogers, a leading contributor to the field, as their main author, the 1971 version differs markedly from the 1962—1964 versions, adding retrospectively a number of publications in 1971, which were not included (but existed already) when the 1962 and 1964 bibliographies were produced.[5] This boundary-problem makes it difficult to compute exact growth rates for a specific area. Part of the development of the area might consist of the later integration of materials published much earlier.

Though the secondary literature in essence describes the primary literature in a given field, in general the description reflects only in a rough way, and with considerable delay, the developments in a discipline. One might wonder whether there are no specific attributes which would make it possible to distinguish in some direct way between the primary literatures of rapidly and slowly growing areas.

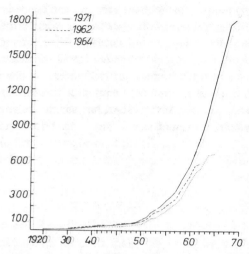

Fig. 7.3. Growth curves for the literature on 'diffusion of innovations', based on E. M. Rogers, *Diffusion of Innovations*, 1962; E. M. Rogers, *Bibliography of Research on the Diffusion of Innovations*, 1964 and E. M. Rogers and F. F. Shoemaker, *Communication of Innovations*, 1971.

The Metabolism of Growth and Primary Literature

Papers in fast growing areas tend to differ from papers in slowly growing areas in several respects. An important characteristic of papers in a fast growing area is that they are predominantly related to recent papers in the area. This should be indicated by the *rate of obsolescence*, or the *rate of decay*, of the publications mentioned in the list of references added to each paper. The rate of obsolescence is based on the publication dates of the referenced texts. These references presumably form a list of links which connects the paper to previous publications in the area. In a fast growing area these references concern mainly recent publications, in slower areas the proportion of references to older publications is significantly higher. Some specific measures have been proposed to express this 'degree of recency' of references. They require an analysis of the distribution of the publication dates of those references connected with a set of papers representative of the area at the time one wants to investigate the growth rate in that area. In general, the histogram of a random sample (e.g. the references in one volume of a journal) suggests a skewed distribution, with a peak in the recent publications of three years ago, and a more or less gradual decline toward the older publications. One measure of obsolescence, *Price's index* (1970) expresses the recency of references in terms of the percentage of references to publications published within the last five years. Another measure is the *half life* of references, expressed in terms of the time span within which the fifty percent most recent references can be found (see Figure 7.4). Price's index of very fast

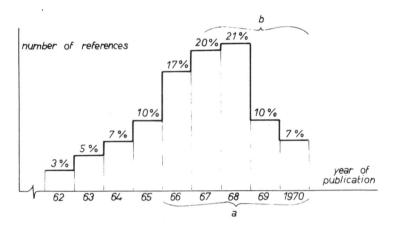

Fig. 7.4. Rate of obsolescence measures for scientific literature illustrated in a hypothetical distribution over time of references taken from a sample of literature published in a given year. *Price's index* corresponds to the percentage of items that refers to materials published within the last five years (a). The *half life* corresponds to the time in which the fifty percent most recent references can be found (b).

growing areas might be as high as 80, and even more. Dormant areas do not exceed 20. Journals which represent fast growing areas in science should, according to Price, attain values of 60 or more.

The recency of references seems to reflect some aspects of the metabolism of growth in the hard sciences, characterized by faster growth, and in the humanities, characterized by slower growth. Price remarks:

with a low index one has a humanistic type of metabolism in which the scholar has to digest all that has gone before, let it mature gently in the cellar of wisdom, and then distill forth new words of wisdom about the same sorts of questions. In hard science the positiveness of the knowledge and its short term permanence enable one to move through the packed down past while a student and then to emerge at the research front where interaction with one's peers is as important as the storehouse of conventional wisdom. The thinner the skin of science the more orderly and crystalline the growth and the more rapid the process (Price, 1970, p. 15).

In this model of the metabolism of scientific growth, Price makes a distinction between, on the one hand, what he calls the *archival body* of scientific literature, and, on the other hand, the *research front* literature. The archival literature is the most substantial part and is compared to the *body* of science, while the research front literature, which might be relatively small in size, is referred to as the *growing skin*. According to this image, science not only grows like a tree, in the sense of branching off in twigs but it also resembles a tree in growing only on the outside — being supported by a lifeless stable structure on the inside. The archival body of literature is rarely used in the construction of new science; the research front literature is, as far as can be determined from the references, heavily used in that respect. Price's statement that "the thinner the skin of science the more orderly and crystalline the growth and the more rapid the process" (see quotation above) refers to the width (expressed in time-units) of that segment of the literature heavily referred to in the new papers continuously added to the area. In fast growing areas this segment is very thin. Papers very soon become obsolete in the sense that their content is readily assimilated and integrated into the body of knowledge so that specific reference to them is superfluous. They become part of the archival body which serves more the purposes of history of science than of science itself. In slowly growing areas, the research front literature is a much wider segment of the literature, and it might be difficult to distinguish between archival and research front literatures. Almost any paper from any time might turn out to be relevant to a current research problem. In Menard's (1971) view, slowly growing areas can expand by periodically dragging up unsolved problems or by questioning semi-accepted solutions, so that time and time again the older literature is brought back as potentially relevant.

The rate of obsolescence of references is clearly related to the 'degree of paradigmaticity' of an area, and might even be considered an operational specification of it. Remembering Kuhn's own operational criteria: use of textbooks and many exercises, no history, no classics, no current research literature during the education in a paradigmatic field, one should notice that this rate of obsolescence is of a similar nature. Applicable to the endeavors of the mature scientist instead of the student, a high rate of obsolescence suggests an orderly step by step progression in a well-defined problem area, where all participants know both what the general problems are as well as what the sub-problems look like; so that, whenever the answer to one set of subproblems is found, the whole community easily and quickly moves on to the next generation of subproblems, ignoring the troubles and struggles with the previous set. A high rate of obsolescence thus appears quite compatible with puzzle-solving-areas. Slower rates of obsolescence indicate that, while the literature accumulates, no substantial progress is made. The literature develops into an impenetrable network where several vaguely related sub-problems obscure the general challenges of the area and every single item of literature remains potentially relevant. The low rate of obsolescence seems to indicate either that no problems are solved or that there is no well-ordered sequence for tackling the subproblems so that participants work crisscross over each other's contributions.

Besides Price's index, several other bibliometric indicators have been studied in connection with the growth characteristics of scientific fields. Among them are: length of papers, number of references, journal-monograph-ratio, self-references of journals, scattering index, secondary versus primary literature ratio, and, when computable, indicators such as rejection rates and speed of information flow. Table III summarizes the major ideas behind each of these measures.

Most of these apply in principle to the set of references derived from the primary literature covering a given area in science. In practice, however, they have been used to characterize journals, and, as such, their indications with respect to growth characteristics depend on how well a given journal represents a given area. Since, on the one hand, it is difficult to unambiguously delineate the literature coinciding with an area and since, on the other hand, many journals cover several areas, these measures are not very appropriate for detecting growth areas in science. They have been used to indicate the degree of connectedness and continuity within disciplines and are invoked to support the distinction between hard science and soft science. But, at best this is only an amplification, or operationalization, of Kuhn's (1962)

Table III. Bibliometric indicators and the metabolism of fast-growing and slow-growing areas

BIBLIOMETRIC INDICATOR (applies to a representative sample of the literature of an area)	CHARACTERISTIC OF FAST-GROWING AREA and corresponding "hard science philosophy"	CHARACTERISTIC OF SLOW-GROWING OR DORMANT AREA and corresponding "soft science philosophy"
CITATION DECAY or OBSOLESCENCE RATE: the relative share of recent and older materials in the list of references as measured by Price's index or *half life* of citations (Price, 1970)	*most references are recent*: progress is stepwise and systematic, any new layer of advances is based on an immediately preceding generation of papers; older literature is considered as definitely acquired and is rarely cited	*references are older*: progress is diffuse and many contributions attempt to bring a new orientation towards old problems by reviving older literature
LENGTH of article as estimated in terms of mean number of words	*short articles*: the problems as well as their possible solutions are well-defined; a short description is sufficient to locate the problem, to describe the method and results.	*long articles*: the problems in the area are ill defined and the author needs an extensive introduction to present his problem representation
NUMBER OF REFERENCES as calculated in terms of the mean number of references per article	*few references*: the problem is precise and easy to locate with respect to the work of other relevant authors who contributed to the actual state of the question: knowledge of the problem is more important than knowledge of the literature	*many references*: the problem is vague and can be related to many items in the literature: a new approach might consist in a new combination of written materials illustrating the erudition of the author rather than his skill in puzzle-solving
SCATTER INDEX: the dispersion of the references over the population of scientific journals measured in terms of the number of journals required to account for fifty percent of the references to journals	*low scatter*: the divisions and subdivisions in the area are clear and are systematically covered by one or a few journals so that progress is mainly "internal" i.e. going back on the same small group of highly relevant journals.	*high scatter*: the references are to many journals both from within and from outside the discipline, illustrating a certain lack of clear internal subdivision and less continuity and less direction
JOURNAL SELF-REFERENCES: the percentage of references to the source journal	*high percentage of self-references*: the same idea as with respect to scatter index	*low percentage of self-references*: idem, cfr. scatter index
JOURNAL-MONOGRAPH RATIO: the dispersion of the references over the population of scientific publications expressed in terms of the relative share of publications in journals compared to the relative share of other publications (mainly books) (Roe, A., 1972, Small & Crane, 1979).	*references are predominantly to journal publications*: almost all references indicate journal publications and only rarely a book-type document is cited; the fast growth is incompatible with time lag involved in the publication of a book	*monographs receive a considerable amount of references*: books and monographs account for a substantial part of the references and the time lag involved in the production of a book does not seem to affect its use as a "research front" item
REJECTION RATES: the proportion of materials presented for publication and refused by the editors of the journal, as compared to the total number of manuscripts sent to the editors (Zuckermann, H., & Merton, R.K., 1971).	*low rejection rate*: relatively few manuscripts are refused for publication; both editors and authors have clear concepts about, and are well acquainted with, the standards governing publication policies in the group; few manuscripts below standard are presented to editors	*high rejection rate*: relatively many manuscripts are refused because neither editors nor authors have a clear picture of the specific problems and methods acceptable to the group
SPEED OF INFORMATION FLOW: the period of time required for the dissemination of a unit of information corresponding to one article (Lin, N., Garvey, W.D., Nelson, C.E., 1970)	*Information disseminates fast and efficiently*: the time between obtaining the results of an investigation and the availability of a report on that investigation in the professional literature is relatively short	*Information disseminates slowly and diffusely*: the time between obtaining the results of an investigation and the publication of a report on it in the literature is longer because of, among other factors, high rejection rates which force the authors into several time-consuming submission procedures
SECONDARY TO PRIMARY LITERATURE RATIO: the proportion of the total literature devoted to the description of the primary literature (Menard, 1971)	*low proportion of secondary literature*: although the information services might be impressive, the number of secondary literature items is only a small proportion compared to the huge primary literature	*high proportion of secondary literature*: slower growth combined with diffuseness requires a substantial part of the research to be devoted to literature searches, resulting in overlapping reviews and bibliographies.

distinction between paradigmatic and non-paradigmatic *disciplines*.[6] From the analysis of the growth of sciences we learn that units of growth should not be looked for at the level of disciplines but at the level of units of a smaller scale, such as scientific *specialties* and *research areas*. Is there any bibliometric method that can isolate those scientific communities which, according to Kuhn (1970), "should be isolated without prior recourse to paradigms"?

Citation Networks

The measures indicated in Table III use references as a kind of aggregate data. They are classified, summed up, and compared in quantities which do no longer contain the information on specific links connecting specific items of scientific literature. Citation analysis, in a more restricted sense than the methods discussed up to now, makes a direct attempt to expose the social structure of science in terms of the links expressed in the references. Within such an approach, specialties are not tracked down by looking for areas of striking growth but by looking for particularly dense patterns of citation links. Within that approach references are dealt with as sociometric[7] data reflecting a scientist's own choices of the colleagues he considers to relate to. The basic assumption is that the list of references added to a paper provides a sociometric profile of the author of the contribution. Several methods have been used to explore the possibilities offered by this approach.

The most straightforward manner for developing citation networks would be multiple step citation analysis. For a given source paper all bibliographic references are traced down. In their turn, all cited items in those documents are retrieved and this cycle is repeated for as many generations of references one feels capable of analyzing. Theoretically, one could hope to discover in this way some stable circular substructures within an immense web of linear relations. However, since the mean number of references for scientific papers is of the order of ten, it is obvious that a few generations of references are enough for entering what in AI is known as the 'combinatorial explosion'. Researchers at the university of Bath exploring this type of methods report that "networks rarely evolve towards such tight concentrations as portrayed in the theoretical cases given, and where concentration can be developed this is often due to the application of quite rigid cut-offs in the data ... "

(DISISS, 1973, pp. 6—7). Specific cut-off criteria can be used for particular purposes such as citation tracing of key publications. However, they are difficult to devise for the detection of interacting groups in science. Other, more selective usages of citation data, prove more useful for that purpose. Two of these are of interest to us: bibliographic coupling and co-citation.

Bibliographic coupling introduced by Kessler in 1963 uses references to determine a degree of affinity between papers. The more references they have in common, the more related papers are supposed to be. The number of shared references is in fact taken as an indicator of the strength of the relationship. Applied to a set of papers of *Physical Review*, Kessler (1965) was able to show that groups formed on the basis of strongly related pairs correlate highly with the subject index composed by the editors of the journal. But though bibliographic coupling seems to capture the relatedness of papers, it is apparently not the optimal tool for detecting specialties. One aspect of its nature is revealed by its tendency to cluster around review papers, obviously because of their high number of references. In general, it has been noticed that "bibliographic coupling should tend to favor repetitive literature" (Weinberg, 1974, p. 193). Thus, while correlating with subject indexes, it does not allow to go significantly beyond them. A more suitable method for locating specialties appears to be co-citation clustering.

Co-citation Clustering

Sometimes, co-citation is discussed as a variation of bibliographic coupling. Like the latter, it is based on some overlap between lists of references, but its peculiar feature is that it is detached from the source documents. Bibliographic coupling uses *cited* documents to measure the relatedness of *citing* documents and other citation analyses focus on the relation between citing and cited documents. Co-citation ignores the citing sources to focus solely on the *frequency with which cited items are cited together*. In principle, it determines for a given sample of papers how many times pairs of papers reappear in the lists of references provided by the papers in the sample. Obviously, as a measure of relatedness, this only makes sense if its value is a whole number exceeding one. All papers in any given list of references are, taken two by two, cited together, i.e. *co-cited*. The point of the measure is to establish, still in principle, for all possible pairs, how many times they can be found in the lists of references of the participating papers. In practice however, the 'combinatorial explosion' involved imposes restrictions on the kind of possibilities which are in fact considered.

Co-citation has been discovered independently by Small (1973) and Marshakova (see Marshakova, 1981). As developed by Small at the Philadelphia Institute for Scientific Information (ISI) it uses the huge data base provided by the *Science Citation Index* (SCI)[8] and the *Social Science Citation Index* (SSCI) with certain thresholds.

The standard procedure involves the following steps (Small, 1975):

(1) Documents cited above some *citation frequency threshold* in an annual SCI or SSCI are selected and listed along with their source documents, i.e. the citing document;

(2) A rearrangement of these data according to citing documents allows to determine *co-citation pairs*, i.e. pairs of highly cited papers which are cited by the same source document, and the number of times they are cited together;

(3) Pairs cited above some *co-citation threshold* are then combined into *clusters* on the basis of a single citation link, i.e. different pairs that have one element in common are connected;

(4) Relationships between clusters are established in terms of *cluster co-citation* scores, i.e. the number of times two clusters are cited together by having one document of a pair in one cluster and the other document of the pair in the other cluster;

(5) The application of this procedure to subsequent collections of citation data is used to arrive at a picture of the structural dynamics of specialties over time.

Structure is defined in terms of the links between co-cited pairs. Depending on the manipulation of the thresholds involved, in particular the co-citation threshold, structural patterns are obtained which make both the microstructure and macro-structure of science manifest. Figure 7.5 provides an example of a *co-citation map*. Small's own analyses (Small, 1977) and other validation studies (Sullivan *et al.*, 1977; Mullins *et al.*, 1977) have supported the idea that co-citation clusters relate strongly to the kind of social and intellectual units which growth studies have identified as specialties. As indicated in point 5 of the procedure, the ultimate goal is to connect series of such maps obtained over several years in order to arrive at a kind of representation comparable to Newell's map for AI (see Figure 7.6) or the map

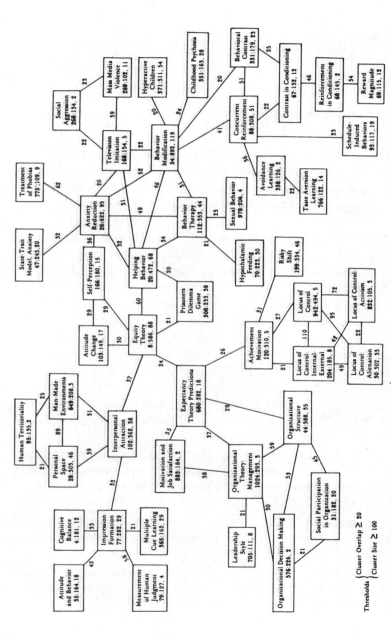

Fig. 7.5. Map based on 1974 SSCI, indicating clusters of highly cited papers in the social sciences. The third number indicates the number of papers in the cluster, the second one indicates the number of source papers. On the links between clusters: the number of co-citations connecting them. (From Garfield, E. *Essays of an Information Scientist*, ISI Press, Philadelphia, 1980, Vol. 2, p. 511; *copyright © 1977 by Eugene Garfield*, reprinted with permission.)

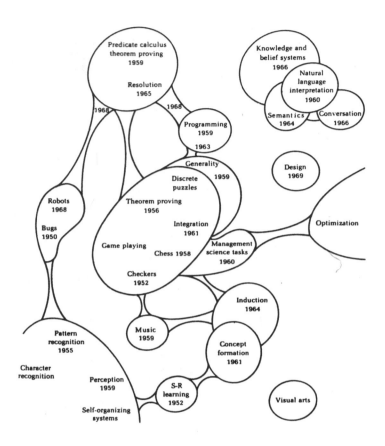

Fig. 7.6. Newell's map of the domain of Artificial Intelligence (Newell, 1973). According to him: "The theoretical structure of science does not consist of a set of isolated entities called theories. Theories derive from each other and are inferred from and built to deal with bodies of data. With our *firmly embedded scholarly tradition of referencing* our progenitors it is usually possible to *trace down* these *genetic relationships*" (p. 25, italics ours). It is not clear in how far this particular map is based on citation analysis. As it expresses the view and the knowledge of a pioneer and expert in the field, it provides an ideal standard for achievements in the algorithmic analysis of citations. (From Newell, A., 'Artificial Intelligence and the Concept of Mind', in Schank, R. C. and Colby, K. M. (eds.), *Computer Models of Thought and Language*, Freeman and Co., San Francisco, 1973, p. 3; *copyright © 1973 by Freeman and Co.*, reprinted with permission.)

of Lachman *et al.* (1979, p. 408) for studies on comprehension. In fact, the largely automatized and large scale production at regular intervals of time of similar types of maps for most areas of science constitutes a major impetus for co-citation studies.

Nevertheless, though Garfield (1979, p. 148) is certainly right in claiming that "there is no doubt that co-citation clustering has uncovered a structure in science", it remains unclear what it is that structure depicts. It provides the kind of image of science that is similar to one obtained from persons by means of X-ray photography or thermography. Globally, with some effort, we recognize the human figure and some major parts, but, in the case of co-citation we apparently do not know whether what we capture represents the supportive skeleton or the highly metabolically active tissue. The problem is manifest in the discussion whether what is captured is social or cognitive.

The highly cited papers in the co-citation clusters are ambiguous units. In terms of their *authors*, they represent individuals, and, socially interpreted, a co-citation cluster could be considered as indicating a *group*. In terms of their *content*, the papers represent contributions of knowledge, and, from the point of view of cognition, the co-citation cluster could be considered to represent a *cognitive structure*. Indeed, according to Small (1977), "highly cited papers are 'markers' or symbols of important concepts and methods in science and represent the current loci of consensus in the scientific community" (p. 141). This assumption seems to allow him to consider social structure and cognitive structure being of the same dimension. It is as if with a finer grain analysis, cognitive structures can be revealed with the same methods that are used to delineate specialties. Small and Greenlee (1980) claim that "we are beyond the point where we simply assert that a 'paradigm' exists in this or that field. If progress is to continue in the study of scientific specialties, systematic methods must be devised for delineating what knowledge elements are involved in a given case" (p. 277). They use citation context analysis in a detailed study of a co-citation cluster on recombinant DNA, allowing them to identify the 'cognitive' contribution of the papers involved. However, the result is not very different from the kind of conventional intellectual history that specialists provide in overviews for non-specialists. Irrespective of the question whether the history is complete or the only possible one, one wonders whether such a structured sequential account can count as *the* 'cognitive structure'. Small and Greenlee describe the results of their study as an exploration of "the 'mental furniture' of a hypothetical and ideal recombinant DNA researcher" (1980, p. 287). But our analysis of the cognitive paradigm in preceding chapters has already indicated that one cannot simply identify cognitive structure with a structured account of the knowledge involved. To go beyond the structural stage, we need to comprehend the kind of partial knowledge that shapes expectations and provides directionality to search. Understanding those aspects of knowledge related to what is sometimes called *tacit* knowledge is essential to understanding what paradigms are. None of it is captured by the methods of Small and Greenlee.

Sociologists of science are intrigued by the relationship between social structure and cognitive structure. Cole and Zuckerman's (1975) remark that "sociologists of science, cognitive psychologists, and philosophers of science focus on different aspects of cognitive structures" (p. 143) might be but an euphemistic way of saying that the issue is confused. The mixture of relative autonomy and mutual dependency between the cognitive and the social explored by Weingart (1974) and Whitley (1974) reminds of Weber but is conceptually not a clear-cut solution. Maybe the distinction is, in Weingart's (1980) terms, one of the "unsatisfactory dichotomies" haunting sociology of science, to be dissolved only by overcoming the segmentation between internal and external.

If there are merits to the co-citation clustering method, we better restrict them to its capacity for detecting 'hot spots' or locations 'where the action is' in science. Its wide scale application and time series representation (Coward, 1980) will probably lead to a more differentiated understanding of the dynamics of scientific specialties and might in the end contribute to the elucidation of the social/cognitive distinction. For the present, we tend to think of co-citation clustering as a method for the detection of specialties rather than for the dissection of paradigms.

Bibliometrics and Scientometrics

Originated in some clever techniques of library scientists, bibliometrics has developed into an almost autonomous endeavor dealing with quantifiable aspects of scientific literature. It occupies a core position in the quantitative study of science which has found an identity and an outlet in the journal *Scientometrics* founded in 1978. Bibliometric indicators have a prominent role in the *Science Indicators*[9] published biennially by the U.S. National Science Board since 1973. International organizations, such as OECD, devote substantial effort to the promotion and utilization of science indicators. It might seem that with bibliometrics and scientometrics, we have finally developed a tenacious hold upon the phenomenon 'science'. However, maybe also partly because of its explosive character in recent years, the quantitative approach to science of science has become very controversial. In particular citation studies constitute the focus of strong debates.[10]

Whereas upholders of the quantitative approach consider references as unobtrusive[11] measures of genuine affinity and intellectual descent, easily tapped and processed, opponents look upon them as highly ambiguous and poorly understood artefacts of the scientific literature which are easily misused. They find it rather simpleminded to think of references which in scholarly papers are provided in footnotes, as footprints indicating the route along which the author has reached his results. Not only are there different ways of citing, such as negative versus positive citations, which mean very different positions toward the cited document, i.e. rejection or support. Many crucial items and influences are not cited at all and many of those that are cited bear only a superficial relation to the work presented in the paper. Furthermore, it is generally questionable whether the scientific literature reports and reflects so faithfully the social and intellectual processes of scientific research. One should only trust that literature and analyze what it reflects when truly conversant with the context of its production, viz. the research laboratory, the university department, etc. Observations and objections like these have led to a rather skeptical attitude regarding quantitative methods in the study of science. The prevailing attitude, e.g. Edge (1979), is not unlike the donkey's attitude toward reading in Orwell's *Animal Farm*. Though he has managed to acquire the skill of reading and has read many things, he has not yet come across a piece of writing sufficiently rewarding to justify the effort of learning to read.

Some earlier adopters of the quantitative approach, such as Whitley (1969) have joined the ranks of the skeptics. But the shift in their position is not solely based on disillusionment with scientometric methods. It is also the expression of enthusiasm for what some consider a complementary method and what others consider an alternative approach: the social study of the *informal* organization of science.

CHAPTER EIGHT

INFORMAL GROUPS AND THE ORIGIN OF NETWORKS

> ... in quest of an education ... , thrown into
> personal contact by that combination of design
> and accident which always governs human
> affairs. – Panofsky, E., 1953, p. 323.

It might seem ridiculous to doubt whether the papers that appear in scientific journals are really intended to be read. Why would scientists engage in reporting their observations and results unless they would want their colleagues to know what they are doing and to communicate what they have found? What would be the point of careful editorial screening unless reliability and readability were the purpose?

Some sociologists of science seriously doubt whether the major use of scientific papers is in the readership they obtain. According to their point of view, scientific papers function primarily as titles to intellectual territory. They provide evidence of a particular piece of research and substantiate a claim for exclusive possession of a particular segment of a domain of knowledge. The publication in a scientific journal certifies them as legitimate requests for intellectual ownership of a fragment of science which is the sum total of 'certified knowledge'. Scientific publications thus function primarily as cadastral registrations for intellectual property, not as genuine communications.

Are there any facts to support this view and is there an explanation that makes it plausible? With respect to data, the fact is that scientists do not read the literature so intensively as one would expect if they were heavily dependent on it for information. In an impressive series of investigations on communication in psychology,[1] APA (American Psychological Association) came to the rather surprising conclusion that papers distributed through several thousands copies of a journal issue were only read by a few hundreds of subscribers. Swanson (1966) showed that only ten percent of the relevant biomedical literature in cancer research is read by the specialists in the area.

132

Don't they have time to read? Hagstrom (1965) quotes an indicative response of an eminent statistician: " . . . a man must decide whether to read or to write . . . " (p. 45). Also de Solla Price in his editorial statement for *Scientometrics* suggests that "journals, after all, are for publishing in rather than for reading" (1978, p. 8). Apparently, reading is not the major use of scientific publications.

What is the explanation? Are scientists sloppy and inefficient, doing things over and over again because they fail to check in the literature what others have done? No, another fact is that they are very keen on priority questions (Merton, 1957) and hate to redo what others have done unless it is for the purpose of replication. If they do not read as much as we expect, the reason might be very simple: they probably know already what is in the paper. An active scientist does not depend solely on the formal channels of communication by means of journals for keeping in touch with his area. The more active he is, the more he will have established informal contacts to monitor what is going on. He will know the researchers in his area from meeting them at conferences or from private correspondence and he will be acquainted with their research and generally with their findings well ahead of the time they become accessible through the journals. He will know what goes on in his area by being a member of the *informal communication network*, the social unit in science which has become popular under the label of *invisible college*.

Invisible Colleges and Specialties

Invisible colleges constitute the informal counterpart of the units of growth introduced in the previous chapter. They can be considered a 'natural' outgrowth of the search for specialties.

Citation networks, including those constructed by means of co-citation clustering, are based on highly cited papers. Citations are unequally distributed over citable authors and they tend to exemplify the mechanism of positive feedback or the law of 'success breeds success' or, with Merton's (1968) preferred term the 'Matthew-principle'.[2] Above a certain 'visibility threshold', a small elite of authors attracts ever more citations while the mass of authors sharing few citations lose the few they have and sink away into total oblivion. Some of the earliest studies of social structure in science combined citation analyses with direct sociometric techniques such as interviews or questionnaires. The results amplify the view suggested by citation studies in terms of a concentric model for the social structure of specialties. Central to the specialty is an *elitist* core group which dominates by number of publications and citations and which coordinates activities. Around the core group one finds a *peripheral* group which is substantially larger in population but which is in many ways, intellectually and socially, dependent upon the members of the exclusive circle of the core group. Crawford's (1971) analysis of the informal communication links between the 218 members of the community of sleep researchers revealed that 11 percent of the community received 54 percent of the choices [3] while 45 percent of its members received no choice at all. The 33 members who received most choices turned out to occupy key positions for the diffusion of information within the local subgroups to which they belonged and they disposed of an informal communication network to keep in touch with each other on the level of the whole group. One can introduce more refined distinctions between social strata in specialties. In his analysis of animal physiology researchers, Whitley (1969) distinguishes between high, low and middle-status groups, thereby discovering a quite active middle-group with an important screening function for the acquisition of new core members.

Whatever their internal structure, these are the kind of units we are looking for. They correspond to those indicated by Kuhn in the *Postscript 1969* as "communities of perhaps one hundred members, occasionally significantly fewer" (1970, p. 178) with "recourse to attendance at special conferences, to the distribution of draft manuscripts or galley proofs prior to publication, and above all to formal and informal communication networks including those discovered in correspondence and in the linkages among citations" (pp. 177–178). The description is not very different from one which Price and Beaver (1966) offered for similar groups and for which Price (1963) reintroduced the label *invisible college*. Originally, this term was used by Boyle around 1646 [4] to designate the precursor of what later on (in 1662) became the Royal Society. In its use then as well as in the more recent use of a closely related term such as *circle*,[5] as, e.g., in Wiener Kreis (Vienna Circle), the connotations are: minimal formal organization and free interaction within a relatively closed community.

Terminological confusion might arise from using indistinctively the notions *invisible college* and *specialty* to designate the productive groups which are behind growth points in science. In line with Crane (1972, pp. 34–35) we could restrict to the two-group distinction with on the one hand the members of the entire specialty, and, on the other hand, the members of an elite sub-group, the invisible college. With his sense for neat mathematical relationships, Price prefers to think of the size of such in-groups as being equal to the square root of the number of people in the specialty. In addition, he holds them responsible for half of the total production. Before considering again such numbers (see Chapter Nine), we will focus upon the structure and origin of specialties and invisible colleges.

Characteristics of Invisible Colleges

Griffith and Mullins (1972) and Mullins (1973) provide a listing of properties of invisible colleges which has resulted from the analysis of various productive and influential informal groups in science. They notice the following distinctive features:

— invisible colleges turn up in the scientific literature as areas of fast growth and high density in terms of the connectivity between items of the literature;

— this expansion in the literature and the concentration of researchers that goes with it is related to a radical *theoretical innovation* or *new method*;

— the utilization of the new idea occurs according to *well-defined procedures* and within well-defined limits. Based on Mullins analysis of molecular biology, the example of the Phage-group is mentioned with its focus on the study of seven bacteriophages on only one host;

— there are a few *geographical centers* which provide the pioneers and 'founding members' of the specialty. For most members at these centers, participation in the area develops into their highest priority;[6]

— there is *intensive interaction* between the members of the group both at professional and at extra-professional level. They meet each other at symposia and congresses. They exchange preprints and keep in touch by mail. At the occasion of meetings, they might also engage in some form of group recreation which might range from a modest dinner in a local restaurant to demanding mountain climbing or risky desert-tours;

— there is an *organizational leader* who takes care of the funding and the management of joint events, the distribution of communications and the relations with official agencies in science and administration;

— there is an *intellectual leader* (occasionally he is also the organizational leader) who has discovered the theoretical innovation or new method which constitutes the center of attraction and the raison d'être of the group. Usually, he has provided a publication or informal document that describes and defends the intellectual innovation. He also serves as the principal critic with respect to the contributions of the other members of the group. In the Phage-group, Max Delbrück is the uncontested leader. In ethnomethodology, another group studied by Mullins, it is Garfinkel who functions as the intellectual leader while Cicourel fulfills the role of organizational leader;

— somewhat surprisingly, the group has a rather *high rate of turnover.* While there is overall growth, that does not mean that everybody who joins will stay for a limitless time. Again taking the Phage-group as an example, Mullins mentions 3,2 years as the mean time of stay within the peripheral group and 6 years for core group members. Occasionally, an intellectual or organizational leader might stay much longer providing a misleading impression of static stability;

— as can be inferred from the previous chapter, the group has a *restricted lifetime.* The leaders and founding members of the first years are the most productive ones. While the number of participants grows, productivity gradually declines and finally the group dissolves. If meanwhile, it has attained some formal status in terms of an official institution, a residual group may remain as a rearguard without intellectual attracting power.

In view of these characteristics, the crucial factors in the inception of such groups seem to localize in the nature of the innovation which functions as point of attraction, the nature of those who join the group during the initial stages and the nature of their informal collaboration during that formative period. Could the latter provide a key to the mechanism of group formation?

Invisible Colleges and Small Groups

The discovery of the small informal group as a major component in formal organizations is almost ritualistically connected with the Hawthorne studies. The name now stands for a series of findings and convictions which stress the importance of informal groups and informal communication channels at the expense of formal components (see Rogers, 1976, p. 38). The real workings of complex social systems are supposed to occur behind the scenes, controlled by small unofficial groups. Certain orientations in sociology of science seem to develop along these lines when they stress the importance of specialties and invisible colleges as units of *social control*. Specialties are considered as reference groups providing norms to their individual members. They are some refined version of the Mertonian norms discussed in Chapter Five. Control on the adherence to the norms is executed by a small number of peers who determine through the distribution of rewards (e.g. citations) who is to be promoted and who is to be sanctioned (by being ignored).

The social control approach has served as background for some pioneering social studies of science. Hagstrom's (1965) study of the informal organization of the scientific community has been a major impetus for the study of 'specialties'. While in his analysis, social control seems more distributed over the whole specialty, the equally important studies of Cole and Cole (1973) emphasize the role of the smaller elitist core. However, the wealth of interesting empirical findings provided by these studies tends to cover up the theoretical difficulties which accompany this approach. Implicitly, individual researchers are viewed as an amorphous mass of cognitive information gatherers upon which a system of social control is superimposed to amplify the activities of some and to induce pressures on the others to align with those selected as stars. In its most crude form, such a social pressure hypothesis would come down to the *conformity model* of Asch (1956). By means of simulated group pressure an individual is brought to the point that he accepts even blatantly false statements on the length of lines in a simple drawing.[7] The social control approach is more complex and more subtle. It is of no use to put scientists in a forced compliance situation. Science does not operate by vote and passive followers do not contribute and do not count very much. To have a grip upon the productive scientist and to control the quality of his products, the social control has to have bearings upon the quality of his products as well as upon his behavior.

The social control approach views the individual scientist as loaded with a set of augmented Mertonian norms and assumes that his scientific success or failure results from the degree to which he succeeds in making these norms the principles that govern his behavior. The norms guarantee that what he produces is scientific while the group keeps him sticking to the rules. In principle, by emphasizing the norms, the social control approach recognizes the need for an active contribution by the subject in the process of knowledge acquisition. However, whether in the form originally proposed by Merton (see Chapter Five) or in the extended version by Barber (1957) or in the list of counter norms discussed by Mitroff (1974), the norms are far from being generic cognitive structures that shape perception and direct investigation. Despite suggestions that the analysis of norms can be extended in the direction of cognitive structure, the gap between cognitive structures and norms seems still immensely wide. The dual process of control, norms on individual and group on individual's sticking to norms, might cover it up but it does not fill it. This is most apparent in the use of the notion of information. Hagstrom's position is that "social control in science is exercised in an exchange system, a system wherein gifts of information are exchanged for recognition from scientific colleagues" (1965, p. 52). This might well be true but how do colleagues recognize a particular information gift as relevant to their trade? It is not too difficult in an established group with a well-defined world model. Social control might be relevant to the continued existence of such groups, but how could it bring such groups into existence? How do the founding members of the group find out about their mutual interests and the potential relevance of each others work? Information is probably prior to social control. To understand how it achieves the status of a well-defined entity in a scientific group, we have to understand the formative communicative processes in such groups.

Communication Patterns and Information Flow

The upsurge of quantitative studies on formal communication channels has been counteracted by the prevalent impression that informal networks in science were significantly more important. Current convictions seem to restore some balance with Latour and Woolgar's (1979) warning that "some care needs to be exercised in interpreting the relative importance of different communication channels" (p. 52) and Knorr's (1981a) emphasis on the heterogenuity in contact and information sources of the laboratory scientist. They bring back to mind Menzel's (1966) findings about the system characteristics of the information channels in science and the synergic interlocking of formal and informal communication.

The study of mass communication offers a simple and celebrated model for studying the relations between formal and informal communication channels. Mass communication is concerned with information through mass media such as newspapers, radio and television. After the one-step *injection* model in that area proved inadequate, the *two-step flow* model became the dominant hypothesis.[8] According to that model, information diffused through mass media does not reach all members of an intended audience directly. It takes two steps to reach the majority of them. In the first step, information transmitted through the media reaches a special category of alert and influential people, the so-called *opinion leaders*. In the second step, those opinion leaders pass on the message in face-to-face contact with members of their local community. The first step makes use of *formal* communication channels: the mass media. The second step involves an *informal* channel: interpersonal communication.

Apparently, the model could fit scientific communication nicely. The scientific journals could be considered as formal channels, while the members of the invisible college could be viewed as informal opinion leaders. They would *control* the flow of information in science. Through their personal influence, some messages on the formal channels would have an impact on certain communities within science while the impact of other items of information, unamplified by them, would remain negligible. Though there are some problems with the validity of the two-step flow hypothesis (see Nan Lin, 1972, p. 202) it does not seem too implausible to fit the diffusion of innovations through the scientific community at large. Not every Ph.D. holder and university or industry employed scientist is an active participant in a specialty. The majority of them has to devote most of their time to duties such as teaching, administration, use of science in local developments and applications, etc. One could say that their reference groups are more local than the cosmopolitian clubs that constitute specialties and invisible colleges. They are probably dependent upon opinion leaders to know what is going on at the 'cutting edge' of areas they relate to. At this point, they are not highly relevant for developments within the specialty.

For diffusion of information within the specialty, the two-step flow model seems not applicable unless we are willing to reexamine the notion of information channel drastically. Obviously, for communication within the specialty, media such as journals are too slow. Remember, however, that Crawford's analysis identified persons with a key position in information distribution as a core group within the sleep-researchers network. Reconsidering the

composition of productive groups in such networks, it is striking to notice
that, basically, they exemplify two types of social relations which correspond
to two types of communication. First, there are the few pioneers, sometimes
working together, but more often working at different places as director of
one of the few geographic centers involved. Secondly, there are the students,
working in close collaboration with one of these pioneers. Obviously, in such
a constellation, pioneers will almost naturally develop into 'opinion leaders',
although within this type of network they will transmit information produced
within the system rather than information handed over from outside. However, we should not restrict our attention to the manifest flow of messages
and the visible points of contact in an established network. This is superficial
compared to the elaboration of a shared world model involved in the first
interactions of pioneers and the first recruitment of students. These two
types of communication are more important in their contribution to the
construction of the cognitive structure of the specialty than in their influence
on the flow pattern of information.

Interdisciplinarity and the Origin of Specialties

Interdisciplinarity is an ambivalent term in science. Usually, it is discussed
in the context of urgent practical problems which manifestly need a team
of various specialists to be dealt with more or less effectively. Engineers,
economists and biologists are involved in solving agricultural problems. Jurists,
sociologists and psychologists sit together to draw up plans to deal with
juvenile delinquency. For practical problems it is considered valid and un-
avoidable but for theoretical purposes in science, interdisciplinarity is handled
with great caution and even with suspicion. While they pay lip service to
the principle, most scientists look upon their own discipline as either too
incomplete or too immature to be coupled to another one. The prevailing
attitude seems to be: first disciplinarity before engaging in interdisciplinarity.
However, when focussing upon the origin of specialties, the reverse position
might turn out to be true with interdisciplinarity preceding rather than
following disciplinarity. Law's (1973) analysis of X-ray protein crystallography
is suggestive in that respect.

Law makes use of two Durkheimian concepts which Downey (1969)
reintroduced for describing social relations in science: *organic* and *mechanical*

solidarity. In his application, mechanical solidarity is "defined as the development and maintenance of relationships which depend on shared standards and exemplars, and hence on a relatively high degree of consensus about theory and method" (p. 279), whereas organic solidarity is "defined as an aspect of the division of labor in which scientists come into relationship with one another because one performs services which the other cannot easily carry out for himself" (p. 279). Although Law distinguishes between different types of specialties (theory-based, method-based and problem-based), his argument tends to be that organic solidarity networks precede mechanical solidarity clusters and specialties. The terms 'network', 'cluster' and 'specialty' are borrowed from Mullins' more differentiated analysis of stages in the formation of a specialty. X-ray protein crystallography resulted from an interaction between X-ray crystallographers and protein researchers in the second quarter of this century. The X-ray protein crystallography community consisted of a method-based group which typified mechanical solidarity. The protein researchers appeared to be a more heterogeneous community interested in a particular subject studied from different points of view. Around a few important members of the X-ray crystallography group, who are 'permitted' participation in the other group as well, arises a new group which is to develop into a specialty on its own.

After a detailed analysis of the development of radio astronomy and a comparison with several other specialty studies, Edge and Mulkay (1976) conclude that a general model of specialty formation and development is still not available. However, a recurrent trait and theme, repeatedly emphasized by Mulkay, is *migration*. Intellectual and social migration of scientists who are in a marginal position in respect to their own group is somehow connected to the innovations which give rise to new specialties. Marginality should be understood in a broad sense. The X-ray crystallographers in Law's study who were allowed to move to the protein community were prestigious members: Astbury and Bernal. Marginality and high reputation are compatible in cases of prestigious scientists using unorthodox ways to innovate. While their prestige allows them to make risky moves, their work remains marginal to the main endeavors of their group of origin (in Law's terms: their work is 'permissible' but 'not preferred'). This is in line with the Homans model for noncomformity used by Mulkay (1972, p. 48) to account for nonconformist behavior in both high and low status categories of scientific communities and conformist behavior in the middle status category. Mullins' (1972) account of the origins of molecular biology provides another example of a migration led by a prestigious scientist. Delbrück, a physics student of Niels Bohr, moved into biology where he organized the Phage-group to tackle 'the secret of life'. In general, though the criterion for success is the disciplinarity exhibited by the coherent and uniform group which is to constitute the specialty, at the origin we find, often as the result of migration, heterogeneity and interdisciplinarity.

Migrations into Psychology: Two Examples

Though there is an extensive literature on migration in science,[9] disappointingly little of it deals with the intellectual transposition and recombination of skills and ideas which lead to the innovations that draw people together and result in new groups. The discipline of psychology provides both in its origin and in its current dynamics some examples of the scope and impact of migration.

The Origin of Psychology

A classical study of migration in science is Ben-David and Collins' (1966) analysis of the origin of psychology. Sometimes new scientific disciplines are considered to be products of geniuses who in an extraordinary act of discovery and vision provide outlines and lay down foundations for an intellectual edifice that then takes many generations of scientists to erect. Several disciplines reinforce this view by developing modest cults in the honor of the founding father(s). Thus, psychology, as a full-grown scientific discipline, arranged its 22nd international congress in 1980 at Leipzig, the place where its founder Wilhelm Wundt set up the first laboratory of experimental psychology in 1879. But what were actually Wundt's merits in the establishment of psychology as a new discipline? Ben-David and Collins show that in addition to possible intellectual innovation, Wundt's case clearly fits a model of *migration* which is familiar to sociologists and which, besides some intellectual 'capital', takes into account institutional opportunities and individual ambitions.

According to Ben-David and Collins, new scientific roles develop through a process of *role-hybridization* which results from an *intellectual migration*. Because of a lack of opportunities through saturation, overpopulation, high competition, etc. in the field in which they have been educated, some scientific workers might move into other fields where competition is less severe and opportunities are less restricted. They bring with them, however, the methods of their original role and use them to give a new identity to the domain they enter, thus creating a new (hybrid) scientific role.

In nineteenth-century Germany, between 1850 and 1870, physiology happened to be a very prestigious fast-growing field. Between 1850 and 1864 fifteen chairs were created which were mostly split off from anatomy. By 1870, however, the German universities were near saturation with respect to physiology chairs and related professions such as extraordinary professor and 'privatdozent'. This made possibilities for promotion extremely scarce for younger physiologists like Wundt who started his career in 1857 when competition for the new chairs was already severe. After having been passed over for a professorship in physiology at Heidelberg in 1871, Wundt apparently selected a new field. In 1874 he becomes professor of philosophy at the university of Zürich and in 1875 he receives the philosophy chair at Leipzig. The transition can pass as a good example for any migration. From physiology which was high in prestige and attractiveness but low in opportunities, Wundt moves to philosophy, a domain strongly attacked and battered by science since about 1830 and accordingly low in prestige and attractiveness but providing more favorable conditions for promotion. To resolve the *identity crisis* induced by this change, Wundt has to *innovate* in order to show that he is reshaping the inferior field of philosophy by preserving the superior methods and standards of the field in which he had been educated. This explains why psychology as a discipline came into being as *physiological philosophy*.

Ben-David and Collins' study attempts to illustrate that a body of problems and ideas does not spontaneously develop into a scientific discipline. Only under certain sociological conditions an intellectual endeavor couples with a social movement and becomes institutionalized as a science. In suggesting a segmentation between the cognitive and the social however, the model fits the social control approach which superimposes upon the cognitive world of ideas the social world of rewards. Wundt does not move into philosophy because he feels inspired by its methods or intrigued by its problems, but because he is in intense need for recognition. This one-sided emphasis on the institutionalized reward system cannot account for all cases of migration. Reducing the motivating forces to a desperate need to become a professor fails to appreciate the sense for intellectual excitement and the cognitive opportunism which induce a scientist to engage in sometimes socially risky moves into foreign areas. The relation between artificial intelligence and psychology is instructive in this respect.

The Field of Psychology and AI

The relationship between AI and psychology is comparable to the relationship between radio astronomy and optical astronomy thoroughly analyzed by Edge and Mulkay (1976). In the way radar techniques and radar equipment, developed during the war, were instrumental in launching radio astronomy, the development and availability of powerful computers around the fifties was a major factor in the launching of AI. And in the way radio astronomy evolved from a separate and autonomous development into a major component of astronomy at large, AI is developing into a major component of the study of cognition at large. However, positions have not yet stabilized in the area of cognitive studies.

In the early pioneering days of AI and computer simulation of cognitive processes, it was expected that AI and cognitive psychology would tend to diverge. Computers were thought to be so fast and so powerful that they would soon excel by orders of magnitude the slow and sloppy procedures of human thinking. As automobiles and airplanes extend the range of human locomotion in such a way as to make it almost incomparable to what is possible with only two legs, so engineers expected that 'thinking machines' would extend in a similar way the range of human brains. In a way, of course, computers have brought such an extension without, however, achieving the superiority that was hoped for. We do not need to repeat the development traced in Chapter One for pattern recognition and language processing. It is typical for the development of AI in general that, while rephrasing its problems from algorithmic into heuristic ones and from general heuristics into domain specific heuristics and representations, it grew in respect and inquisitiveness for human cognitive processes, which, despite apparent weaknesses, are still vastly superior to computers in generality and flexibility. Barbara Hayes-Roth (1978) concludes a discussion of artificial knowledge systems with the remark that " . . .the most promising avenue of approach toward the development of artificial knowledge systems may be to investigate and model human cognition as closely as possible" (p. 346). Chandrasekaran (1975) plainly states, in a review called *Artificial Intelligence — The Past Decade* that "AI can be legitimately looked upon as falling squarely within the field of psychology" (p. 174). Also according to Newell, who, in 1980, became the first president of AAAI (the American Association for Artificial Intelligence), the endeavor could as well be called *theoretical psychology* (1973, p. 25), though he had previously indicated that "scientific fields emerge as the concerns of scientists congeal around various phenomena. Sciences are not defined, they are recognized" (1973, p. 1). Only if we consider AI the product of a migration from computer science to psychology, a disputable interpretation, the direction of the migration is in agreement with what Ben-David and Collins would predict: from prestigious fields to less prestigious fields. The other professional conditions seem not fulfilled: there is apparently no lack of opportunities in computer science and no abundance of easily accessible positions in psychology.[10] If there is a movement towards the study of human cognitive processes, it is because of cognitive demand rather than because of social pressure. The AI community has developed an urgent need to understand how human cognition operates. If it is available in psychology, they seek it in psychology. If it is not available in psychology, they construct it, in collaboration with psychologists or with other disciplines or on their own, irrespective of the disciplinary turmoil their moves might cause. Cognitive interest backed by the computer as an increasingly powerful new instrument is the driving force here.

Innovation and Discipleship

It should not escape our attention that in the migration-model applied to science, immigrants are not leaving the specialty of their education without cognitive resources. Jean Piaget is also an illustrative case of an innovator in psychology and epistemology who does not hesitate to stress the importance of his training as a biologist. Sometimes the success of the immigrant is attributed to naivety. He enters the new domain equipped with skills and methods that are often not familiar to the autochtones but which might prove very valuable for the area. In addition the immigrant is not burdened by the accumulated a priori beliefs that might trouble those who have been working in the area for a long time. His ignorance permits him fresh approaches which insiders would not dare to consider because the product of their long experience contains a high dose of skepticism. Therefore, from time to time the apparently contradictory statement that "he realized it because he did not know it was impossible" applies to that immigrant. But eye-catching advantages of naivety should not lead us into underestimating the importance of knowledge and expertise carried along.

Though innovation might seem to be coupled to erratic and unpredictable movement, there are also impressive chains of continuity in science through master-apprentice relationships. Zuckerman (1977) has traced these relations for a sample of Nobel prize laureates and stressed their importance. Apprentices need not to remain in the area of their master to exemplify this continuity. Delbrück, who innovated in molecular biology, has already been mentioned as Bohr's student. Einstein is notorious for saying that he used Minkowski's geometry because that happened to be the one he had been taught by his teacher Minkowski. The conservation and transmission of a tradition of scientific leadership and innovativeness has again both sociological and cognitive aspects. Through their position in the social organization of science, masters can introduce their protégés more effectively into the leading circles and innovative areas. But again, this aspect should not induce us to neglect the substantive cognitive skills that are transmitted in high quality teaching. Such teaching does not necessarily include very didactic and spectacular courses that could have the indoctrinating effect Kuhn is pointing at. We do not mean the kind of good teaching that is only needed for the weaker students.[11] It is the kind of teaching Ravetz (1971) refers to when he emphasizes that good science has much in common with medieval crafts and guilds in which skills are transmitted from artisan to apprentice "largely through a close personal association of master and pupil" (p. 103). Such teaching is probably the most effective way of installing cognitive structures which remain intact when the apprentice migrates to other fields or encounters revolutionary turmoil. Modern mass media courses might install a disciplinary paradigm but mastery teaching in the older fashion establishes the more solid skills that are constitutive to specialties and that survive paradigms.

In late applications of concepts developed in other areas of sociology and social psychology, sociology of science has tended to focus upon the role of the small group and its complex internal pattern of interaction. The ultimate goal, however, is to reveal, in the tradition of sociology of knowlege, the detailed links between social structure and cognitive content. Some current trends attempt to retrieve this connection from a minute analysis of the communicative interactions on the location where science is supposed to be constructed: the laboratory (Latour and Woolgar, 1979, Knorr, 1981b). An emphasis on the 'structural contextuality' of the construction of science need not be in contradiction with a recognition of the importance of historical antecedents brought in by education. However, this kind of imprinting and cognitive transmission of skills has attracted less attention from sociologists of science than the transmission of information between colleagues. Focussing upon the master-apprentice relation does not offer an escape from the pressing need of disentangling the connection between the cognitive and the social but it might provide another vantage point.

Both quantitative and qualitative sociology of science have discovered the 'small groups' in science. The hundreds of specialties that replace the few tens of disciplines as units of growth might harbor a comparable number of cognitive units. In his search for "the new approach to the study of history of science that has been emerging, one that looks for fruitful ideas in fields ranging from the philosophy and sociology of science to psychology and aesthetics", Holton (1973, p. 11) introduced the notion 'themata' to designate basic units that keep recurring under various forms and recombinations. After having witnessed the recent flourishing of sociology of science, which readapted the older 'themes' of *small groups, mass communication, diffusion of innovation*, etc., this approach obviously has some appeal. As paradigms in Kuhn's (1962) sense, such themes can both be taught and function as the center of attraction for a group, but in addition, they can account for the basic continuity that is underneath innovations brought by migration. The suggestion of both more continuity and more innovation than Kuhn's original notion of paradigm does not make the search for generic cognitive schemes less intriguing.

CHAPTER NINE

THE LIFE CYCLE OF SCIENTIFIC SPECIALTIES

> Steady advance implies exact determination of
> every previous step. – Sarton, G., 1962, p. 66.[1]

In some applications of the paradigm concept, the difference between implicit knowledge and expectations on the one hand and explicit and established knowledge on the other, is accounted for in terms of a trade off relation maintained over the paradigm life cycle. In its initial phase the paradigm is a promising source of ideas evoking possible investigations and generating expectations about their outcome. As the suggested research is gradually realized and solid data are assembled, the knowledge involved gradually becomes more articulated and the share of programmatic principles diminishes. Finally, the suggestive power of the paradigm seems used up and the scientific community is left with a more or less stabilized body of knowledge which can either be added to the store of certified knowledge or classified as a dead end. Viewed on the level of specialties, in Yellin's (1972) terminology, the specialty enters the 'post developed state'.

Several attempts have been made to trace paradigms in terms of a developmental model of specialties. Some of them distinguish four stages within paradigm life cycles. Goffman (1971) introduces a sequence of stage 1: insufficient and unordered information, stage 2: insufficient but ordered information, stage 3: sufficient but unordered information and stage 4: sufficient and ordered information. Crane (1972) couples the four stages of knowledge development to four developmental stages for scientific communication. In stage 1, when the paradigm appears, there is no developed social organization; in stage 2 when normal science flourishes, invisible colleges appear; in stage 3 with the major problem solved and anomalies turning up, social splitting occurs; in stage 4 with the paradigm exhausted, the number of participants decreases. With a sequence of stage 1: exploration, stage 2: unification and stage 3: decline or displacement, Mulkay *et al.* (1975) restrict to three stages. However, all these models attempt to cover the whole

148

life cycle, i.e. from birth to death. Mullins (1973) four-stage model with a sequence of normal stage, network stage, cluster stage and specialty stage focusses more on the earlier formative period. Others, such as Clark's (1968) focus on later stages such as institutionalization. Most of these approaches suggest specific links between the 'cognitive state' of a specialty and its social structure in any particular stage.

The Stages of the Specialty Life Cycle

Figure 9.1 provides an augmented table of stage-characteristics super-imposed upon Crane's (1972) segmentation of the logistic growth curve for specialties.

The table combines features collected from various stage-models, including those which Radnitzky (1973) uses to characterize the dominant methodology at each stage. Among the most influential critics of Kuhn, Popper (1970) stands out as the strongest defender of a view which sees science as the product of an inexorable methodology. Whatever the origin of ideas, whatever their appeal or apparent soundness, there is an ultimate criterion to decide whether what is proposed is scientifically tenable or not. Methodological criteria range from all-embracing principles to long lists of methodological rules, such as, e.g., Bunge's (1963, pp. 100—107). The intriguing observation to be made with respect to specialties is that, while they are relatively shortlived, the methodological principles that govern the behavior of the participating scientists vary substantially over time. This is not solely important for arriving at a more dynamic philosophy of science. As we have repeatedly seen, sociologists are tempted by the idea of considering norms as the cognitive reflection of social structure or stratification. The changing content of methodological norms through the various stages of the specialty life cycle, traced together with cognitive content and sociometrically accessible aspects of structure, might give us a hint whether such norms are indeed social or cognitive, or both. Guided by Radnitzky's suggestive labels, we will now focus on the methodological norms in a global characterization of each stage.

The Pioneering Stage

The intellectual innovation which constitutes the foundation for further

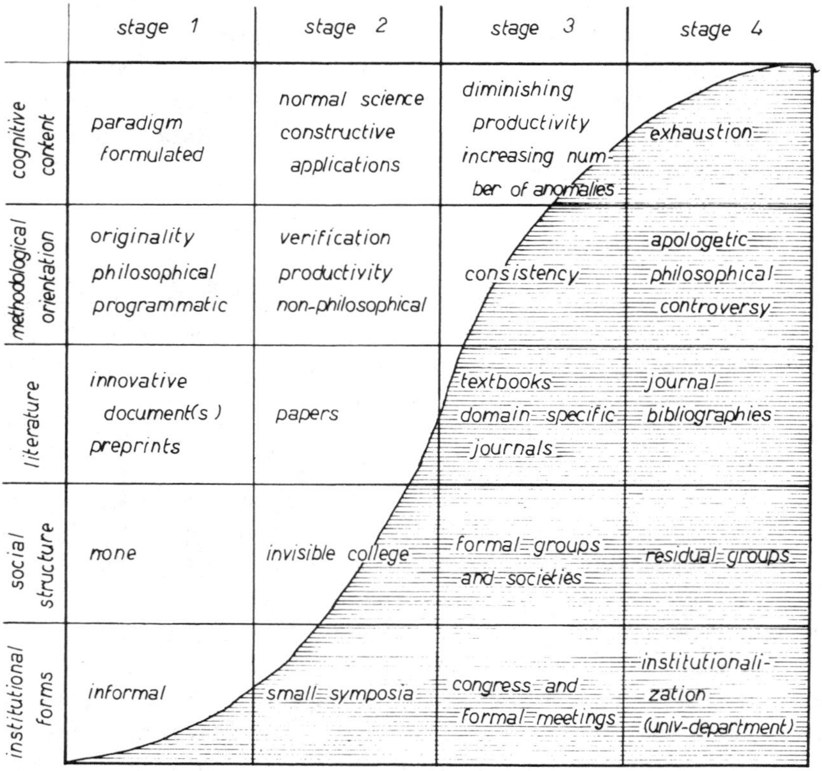

	stage 1	stage 2	stage 3	stage 4
cognitive content	paradigm formulated	normal science constructive applications	diminishing productivity increasing number of anomalies	exhaustion
methodological orientation	originality philosophical programmatic	verification productivity non-philosophical	consistency	apologetic philosophical controversy
literature	innovative document(s) preprints	papers	textbooks domain-specific journals	journal bibliographies
social structure	none	invisible college	formal groups and societies	residual groups
institutional forms	informal	small symposia	congress and formal meetings	institutionalization (univ-department)

Fig. 9.1. Characteristics of the life cycle of scientific specialties in relation to the various stages superimposed upon a logistic growth curve.

development is formulated at this stage. As indicated in the previous chapter, this might be done in a rather covered way by a very restricted number of pioneers who hide from publicity both for fear of piracy and for fear of severe methodological critique at an inopportune time.

In line with Max Delbruck's *'principle of limited sloppiness'* (reported by Mullins, 1972), Radnitzky notices a "high willingness to take risks", "openess to new points of view" and an emphasis on *creativity, originality* and *richness of ideas*. The dominant attitude can be called philosophic in the sense that quite general beliefs are questioned and replaced by programmatic alternatives based on speculation. The end product of this stage might be a programmatic text circulated through informal channels. It might be made available through the formal communication system as well. Minsky's *Frame paper* (1975) can be considered a published program text for the M.I.T. version of cognitive science. However, many of its major ideas were earlier accessible through a 'progress report' (Minsky and Papert, 1972), that circulated in AI circles.

Minsky's (1975) paper provides some representative statements. General and programmatic is:

It seems to me that the ingredients of most theories both in artificial intelligence and in psychology have been on the whole too minute, local, and unstructured to account – either practically or phenomenologically – for the effectiveness of common-sense thought. The 'chunks' of reasoning, language, memory, and 'perception' ought to be larger and more structured, and their factual and procedural contents must be more intimately connected in order to explain the apparent power and speed of mental activities (Minsky, 1975, p. 211).

A version of the 'principle of limited sloppiness' is presented in terms of a rejection of parsimony at this stage combined with what Radnitzky calls "a polemic edge directed against some older competing tradition" (p. 33):

Workers from psychology inherit stronger desires to minimize the variety of assumed mechanisms. I believe this leads to attempts to extract more performance from fewer 'basic mechanisms' than is reasonable. Such theories especially neglect mechanisms of procedure control and explicit representations of processes. On the other side, workers in artificial intelligence have perhaps focused too sharply on just such questions. Neither have given enough attention to the structure of knowledge, especially procedural knowl-edge . . . . It is understandable why psychologists are uncomfortable with complex pro-posals not based on well established mechanisms. But I believe that *parsimony* is still *inappropriate at this stage*, valuable as it may be in later phases of every science. There is room in the anatomy and genetics of the brain for much more mechanism than anyone today is prepared to propose, and we should concentrate for a while more on *sufficiency* and *efficiency* rather than on *necessity*. (Minsky, 1975, p. 215; italics ours).

The Building Stage

Once the pioneers have formed a minimal network and have recruited a first generation of students, the thrust of the activity is no longer directed towards promising programs but to productive investigations that turn the first stage programmatic statements into solid knowledge supported by *empirical evidence*. It is the stage during which rapid growth becomes manifest as new participants join and contribute their personal piece of empirical support. Standard research methods are developed and widely used. Around the beginning of the sixties, Chomsky-inspired psycholinguistics adapted some older techniques of mentalistic psychology to study the processing of isolated sentences. They constituted a framework for many experiments in the area and contributed much to the conceptual expansion of the specialty and to its growth in numbers of experimental papers and participants. However, the second-stage participants tend to be highly productive so that a high increase in papers is accomplished by a relatively small group of members. In general, there is a tendency to agree easily on the supportive nature of the empirical evidence provided. Phrasing their remarks in statistical terms, some authors have indicated this tendency as a high risk for type I error: the unjustified acceptance of a hypothesis (i.e. unjustified because of inconclusive evidence to reject the null hypothesis). Philosophical discussion is strongly rejected. Apparently, it might endanger specific commitments which are no longer considered to be disputable. To take again the example of psycholinguistics: one of the hallmarks of Chomsky's doctrine is the competence-performance distinction (see Chapter Four). Second stage psycholinguistics did not allow questioning of this principle. It was part of the rules of the game and those who wanted to join had to accept it or at least to respect it by refraining from open critical discussion.[2]

Stage of Internal Criticism

The third stage has a Janus face. At the surface everything looks great and the paradigm is highly successful, but internally the first signs of disintegration are perceptible. Intellectually, the group has obviously conquered a new territory and the major task is now to organize it in a well-ordered fashion. The empirical data which have accumulated in a rather disorderly way during the building stage have to be classified and assimilated into the conceptual system without endangering its coherence. The dominating methodological values are, in Radnitzky's terms: *rigor, precision* and *transparency*. The typical product of this stage is the *standard text* which presents a coherent picture of the specialty and introduces its history as a success story. Levelt (1973) reports how, at about the same time (1966), such texts became available for psycholinguistics. However, he also reports that several of them remained in the drawer as manuscripts without ever reaching the stage of published books. The reason is that, while the specialty is very successful when looked upon from the outside, insiders increasingly become aware of the scope and importance of resistant problems left to be solved before the paradigm can really be claimed to work. The aspirant text writer is apt to experience such a tension between success and distress most dramatically and it is understandable that pessimists among them abandon.

Paradigm-confirming research which supports the intellectual innovation is no longer easy and incompatible findings which could be overlooked during the euphoria of the building phase have to be confronted seriously. Two alternative reactions are registered. Some participants seek a solution in methodological sophistication in the hope that well-designed experiments and refined measurement will allow to settle the issues. Such an increase in methodological sensitivity should be apparent from a greater vulnerability to statistical error of type II: the unjustified rejection of an hypothesis as the expression of an overcautious critical attitude. Other participants start questioning the validity of the research methods of stage two or even the basic conceptions of the paradigm itself. Again, the distinction between competence and performance in psycholinguistics is an example of a basic distinction which, at this stage, is no longer an uncontested principle. An example which shows how doubt is casted upon the research previously done is the changing attitude, also in psycholinguistics, with respect to research based on isolated sentences. What were considered elegant and convincing experiments in stage two are now looked upon as doubtful artefacts which are of little relevance to the crucial issues. Wason (1972) documents this development in a review of his own work on negation and some related studies. In Chapter Ten we will refer to similar remarks of Moray and Fitter (1973) with respect to research on attention.

The Stage of External Criticism

With his distinction between internal and external criticism, Radnitzky wants to indicate whether the participant offers his criticism from within or from outside the specialty. While in stage three he can still be considered a loyal adherent, in stage four he is dissociating himself from the paradigm that constituted the innovation. Criticism is now based on the same general philosophical grounds that were used for the promotion of the paradigm in the programmatic first stage. Though the suggestive power seems used up and the paradigm appears intellectually dead, social activity might still be important. If the specialty has achieved an institutionalized form in terms of research centers or university chairs, it can produce its own supply of participants and be kept alive for quite some time. But the sense of unity seems gone with the intellectual challenges fragmented and the social structure is less tight.

As a consequence of the intellectual attraction and excitement produced by the Chomskyan 'revolution', professorships and educations have been established in psycholinguistics, although, meanwhile, the specialty has lost much of its original appeal.[3] While this has allowed the field to grow in terms of researchers and reports, complaints have arisen about the a-theoretical and 'dataistic' nature of the work in the area. With respect to journals such as *Journal of Verbal Learning and Verbal Behavior*, Norman (1974) makes the remark that "these journals are now filled with paradigms that are neatly and carefully done, but that often seem to answer no questions" (p. 176): typically the non-paradigmatic state in Kuhn's (1962) sense. But the methodological values with which Radnitzky characterizes this particular stage do not apply to those researchers who find themselves in a specialty or field without a sense of direction. *Dogmatization, immunization, trivialization*, etc., are typical for the over-defensive orientation exhibited by a rearguard of active defenders and cultivators of the specialty. By closing it off from external influence, they turn it into a sect. One does not need to go back in history as far as astrology and alchemy to find fossilized sciences. The values listed by Radnitzky can be pursued in today's universities as well, leading into a pathological form of isolation. According to Krantz (1972), Skinnerian behaviorism demonstrates such a defensive isolationism in contemporary psychology.

Regulative Mechanisms and Growth

The notion of specialty life cycle is nothing but a scheme for developing a grip upon the evolution of growth points in science. Radnitzky uses Weber's notion of *ideal-type* to characterize the epistemological status of the concept.

It should not be thought of as fitting in detail all specialties encountered. With respect to specialties detected by means of co-citation clustering, Coward (1980) noticed that there is a "high rate of infant mortality". Many clusters of authors and ideas which appear promising, even in citation scores, fail to maintain the kind of metabolism that leads into the full-scale developmental pattern suggested by the life cycle scheme. Also, clusters of authors and ideas that do enter into a full-scale deployment of specialty features may vary widely in pattern and profile so that they correspond only weakly to the life cycle model that we presented. Further on we will discuss some important variations. But whatever the list of features and the sequence of stages, the dynamics of the life cycle seem to derive from the interaction between the ever-recurring cognitive and social factors. Attempts to arrive at a more theoretical model of the life cycle have focussed on this interaction.

Life cycle analysis is well known in science. Among the recurrent themes that Holton (1973) presumes to be sources of inspiration, it is undoubtedly a prevalent one. Familiar from biology in terms of birth, growth, maturation and death of organisms, it has been extended to include the short-lived existence of sub-atomic particles as well as the astronomical lifetime of stars. In biology, the phenomenon of growth and life cycle have been closely linked. D'Arcy Thompson's *Growth and Form* has shown how changing size is coupled to requirements of internal reorganization manifested through a series of stages characteristic of the life cycle. The impressive elegance of his demonstrations stems from the fact that forms are not based on mysterious vitalistic forces but on such awfully simple principles as Galileo's square-cube law on the relation between size and structural stability, explaining "the impossibility of increasing the size of structures to vast dimensions either in art or in nature" (Galileo, 1638, p. 187). Researchers who have realized that a similar relation might obtain between changes in the size of a group and the nature of communication between its members are tempted to look for similarly simple formulas.

Holton's (1962) *exploitation model* derives a logistic growth curve from a combination of two factors: one related to utilizable secondary ideas and one related to population size. He assumes that each major innovation or discovery which starts a specialty contains a finite potential in second order discoveries or derived applications. The exploitation of such an original discovery is comparable with the exploitation of a new lode for someone

digging ore. When a researcher is working at a steady pace, the number of
undiscovered derivations or applications decreases in a linear fashion (see
Figure 9.2). The curve P for the population of researchers who participate in
this exploitation has an inverted U-shape, somewhat skewed in the direction
of the later stages. A curve I', based on the curve for derivable ideas and the

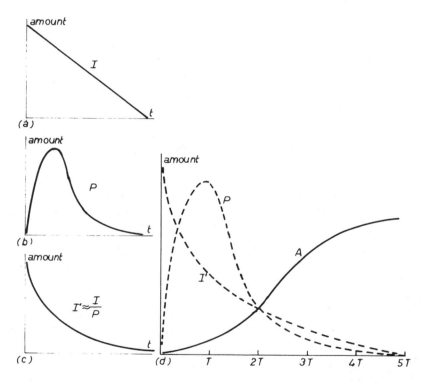

Fig. 9.2. In Holton's model of the exploitation of a major discovery the curve I expresses
the linearly decreasing number of derivable secondary discoveries for a fixed amount
of researchers (see (a)). The curve P indicates the actual population of participating
researchers (see (b)). The two curves combined yield I', indicating the actual rate of
exploitation in terms of a decreasing reservoir of derivable discoveries (see (c)). Curve A,
expressing the applications, is inversely related to I' (see (d)). (After Holton, G., 'Models
for Understanding the Growth and Excellence of Scientific Research', in Graubard and
Holton (eds.), *Excellence and Leadership in a Democracy*, Columbia University Press,
New York, 1962, pp. 122–123; reprinted with permission.)

curve of the population, illustrates the pace of exploitation at the global level
of the specialty. When accounted for in terms of an increase in applications,
this curve is inverted and yields the familiar shape of the logistic curve,
indicating this time the 'already known and applied ideas'. Holton also
introduces a period T which denotes the 'half life' of the secondary discovery
reservoir. The point of time at which half of these applications have been
discovered coincides with the point of time at which a maximum number
of researchers participates in the specialty. Though Holton indicates some
implications for university education based on an estimation of T as between
five and fifteen years, his approach also suggests consequences for individuals
who participate since it is quite clear that opportunities for discovery differ
according to the stage at which one enters the specialty. However, there are
problems with Holton's elegant mathematical simplifications. Besides some
ambiguity and confusion inherent in counting 'interesting basic ideas' and
'applied ideas' as basically identical, his two fundamental assumptions also
seem questionable. The first assumption is that the exploitation of a fundamental discovery consists in unearthing the secondary discoveries which
it entails. The second assumption is that each participant in a specialty
contributes an equal share of secondary discoveries. Further on we will have
an opportunity to discuss the first assumption. Let us first consider the
second one.

Holton's second assumption expresses what Yellin (1972) calls a *populistic*
view on scientific workers. Others have labeled it as the *Ortega hypothesis*
(Cole and Cole, 1972). It assumes that whoever participates in scientific
research makes a contribution and that the whole of science is the sum total
of the contributions of all scientists. Opposed to a populistic view is an
elitistic view that considers science to be the work of an elite whose superior
products survive the noise in which the mass of weaker contributions drowns.
Even within a populistic approach, a strict linear relationship between exploitation of an idea and the number of participating researchers cannot be
maintained. It would be like the fascinating extension of linearity demonstrated by primary school children who, once they have learned that four
persons can build a house in two weeks and that three persons accomplish it
in two weeks and four days, are eager to calculate how many thousands of
persons are required to build a house in one minute.

Because of the nature of the communication and coordination involved in

cooperation, there is an optimal number of participants above which a further
increase in population becomes detrimental. Authors who have used epidemiological models to study the spread of ideas and the growth of specialties
(Goffman, 1966; Goffman and Harmon, 1971; Goffman and Warren, 1980;
Nowakowska, 1973) have taken this factor into account. Also Yellin (1972)
in his mathematical "model for research problem allocation among members
of a scientific community" has pointed out the relevance of population size.
In particular, his suggestions for science policy applications warn for the
negative effects of a population growth as the result of external support:
"external support should be structured in such a way that its total effect
on community population growth is minimal" (Yellin, p. 35). The size of
a community obviously affects the informal contacts between members.
To keep track of the work of a growing number of colleagues increases the
cognitive load on each individual member. It seems plausible to consider
the methodological norms as expressing the relationship between cognitive
load and population size. In the earlier stages, the contributions are few in
number and can easily be tracked. No information overload is induced by
new members joining and contributing. In the later stages, informal communication channels are incapable of keeping control over the productions of all
participants. Formal procedures of control have to be installed and the risk
for a bureaucratic adherence to methodological rules (see Norman's remark
quoted above) increases as the community grows larger and larger.

On the whole, the scenario is not very different from the scheme management science provides for keeping control over a growing business. Greiner's
(1972) model indicates a series of stages which develop according to a Kuhnian type of alternation between periods of construction and periods of crisis [4]
(see Figure 9.3). Life cycle models for scientific specialties have not been
made so explicit as to locate critical points in the organization between each
phase. But, in general, they have followed the same line by locating a rather
underspecified creative event at the beginning followed by a series of social
installments of increasing complexity and rigidity. But while the changing
methodological norms give us a hint with respect to the relation between
optimal size and cognitive load, they fail to indicate the nature and the
guiding force of the original innovation. Theoretical models on the relation
between growth and structure are suggestive with respect to changes in
social structure. But what about *cognitive* structure? How does the creative

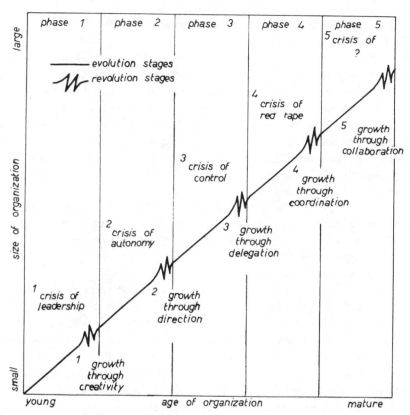

Fig. 9.3. Greiner's five phases of organizational growth should apply as well to growing scientific specialties as to business firms. They link the creative informal group to the big business company with an increasing risk of bureaucracy. (Reprinted by permission of the *Harvard Business Review*. An exhibit from 'Evolution and Revolution as Organizations Grow' by Larry E. Greiner (*HBR*, July/August 1972, p. 41). Copyright © 1972 by the President and Fellows of Harvard College; All Rights Reserved.)

contribution which is supposed to ignite the specialty relate to the secondary discoveries that are supposed to follow from it?

Finalization: Cognitive and Social in Sequence?

The *Finalizierungs*-model of Böhme *et al.* (1973), sometimes called the

Starnberg-model, is also a stage-model for the development of specialties. It is particularly focussed on the currently rediscussed value of *social relevance* of science which, in a sense, can be considered to belong to an extended list of methodological norms. But it also makes explicit provisions for a self-contained cognitive development within specialties.

The model is a three stage scheme similar to Mulkay, Gilbert and Woolgar's (1975) mentioned above. To the phases of *exploration* and *unification* in their model correspond similar phases in Böhme, van den Daele and Krohn. However, the second phase is in the Starnberg-terminology explicitly called the *paradigmatic* phase. The third phase is what they introduce as the stage of *finalization* (*Finalizierung*). Though it takes place after the maturation of the specialty, it does not have the negative connotation with which Yellin (1972) describes the postdevelopmental stage or which is suggested by the use of labels such as 'deterioration' and 'decline' as in Mulkay *et al.* (1975). Finalization means the application or utilization of developed science for *externally* supplied goals, i.e. goals provided by society at large or by specific groups or bodies outside the scientific community.

In their presentation of the finalization-model, Böhme and his colleagues keep close to Kuhn's original scheme. The explorative phase is described in terms of the pre-paradigmatic sciences exemplified by chemistry before Lavoisier or particle physics nowadays. The paradigmatic phase is brought about by a theoretical achievement which imposes a well-defined conceptual structure upon the area. For molecular genetics, Watson and Crick's assumptions with respect to the role of DNA for the transmission of genetic information constituted a paradigm. Further development of the field becomes dependent upon an internal logic. At that stage, the area or specialty is immune to external orientation or influences. Once the development imposed by the internal logic has been carried through, the specialty has reached a conceptual maturity and completeness which makes it available for external use.

Though the finalization-model sequentializes openess and closure in line with Kuhn (1962) and might thus seem applicable throughout the whole history of science, the authors consider it particularly apt to characterize the twentieth-century link between science and society. Central to their argument is that advanced theoretical models such as those found in physics and chemistry are brought to bear upon scientific troubles, e.g. medical problems, and give use to new and extensive research programs. To the layman, these programs might seem as esoteric as pure or basic research, but upon closer examination, one discovers that societal needs are definitely embodied in the program. Thus, the third phase is not simple straightforward application but scientific development shaped by external goals.

Obviously, the crucial distinction is between the internal logic of phase two and the shaping force of the external goals of phase three. How does one distinguish between self-contained internal logic and logic of external goals? Three types of activity characterize the paradigmatic phase. Either the paradigm is better founded (*Begründung*) or its domain of application and validity is enlarged (*Geltungsbereich erweitern*) or it is articulated for specific cases where it is already known to be valid (*Spezification*). In particular with respect to the second type, it is difficult to see a basic difference between an extension in the paradigmatic sense and an application in the finalization sense. Both involve the transfer of concepts with creative reorganization and ingenious adaptation. They seem to use the same cognitive resources. To qualify the paradigmatic phase as more 'closed' or 'autonomous' cannot mean a restriction on the kind of concepts or the domain of ideas invoked to contribute to the formulation of a problem or to its solution. The Starnberg-school refers to Heisenberg's concept of 'closed theory'. But Forman's (1971) analysis of Heisenberg's work in quantum mechanics shows how even concepts which reflect societal situations can eventually penetrate into apparently highly esoteric science. One can question the imperviousness of a socially immune 'cognitive' phase. While recognizing the need for a better understanding of the nature of 'closure' which gives a conceptual system coherence and integrity, opposing 'cognitive autonomy' to 'social utility' does not seem to solve very much.[5]

Sometimes a distinction has been made between internal and external sociology of science. With the finalization-model we have shifted from our previous preoccupation with the structure of the scientific community — internal sociology of science, to an inclusion of societal problems at large — external sociology of science. It did not bring us nearer to a solution of the cognitive-social issue, though it certainly amplified the need for clarification. However, what the finalization-model definitely has shown is that the life cycle of specialties need not be restricted to a single sequence of growth, maturity and decline. Mature specialties can have quite distinct phases of deployment and their line of development can seem somewhat erratic. A discussion of the diffusion of innovations model will illustrate the importance of that point. Notice also that this is in agreement with Starnbergers' critical correction of Kuhn, who they blame for a one-shot model of paradigms. According to them, Kuhn overgeneralizes the phlogiston-case. In general, replaced paradigms are not totally discarded from science. They remain available for application and utilization in other schemes. Recognizing this type of continuity, we witness for some specialties a substantial increase of the span of their life cycle.

Escalatory Expansion of Diffusion Studies

Diffusion studies deal with the adoption of innovations in communities. In Chapter Seven we presented the expansion of that area to illustrate some aspects of growth in science (see Figure 7.3). In Chapter Eight we suggested that some recent developments in sociology of science are late applications of the diffusion scheme. Indeed, while originally developed to trace the adoption of new kinds of sowing-seeds in agricultural groups, it has proven to be equally relevant for tracing the adoption of new ideas in scientific communities. Standard conceptual tools of the diffusion studies [6] include a stage model for adoption and a classification of different types of adopters. The sequence of stages which accompanies each adoption process, i.e. *awareness, interest, evaluation, trial* and *adoption*, combines with the two-step-flow-model of communication, indicating formal channels for stage one and informal ones for stage three (opinion leaders). The classification of adopters into *innovators, early adopters, early majority, late majority* and *laggards* even offers a way out in our attempt to account for changing methodological attitudes during the specialty life cycle. Different methodological attitudes could be linked with different types of scientists who join a specialty in a way similar to that in which different categories of a population adopt an innovation.

In an effort to check the accuracy of the phase-characteristics in a specific case of specialty development, we made some analyses on the diffusion of innovations paradigm. Rogers and Shoemaker (1971) contains a bibliography which covers the development of that area from its very beginning and which distinguishes between empirical studies and publications that do not report empirical results. Furthermore, they provide, for a detailed list of hypotheses which express the diffusion paradigm, a listing of the bibliographic items which either support or do not support the hypothesis studied. In principle, this should permit to verify some of the stage characteristics indicated in Figure 9.1. Non-empirical or programmatic studies should be retrieved mainly among the items of literature of the earlier phases, while negative findings or anomalies should constitute the most substantial portion of the literature reported in later phases.

Figure 9.4a and b shows the distribution of items, split up in *empirical* versus *non-empirical* and *supporting* versus *non-supporting* respectively. With respect to empirical versus non-empirical, it is clear that the non-empirical literature seems to develop as an almost constant proportion of the whole literature, roughly one third, rather than as a category mainly to be found in the earlier stages. With respect to supporting versus non-supporting studies, the proportion of non-supporting literature seems to increase but is manifestly not the category that dominates the later stages.

Are we to conclude from this that the stage characteristics are false or, at least, do not apply to diffusion of innovations as a paradigm?

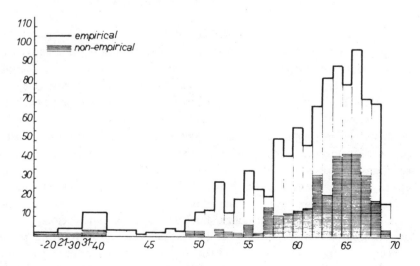

Fig. 9.4a. Distribution of empirical versus non-empirical items in the bibliography on diffusion of innovations (Rogers and Shoemaker, 1971).

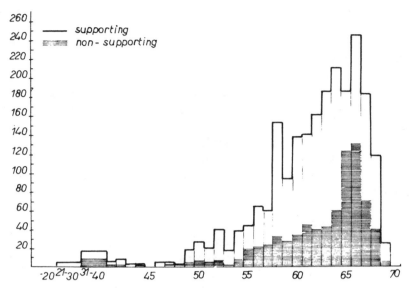

Fig. 9.4b. Distribution of supporting versus non-supporting items in the bibliography on diffusion of innovations (Rogers and Shoemaker, 1971).

Rogers and Shoemaker's bibliography contains a clue to the clarification of this puzzling result. In addition to the classification into empirical versus non-empirical and supporting versus non-supporting items of literature, the bibliography also provides a code which relates each item to certain research traditions. These range from anthropology, via education and rural sociology to marketing, public administration and statistics. Twenty such categories are included. This distinction between several research traditions in the discussion of a single specialty typifies an important characteristic of successful paradigms. Apparently, diffusion of innovations has penetrated into many different areas. Looking at its growth in more detail, one realizes that a substantial part of its expansion has been maintained by spreading into new areas of relevance and applicability. First developed in rural sociology with the diffusion of hybrid corn, applied in medical sociology with the adoption of new drugs by physicians, then marketing ... up to the adoption of new ideas by scientists. In each of these areas, the diffusion scheme might go through the specialty life cycle on its own. Conflating the whole literature into one bibliography results in mixing up late literature of one sub-paradigm with early literature of another one.

This pattern of prolongated growth through a sequential opening up of new apposite areas is not restricted to wandering themes which run through science as superficial straw fires. Edge and Mulkay (1976), as well as Woolgar (1978), have illustrated how radio astronomy in its initial phases gathered momentum by appealing to different audiences at different stages in its development. It is interesting to see how, taking advantage of gravitational pull provided by such audiences, specialties can make a tour through a specific configuration of areas before achieving a stable position in a consolidated discipline. In several papers, Whitley (1975, 1977, 1980) has explored the importance of 'work organization' for tracing such complex paths of development. He considers the specific conditions of scientific activity as the locus of interaction between cognitive and social factors and notices that significant gravitational pull might stem from 'audiences' outside university academia which impose a specific working pattern upon the area. He warns against uncritical adoption of one single pattern of development: "The diversity and plurality of patterns of change and development in the sciences necessitate taking the domination of a particular pattern of social organisation and knowledge development such as that of Physics, as historically contingent and sociologically problematic rather than the inevitable outcome of epistemological assertions" (Whitley, 1981, in press).

Forms of Specialties and Patterns of Life Cycles

Gallant and Prothero (1972) base some considerations on university science policy upon an exploration of the biological relation between size and structure discussed above. They point to the fact that, while animals can differ enormously in size, the range of cell size is rather restricted: they are of the same order of magnitude in all animals. Obviously, the metabolism of cells imposes this kind of restriction on size.

The analysis of specialties suggests a similar state of affairs for scientific enterprises. The vital unit behind all processes is the dynamic social entity which we came to know while learning about specialties and invisible colleges. Because of a delicate balance between the formal and informal relations that connect members, the size of such units can vary only within a restricted range and their overall characteristics coincide with those of a *small group*. In various states of development and in various combinations, these units can account for a much larger diversity of forms and life cycle patterns. Wasson (1978) provides some variations of the familiar logistic curve life cycle for commercial products. They seem equally applicable to a variety of forms of specialty life cycles (see Figure 9.5).

The *pyramided* cycle corresponds to the pattern discussed for diffusion of innovations and for radio astronomy. It contains a sequence of regrowth periods due to the presentation and acceptation of the augmented paradigm in new audiences. It is the kind of evolution that leads into macro-development and might result in new disciplines.

At the other end of the scale we find what Wasson calls *instant busts*. Rather than being the exception, they might be the rule if we accept Coward's (1980) remark about the high rate of infant mortality in specialties discovered by means of co-citation clustering. Many specialties are probably very short-lived and do never get beyond the pioneering stage.

In between are corollaries of *fads* and *market specialties*. An example of a fad is provided in Cartwright's (1973) analysis of research on the *risky shift*. This intriguing 'line of research' (Cartwright) developed in social psychology around the notion that groups tend to take more risk than individuals. But despite an inspiring and impressive start, Cartwright reports:

As time went by, ... , it gradually became clear that the cumulative impact of ... findings was quite different from what had been expected by those who produced them. Instead of providing an explanation of why 'groups are riskier than individuals', they in fact cast serious doubt on the validity of the proposition itself (p. 225).

Cartwright feels compelled to consider the "real possibility that interest in the topic will gradually subside and that the accumulated findings will simply fade into the history of social psychology" (p. 230). He also adds the remark that "such an outcome would not be unprecedented, for this has been the fate of many quite popular research enterprises of the past" (p. 230).[7]

Corollaries of market specialties can be found either in technical laboratory procedures or mathematical techniques. They provide specific methods experienced by certain scientists as advances which require cultivation and refinement in terms of a specialty. They might compare to a fad with a significant residual market when they are the result of methodological innovations

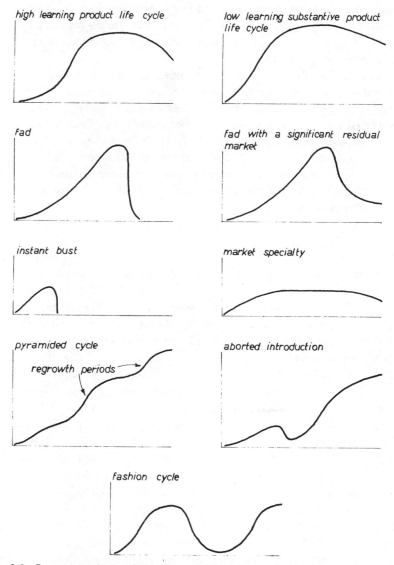

Fig. 9.5. Common variants of the product life cycle applicable to specialty growth patterns. (After Wasson, C. R., *Dynamic Competitive Strategy and Product Life Cycles*, 3rd ed., Austin Press, Austin, Texas, 1978, p. 12; reproduced by permission of the copyrightholder Dr. Chester R. Wasson.)

that, at first, are widely explored and put to test by a large segment of the scientific community and afterwards remain the major preoccupation of a selected group of specialists. Certain statistical techniques in the social science manifest this pattern.

Undoubtedly, the list could be extended and careful analysis of more specialties would reveal a greater variety of patterns. Even the fashion cycle could be made to fit science, although, as Wasson indicates, it is not to be considered a variant of the life cycle. For commercial products, fashion is a superimposed oscillatory pattern of change in fringe attributes which do not affect basic qualities. Similarly, alterations between such trends, as empiricism and rationalism, might be superimposed changes which affect less deeply than it seems the more substantial contributions offered by specific paradigms and specialties.[8]

Social Studies of Science

The social studies of science of the last decades have been looked upon as a very promising strategy in the hunt for the paradigm. This double-faced entity with both a social side and a cognitive side seemed easier to grasp through the glitter of social success than through the inspiring forces of its cognitive qualities. However, started with the hope to get to the cognitive side as well, some basic presuppositions had to be adapted to new findings and some of the earlier ambitions had to be given up during the course of the search. Now one wonders whether the vivid hopes have been fulfilled or whether we did lose more than we have gained.

Price's (1971) impression of invisible college research might well be extendable to the social search for the paradigm in general. "Invisible college research" is, according to him, "rather like hunting the unicorn. Somehow, we got word of the existence of such a beast, and we set out to hunt it in various ways appropriate to hunting something which would be like a sociological clique or peer group". Further on: " ... having caught the unicorn, we looked at it and found that it was a perfectly ordinary animal, a normal sociological clique ... " (Price, p. 5). Having hoped to finally lay hands on the mysterious paradigm, we might feel disillusioned when we are left with nothing but applications of sociological and social psychological concepts which have been discovered and developed in other areas.

Who can remain enthusiastic about diffusion on innovations in science once we realize that soap has been sold to us according to marketing schemes derived from diffusion research long before we realized that it might be applicable to science? Now that we have found the groups, where is the paradigm?

Unfulfilled expectations should not cause us to loose sight of the vast changes in our image of science that have resulted from developments in the social studies of science during the last decades. We used to have a static eternistic model of science. Science was a complicated machine taking massive amounts of data as input and producing solid scientific knowledge as output: everlasting knowledge. The mechanism of this machine: the scientific method, guaranteed truth and scientifically established truth seemed immutable. Kuhn (1962) has become the symbol of a generation that has cast serious doubt upon this image of science. Science has acquired a dynamic and discontinuous nature and became non-cumulative in character. Herewith, the monolithic motionless body of truths crumbled and left room for an image of action, change and diversity. Social studies of science have amplified this picture. We have discovered a science which is in vehement motion with hundreds of turbulent areas: specialties in various stages of development. We have learned how within these areas a feverish level of activity induces a fast rate of turnover for participating scientists. We have located the communities we were looking for but their composition changes faster than we first dared to expect. We had associated intellectual innovations with specific groups but we have seen that successful systems of ideas might embark on trajectories through several groups, thereby gaining substance and momentum. The finalization discussion has increased the diversity and possibilities for heterogeneity by stressing the persistent availability of surpassed paradigms. Although this last characteristic implies more conservation, sometimes of even fragmented achievements, the other characteristics mean more motion and higher rates of change than Kuhn (1962) suggested. Quite some distance has been covered since we left the orthodox view. From the illusion that we knew what science was, we have moved to a position where we face an enormous diversity of social endeavors in knowledge production, utilization and even manipulation.[9] But the most basic questions have not been answered: what is knowledge and how is it generated? What is the source and the structure of this product of intense social activity? Cognitive structures, despite the importance attached to them by sociologists of science, have remained elusive. Would they be easier to catch within the confinements of the individual mind than within the confinements of the group?

PART THREE

COGNITIVE STRUCTURE AND DYNAMICS OF
SCIENCE

PARADIGMS AND THE PSYCHOLOGY OF ATTENTION AND PERCEPTION

> The paradigm observer is not the man who sees and reports what all normal observers see and report, but the man who sees in familiar objects what no one else has seen before. – Hanson, N., 1958, p. 30.

Among the celebrated authors of cognitive science, Sir Frederic Bartlett should stand out as particularly instructive for understanding the combination of social and psychological aspects that typifies paradigms. As indicated in Chapter Six, a paradigm is a *social* kind of entity in that it is supposed to provide *inter individual cohesion* within a group of scientists. At the same time, it is *psychological* in that it is supposed to account for *intra individual coherence* and directionality in the cognitive processes that guide the research activity of the individual scientist. The relationship between the social and the psychological is not simply a matter of juxtaposition. In his remarkable study of memory, Bartlett (1932) indicates how there is an intricate relationship between the basis provided by socially negotiated schemes of interpretation and the variation introduced by individual applications of such schemes. In one of his elegantly simple experiments, the task involves the serial reproduction of a picture by a group. Each participant transmits the picture to his, say, left-sided neighbor, by redrawing it for him from what he remembers from what his right-sided neighbor has shown to him. Some results are reproduced in Figure 10.1. They illustrate how reproduction, kept faithful for a while by means of one plausible scheme of interpretation (a *bird*) can suddenly go astray when an unimportant detail triggers another totally new but equally plausible interpretation (a *cat*). Bartlett identifies this as the outcome of a tendency to assimilate the picture as a whole in terms of accepted standard representations counteracted by a tendency to preserve some detached detail, whereby that detail might constitute the explosive material that can offset the interpretation of the

173

Fig. 10.1. Serial reproduction of pictorial material by the members of a group. Each participant redraws from memory the picture produced by his predecessor. Notice the similarity of this type of development with the degeneration of hand-copied illustrations before the advent of print as signalled by Eisenstein (1979, Note 51, p. 470). (From Bartlett, F. C., *Remembering. A Study in Experimental and Social Psychology*, Cambridge University Press, Cambridge, 1932; paperback edition 1967, pp. 180–181; reprinted with permission.)

whole. Further on we will have to develop this important relationship between whole and detail as belonging to the heart of the matter in intra-individual analysis of cognition. Here it is sufficient to notice that this simple scheme applies to scientific communities when we consider them as composed of similar groups whereby members pass on to each other something comparable to a picture. On the one hand, they strive to adhere more and more to the socially negotiated standard representation, while on the other hand they preserve individuality by the selection of a particular detail. Though the shared interpretative scheme is cultivated by the group and as such a group characteristic, its use by the individual member surpasses mere reiteration of the scheme. Individual use constitutes a source of fluctuations and modifications that changes a static interpretative scheme into a dynamic entity. Individual use is a major factor in the dynamics of the scientific specialties which we have discussed in the previous chapter as social entities with a life cycle of their own. Therefore, it is now necessary to study what this cognitive scheme is that scientific groups install in the brain of new adepts. How does it operate in normal science mode and how might it become the locus of revolutionary change? How does it establish its hold on the individual mind and why, at a given time, might it loose its grip and disintegrate?

Gestalt Perception and Gestalt Switch as Exemplars

An example recurrently invoked by Kuhn to illustrate both what paradigms are and how they change is the rabbit or duck gestalt picture designed by Jastrow. Notice that Bartlett's experiment mentioned above also exemplifies a classical gestalt switch. The change from bird to cat in Figure 10.1 exemplifies, in a gradual fashion, a reversal similar to the transformation from duck to rabbit. Understanding this kind of phenomenon involves an answer to two separate questions. First, how do we see something, either an object or a picture,[1] as something specific? How do we recognize a set of curves and dashes as a duck? Secondly, what is it that makes us see the rabbit in something that we have previously identified as a duck? Kuhn compares the first kind of achievement with the ability to "find the animal shapes or faces hidden in the drawing of shrubbery or clouds", tasks which abound in children's puzzles. It is another type of exemplar — embedded figures — which should indicate the nature of the guidance provided by a paradigm. A paradigm is what allows to locate the hidden figure, to identify the meaningful pattern in a set of otherwise unrelated and meaningless lines.

In the spirit of some solid AI traditions, we should first attempt to solve these apparently simple cases of embedded figures and gestalt switch in the hope that solutions eventually found for these toy-problems will prove of substantial significance for analyzing paradigms and revolutions in the more complicated cases of mature science.

Perception and Selective Attention

The note with which Boring introduced Hill's picture *My Wife and My Mother-in-Law* (1915) into psychology (Boring, 1930) contains a few lines of history of ambiguous figures in psychology and relates it to the problem of attention. He compares it to Rubin's (1915) goblet profile and mentions how Titchener, the leading proponent of the psychology of attention, used such puzzle-pictures in his own experiments. Referring to some of Köhler's illustrations of sensory organization in *Gestalt Psychology* (1929), he indicates that: "'attention' and 'sensory organization' are simply different names for the same phenomenal fact that such figures illustrate" (p. 444). Recent discussions of similar issues reemphasize the role of attention. Dealing with hidden figures in their *Language and Perception*, Miller and Johnson-Laird point out that such task "is almost *purely a matter of controlling attention* to particular features" (1976, p. 137, italics ours). Given this importance attributed to it, one wonders what attention really is.

While attention was considered of central importance to psychology before the era dominated by behaviorism, the latter school looked upon the notion as a confusing mentalistic term and did away with it. However, shortly after Shannon's major publication on information theory (1949), Colin Cherry's investigations on the *cocktail-party-phenomenon* provoked a spectacular revival of the study of selective attention. Again the problem was phrased in terms of perception, the central idea being that, at any time, a multitude of stimuli impinge upon our eyes, ears and skin and that out of that multitude, we select a small portion to which we pay attention. At cocktail parties we succeed in participating in a discussion within the group we join while, at the same time, we manage to disregard the noisy group next to us. Coupled to the prestige of information theory and aided by some newly available equipment such as tape-recorders,[2] this prototypical case of information processing induced a fast growing area of attention studies. It has now developed into a specialty by itself that has reached the stage of a full-fledged professional society: the 'International Association for the Study of Attention and Performance'.[3] The history of this rediscovery of attention is particularly revealing in that it is quite similar to the development of the cognitive view dealt with in Chapter One. Attention reappears as the study of the subject's contribution in perception. In parallel with the presentation of the cognitive view, we will review four stages in thinking on attention in order to see whether they lead up to an explanation of gestalt perception and gestalt switch.

Discrimination Learning Without Attention

Ambiguity being present everywhere, problems of attention can turn up in even apparently simple problems of animal learning. Suppose you train an animal using a stimulus with a low degree of complexity, e.g. a red square. Does the animal respond to the redness of the stimulus or to the squareness or to both? The orthodox behaviorist view sees no need for attention theories which have the animal proceeding through some kind of hypothesis testing, focussing on one attribute in some trials and on another attribute or some combination in other trials. According to him, an animal is not confronted with a confusing cocktail of attributes which force it to make a selection. The central idea is that with every learning trial, the conditioned response is coupled to all aspects of the stimulus situation through what Pavlov calls stimulus generalization. Over a series of such trials, the differential gain in association strength of the relevant attributes over irrelevant ones brings automatically about the discrimination of the effective stimulus. In spite of impressive achievements of Spence's (1937) model for discrimination learning, it has not been able to prevent the development of a lively controversy about the role of attention in animal discrimination learning (see Gilbert and Sutherland, 1969; Sutherland and Mackintosh, 1971). Would we attempt to extrapolate the behaviorist position towards the rabbit or duck figure, it should predict that the subject will come up with the response that has been reinforced most frequently.

Limited Channel Capacity

As indicated above, the major revival of interest in attention was brought about by Colin Cherry's (1953) analysis of the cocktail party phenomenon. To study this at MIT Research Laboratory of Electronics, he devised the technique of *dichotic listening*. Through earphones a subject is presented simultaneously with two different messages: one to the right ear, the other to the left ear. He is instructed to *shadow* one channel, i.e. to repeat instantaneously what he hears in one ear. Once the task is completed, he is interviewed with the purpose of finding out what he might have picked up from the other — rejected — channel. At first Cherry's results seemed to indicate that we can close an ear almost as easily as we can close an eye. Only gross physical characteristics of the rejected channel, e.g. male versus female voice, might be noticed. However, Broadbent (1958) discovered that, depending on the characteristics of the message, similar results could be obtained with two messages presented to one ear or both messages presented to both ears. Selection did not appear so peripheral as switching on and off receptors. Broadbent postulated that all inputs are registered by the senses and temporarily held in a short-term memory while a *selective filter* controlled by a central processor operates on the elements in the short-term memory store. The further story of this line of research is the story of ingenious attempts to locate a selection device somewhere along the information processing channel from stimulus to response. A discovery of Moray (1959) that some particular messages such as, e.g., the subject's own name, were regularly detected on the rejected channel, was used by Treisman (1960) to develop a *filter-attenuation theory*. She introduced different thresholds for various signals according to their degree of importance to the subject (own name always being a high priority item to detect). Deutch and Deutch (1963) went further in the direction of the response by having a selection dependent on temporary interests of the subject. By that move, the selective attention line of research came very close to the third approach which focusses on the subjects' involvement and alertness.

Activation and Motivational Selection

Selective attention studies tend to be one-way information-processing approaches emphasizing the complexity of the stimulus situation and looking for mechanisms that reduce and combine information elements into coherent messages. The multitude of stimuli which impinge upon the senses has, however, an internal counterpart in terms of a multiplicity of states that affect the reactivity and a multiplicity of drives or goals that activate an organism. *Activation* denotes such states as sleep, drowsiness and alertness which are coupled to well-defined brain wave patterns. Though these kinds of state might be influenced by external events (general activation of the cortex via reticular formation as described by Moruzzi and Magoun, 1949), it is also obvious that a rhythmic component is basic to their selection as, e.g., in the diurnal cycle of sleeping and waking. Rhythms are internal selectional devices that operate relatively independent of current input. They apply as well to the selection of specific drives and goals as to the switching of general states of alertness and receptivity. At regular times, independent of external events around us, we become hungry and when in an unknown street, that makes us more apt to spot a restaurant. A great variety of investigations, ranging from experimental studies on P.R.P. ('psychological refractory period' as some fundamental rhythm for information processing; see Bertelson, 1966) to research on set connects to this category. They all share a preoccupation with selective mechanisms imposed by complexity localized in the organism rather than by complexity localized in the incoming information.

Confronted again with great variety and diversity, it is tempting at this point to venture a reformulation in terms of the phases discussed in Chapter One.

While the behavioristic approach could be called monadic in the sense that there is just a one to one mapping of energetic events onto the receptors without any selection or reduction, the selective attention approach introduces a structural bottleneck for incoming information. Due to the limited capacity of the central processor, only certain combinations of attributes are allowed to enter as information. The activation and motivational approach is similar to the contextual stage in that it illustrates how, in addition to structural aspects of the stimulus, various other factors can be proven relevant to human information processing. Though the information is thought to be in the stimulus, this orientation acknowledges the need for some active contribution by the subject in order to have that information assimilated. The great diversity of opinions and suggestions indicated above is typical for contextual factors. The major obstacle seems to stem from the idea that because the 'genuine' information comes from the outside world, the contribution of the subject cannot be very specific and, in cases where it becomes specific, it has to be subjective in the sense of modifying or distorting. Therefore, cautious experimentalists restrict to momentary variations in alertness and receptivity for incoming information while some clinicians explore Rohrschach-type[4] hypotheses in the conviction that if the subject has to contribute something it cannot be but elements revealing his personality. The latter position is in line with some so-called 'New Look'-findings, e.g. that poor children overestimate the size of coins compared to children from wealthy backgrounds. Bruner and Goodman's (1947) *New Look*-paper leaves no doubt about this orientation, being entitled 'Value and Need as Organizing Factors in Perception'. While these approaches might suggest that we see either a duck or a rabbit according to personal taste or other subjective preferences, they fail to explain the more basic problem of how we perceive. How do boys, poor and rich alike, identify a piece of metal as a coin? Somehow, the contribution of the subject in the recognition of an object should be more specific than either some preparatory alertness or some subtle subjective deformation.

Task Demands: The Cognitive Orientation

One of the early pioneers of the revival of attention in psychology, Moray, has been among the first to express the suspicion that things were going fundamentally wrong with the selective attention orientation. In Moray and Fitter (1973), he complains with Fitter that "despite the large amount of research that had been done in the area of selective listening, theory in that area is remarkably poor". However, in the same paper in which he expresses this disillusionment, he also points to a discovery which, in his appreciation, might revolutionize the study of attention. The discovery is Senders' (1964) model for monitoring complex visual displays. In his characterization of the model, Moray emphasizes two aspects which are highly significant. First, he indicates that in Senders' approach, the subject of the experiment "has constructed an internal model of the environment to which he must pay attention, and that model controls his attention" (p. 4). Secondly, with respect to previous research on attention, he points out that "the voluntary direction of attention we usually see in laboratory experiments is merely the early, inefficient stage of acquiring a model that will eventually control behavior without the observer being aware of it" (p. 5). The second remark amplifies the manifest cognitive orientation expressed by the first. It suggests that experimenters have been analyzing the wrong member of a duo. They have manipulated the stimulus environment while attention resides in knowledge brought in by the subject. *Attention* mechanisms are *embodied in* the *expertise acquired* in a given skill. If there is something to attention, it is in the subject that we should look for it. Again we witness the typical shift from an emphasis on the object to an emphasis on the subject.

The trend exemplified by Moray is also indicated by others. In a discussion of perceptual development, E. Gibson remarks: "I see the need for reexamining the concept of attention, and weighting heavily the role of the task assigned . . . . Perhaps *attention is* simply *perceiving that information which is coincident with task demands*" (Gibson, E., 1977, p. 170, italics ours). If we admit that task demands have a representation in the subject, then attention reflects the knowledge contributed by the subject in an act of perception.[5]

In general however, though Moray's appreciation of Senders' model suggested a full-scale revolution, it has not produced the concomitant turmoil and reorientation one would expect to occur in the community of attention researchers. Probably there are several plausible reasons. One might be that Senders' model is based on classical information theory concepts so that, to some, it might appear as a regression. But the basic reason might well be that the cognitive orientation inevitably, in attention studies as well as in picture processing or language processing, leads to the cognitive paradox. According to Neisser (1976): "*Attention is nothing but perception*: we choose what we will see by anticipating the structured information it will provide" (p. 87, italics ours). The problem is that, when perceptions depend on anticipations, it is difficult to see what the further point is of perception. Furthermore, it becomes impossible to conceive of the perception of the unexpected. Nothing seems solved by moving massively into the subject all the information that was previously localized in the object.

According to a cognitive view, confronted with an ambiguous figure, we should perceive the version which our knowledge predicts. Leeper (1935), however, discovered that verbal knowledge not necessarily translates automatically into the kind of expectancies which allow for an interpretation of an ambiguous figure which is consonant with the knowledge provided. Having been instructed to look for an old woman in the *wife* and *mother-in-law figure* does not guarantee that the mother-in-law will be seen at once or that she will be more easily detected. Although attention should be oriented towards data that allow an interpretation in terms of an old woman, the available data have an organization of their own and do not yield automatically or easily to the imposed knowledge. Quite to the contrary: despite the knowledge, one might find oneself trapped in the young woman version. The problem obviously is not solely, as a one-sided cognitive orientation might suggest, one of translating ideas into tangible forms of perception. Tangible forms of perception and ideas interact in complicated ways and the intriguing question is the mechanism of this interaction rather than one way translations or reductions. To understand an eventual steering function of knowledge, we should first try to disentangle this interaction. The study of attention reveals the same paradox which is characteristic for the cognitive approach in general but it does not go beyond it. Perception is the key issue. As Neisser says: "perception is where cognition and reality meet" (1976, p. 9). It is perception we should tackle straightforwardly.

A Stratified Model of Perception

The observation which is recurrently mentioned by students of perception as well as by students of attention is the enormous richness with which the perceptual world floods our perceptual systems. The character of *vividness* with which Hume distinguishes things seen from things remembered is the apparent abundance of details in real perception contrasted with the skeletal nature of remnants in memory. Perception seems to provide flesh to bony structures of memory. In our problem of determining the respective contributions of the perceiver and the world in any act of observation, the flesh and bone metaphor leads to a highly suggestive model. The perceiver is to be conceived of as providing a wirework structure which the world decorates with numerous colorful elements, thus bringing vividness to what otherwise remains but a shade. Perception results from dipping a skeletal memory scheme into the ocean of observational elements. This conviction that perception results from elements which enter through the senses and which are organized by the mind along the lines of remnants of previous encounters with the same object is deeply ingrained in both classical empiricist views and common-sense thinking. Notice that such an approach reduces perceptual entities to the role of bottom elements in a conceptual pyramid. This is, however, basically misleading.

Of course, a two-component model with scheme and data is superior to a crude uni-directional causal chain model of perception. However, how are we to conceive of the link or interaction between scheme and data? How do they ever meet? Should we think of the subject as a source of schemes radiated as expectations in all directions, while the world floods the subject with data? Should we reduce percepts to fortuitous collisions between outward bound schemes and inward bound data? We will see that unless we abandon thinking about data in terms of bottom elements in a conceptual hierarchy, *scheme* and *datum* will remain ill-sorted components for any interactive model of perception.

What We Notice When We Read

As we have seen in Chapter One, a genuine exemplar for the cognitive view is the misreading of words on the basis of expectations. The fact that we miss misprints while reading proofs illustrates, on the one hand, that it is possible to *see* things that are *not there* while, on the other hand, it is obvious that such illusions go back on *something really seen*. What do we see when we read? How can we miss an error such as, e.g., *content* on a location where there should be *context*? The common-sense interpretation is simple. Since it is only a one character difference, we have apparently overlooked the character in the position of the *x*. If that common-sense interpretation is right, it has important implications which are rarely faced. Obviously, seeing is not an all-or-none process: you can see a word while not seeing all its constituent characters. With respect to the psychology of reading, this can hardly be called anything new. Long before Gestalt psychology, authors such as Cattell (1866) and Huey (1908) discussed the question whether reading involved either the recognition of *global configurations* (the gestalt) or the recognition of specific *arrangements of letters* (the elements). Ever since, it has been a major problem in the psychology of reading whether it is the *whole-word* view or the *letter-by-letter* view or some intermediate *letter-cluster* view that can best account for the data in the reading process. It still seems a basically unresolved issue (see, e.g., McClelland, 1977).

Both the common-sense view and the etymology of the verb *to overlook* suggest, however, some kind of stratified view on the process of reading. Depending on contextual requirements, sometimes words might be identified by their global shape alone, sometimes by means of particular letter clusters, and sometimes by scrutinizing specific letter arrangements. There are occasions where it might even be necessary to decompose a single letter into smaller units, as, e.g., when a parent is deciphering the hand-written homework of his seven-year old son. Apparently a word is a multilevel kind of entity which consists of a global shape and of constituent parts (letters) which are themselves composed of even smaller parts (features). The reading-process shifts back and forth between these levels and it might make more sense for psychologists to find out how these levels combine rather than to persevere in attempts to reduce reading to a single level event.[6]

For our present purposes, the important point is a notion of *global* aspects of words which are *directly accessible to perception independent* of the *perception* of constituent *elements*. This allows indeed for a plausible model of misreading. *Overlooking* is the genuine perception of the *global* shape of a higher order entity and of *some* components, illusionary extended to the perception of the particular shapes of *all* components. Would a similar approach be applicable to the perception of objects and scenes?

What We Notice When We See

The multi-level arrangement encountered in reading turns out to be equally suitable for the description of our perception of simple objects and scenes. Although we walk around most of the time with our eyes wide open, we should realize that we overlook many things which our environment allows us to see. Probably we miss these aspects of our environment in the same way as we miss misprints in our proofs. A man might come home and mention to his wife that he talked to a common acquaintance. She might ask "What did she wear? How was her hair?". To his astonishment, the man realizes that he does not know. It is not like he has remembered it for some time and has forgotten it by now. Apparently, he has never perceived those particular aspects of the meeting with the other person. He might recall the topic of the conversation, the location of the meeting and a few other things, but, obviously a lot of aspects have remained unnoticed. If we would not have any suggestion derived from the model of reading, we could feel confronted with the same situation that ignited the revival of selective attention research. If, however, the structure of the perceptual world can be seen as similar to the structure of a text which is composed of nested entities (letters, words, phrases, etc.), we can try the powerful selective mechanism that seems to operate in reading. Are there several perceptual levels in the accidental encounter that we have taken as an example?

Apparently there are different levels of perceptual analysis for such simple events. A passer-by who crosses my path on the sidewalk will probably be identified in terms of the shape of a human body and a very rough localization of major parts such as head, arms and legs. When I meet a friend in the street by recognizing his typical silhouette, I zoom in on his face to fixate his eyes and the several details of the face acquire perceptual significance. When, however, I look at an anonymous crowd, the human bodies are just textural elements in an otherwise amorphous object. An object such as the human body seems to fit into a pile of part-whole relationships at various levels. It can function as either a whole or a part or a textual element. Depending on his particular purposes, the observer may restrict himself to noticing the general shape only and, as in reading, this might occasionally lead him to the illusionary recognition of an acquaintance with attribution of perceptual detail to an object which afterwards turns out not to be the acquaintance but a stranger.

A revealing demonstration by Palmer (1975) shows how a multi-level arrangement could be involved in almost any act of perception. Moreover, it illustrates how the identity of a perceptual object can derive from a specific selection of levels of analysis.

The Stratified Structure of Simple Objects

Palmer's illustration is based on pictures. We should take seriously Gibson's (1979) warning against the notion that perception of the world is comparable to perceiving pictures. In the study of attention we have seen that the sensory system is commonly regarded as flooding us with an overwhelming richness of data for which selection is considered the most pressing exigency. In pictures, that selection has already been made by the draftsman and our perception of the picture might be quite different from our perception of the object represented by the picture. Palmer (1975) is, however, relevant because he indicates how for a picture a minimum number of levels of analysis seems required in order to qualify as the representation of an object. In an indirect way this tells us something about the way human perceivers arrive at the perception of an object. Consider the recognition of faces. An elliptical structure on its own (see Figure 10.2a) might not directly stand for a human face or head. It could represent many things. Neither does an unorganized set of dashes, angles or smaller ellipses evoke the notions of eyes, nose, mouth and ears. However, combined in proper proportions, the two categories of figures represent a human face in an unambiguous way (see Figure 10.2b). If the larger ellipse is made part of a figure representing the human body (see Figure 10.2c), it can be recognized as a face without the additional details of eyes, nose, mouth, etc. These simple illustrations [7] demonstrate that object identification can be quite accurate whenever structural information is revealed in the combination of two or more levels of analysis.

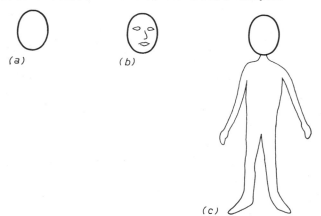

(a)   (b)

(c)

Fig. 10.2. An ellipse (a) on its own does not necessarily represent a human face or head. When filled in with the appropriate details (b), it represents a face in an unambiguous way. If the ellipse is made part of a figure representing the human body (c), it can be recognized as a head without any detail. (After Palmer, S. E., 'Visual Perception and World Knowledge: Notes on a Model of Sensory-cognitive Interaction', in Norman, D. A. and Rumelhart, D. E. (eds.), *Explorations in Cognition*, Freeman and Co., San Francisco, 1975, p. 296; *copyright © 1975 by Freeman and Co.*, reprinted with permission.)

In a generalization of his scheme, Palmer (1977) following the practice in AI,[8] distinguishes between three levels of structural units for pictorial representation of an object:
— the *whole* figure;
— the multisegment *parts*, and
— the individual *line segments*.
We should notice that these are perceptual traits which are partly independent of each other, accessible for perception on their own:
— the whole figure can be represented by the *outline* or *contour* which, like a silhouette, is derived from the border, the continuous contrastline of the figure with the background, and which might reveal the global *shape* of the object;
— the multisegment parts derived from *internal contours* or edges which allow for the identification of a structural skeleton of the major *parts*;
— individual line segments which correspond to *textural* elements, microunits that reveal the nature of the external subparts.
To grasp the power of this scheme, we should realize how radically it differs from our common-sense notions on perception. The common-sense notion is this. Every object we readily see can in fact be decomposed into a number of constituent parts. A human body is indeed a head, a trunk, arms and legs organized in a specific way. These parts in turn can be decomposed into smaller parts. An arm consists of upper arm, forearm and hand. And again, a hand can be decomposed into palm and fingers. The idea is that, though we can do this a number of times, we cannot go on endlessly with such a decomposition. Ultimately, we will reach a bottom level. That bottom level is thought of as the level of elementary sensations. Conceptually, an object can be decomposed into constituent parts and this can be reiterated several times, but in the end, perceptually, there is only a huge collection of sensations. Different levels of aggregation correspond to specific ways of conceptually grouping these sensations. In this way we arrive at the more or less elaborate conceptual pyramid with, at the bottom, the elements of perception. The bottom layer of the conceptual pyramid is supposed to constitute the interface between knowledge and reality. As indicated previously, it is the heritage of empiricism: reality provides the elements, the mind does the organizing.
The 90-degree turn in the new scheme about perception is literally that in

this pyramidal model we detach the bottom layer of perceptual elements and bring it in an upright position along the hierarchically organized conceptual system. This is what we have done in the example of the *face* (see Figure 10.3). Indeed, whenever we accept that the global shape of an object is

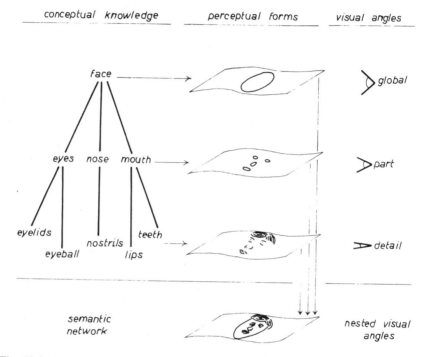

Fig. 10.3. The proposed theory of vision presupposes a stratified structure of percepts corresponding to the hierarchical structure in the semantic network which represents the knowledge of the perceiver. Perceptual features are available at several levels of aggregation of the object. The percept is like a combination of overlays which correspond to levels in the conceptual network. The perceptual features which specify shapes at the different levels can be the same (elementary categories of form). What varies is, in a sense, visual angle. Object perception involves 'zooming' back and forth between global shapes and specific details. Veracity of object perception is dependent on a minimal number of levels on which perceptual features are registered and between which part-whole relations obtain. This is in line with Gombrich's remark that "perceptual skill . . . comes with familiarity and allows us to proceed from the overall form to the detail" (1979, p. 102). The key concept in this approach is the relationship between overall form and detail.

accessible to perception independently of the way in which constituent parts are perceived, we introduce an hierarchical dimension in perception which parallels the one encountered in the conceptual pyramid.

Having sketched a new way for looking at perception, we should now come back to our original question and investigate whether along the lines of this scheme, a model can be developed that handles the classical ambiguous figures and their gestalt switch.

Interactive and Integrative Processes in Perception

Despite its apparent simplicity, perception involves the extraction of salient features at various levels of aggregation of an object or a scene. The general impression is that in real life situations there is a continuous shifting back and forth between global views and close-ups of details. It extends over many levels but at each level the features can be specified in the same kind of units. Though the subjective assessment of an object of perception as 'real' seems to involve its assimilation at two or more levels, there are always both more global and more detailed features that are ignored in any situation.

Top Down and Bottom Up Versus Concept Driven and Data Driven

Identification of an object might proceed from recognition of a significant detail (e.g. a textural element) towards recognition of the shape of the whole as well as from the recognition of the whole to the identification of the relevant detail. This means that the connection between levels can be established in both directions, i.e. *top down* as well as *bottom up*. With respect to this terminology, the notions 'top' and 'bottom' might be misleading. In this gliding conceptual hierarchy, there is no real top or bottom. It seems better to use Duncker's terminology, i.e. 'from above' which means 'from the whole to the parts' and 'from below' meaning 'from the parts to the whole'. Furthermore, the notion of bottom might suggest in a misleading way: the bottom of sensations at the base of the conceptual pyramid. In the approach we have presented, the bottom is not in any way nearer perceptual elements than the top! Data are orthogonal to this dimension.[9]

In a hypothetical reconstruction of the *duck-rabbit*, we suppose that the drawing is encoded at various angles in terms of a basic vocabulary of forms. These constitute *data*. Some of these, the most salient ones, evoke a conceptual interpretation, e.g. the large protruding structure is interpreted as representing a beak. Thus far, we would consider this *data driven*, i.e. from pictorial form to conceptual element. The interpretation of part of the figure as a beak allows the perceiver to invoke a conceptual network in which beak is part of bird. A conceptual network of this kind provides a kind of ladder for conceptual *top down* or *bottom up* explorations. Ascending the conceptual ladder from beak up to the level of bird is a bottom up conceptual process. Transforming the concept of bird into its standard pictorial representation so that the perceiver will be on the look out for the outlines of a bird should then be considered *concept driven*. [10] It is a higher order activated concept that generates specific expectations in terms of pictorial elements. Only when top down and bottom up processes are clearly separated from data driven and concept driven processes can one conceive of a genuine interaction between conceptual knowledge and data in perception. Otherwise one ends up with the collision model whereby perception would only result from clashes of outgoing expectations (concept driven/top down) with incoming data (data driven/bottom up). An analysis with two versions of an ambiguous figure specified in hypothetical data driven and concept driven interpretations is given for the *rat-man* figure in Figure 10.4.

We have now sketched a model of perception in which the contributions of the subject do no longer need to collide with the contributions of the object. In line with some developments in current cognitive psychology and AI, perception has been introduced as an *integrative* process which combines perceptual features selected at various levels of resolution. A perceived object is an amalgam of nested forms. Its coherence is solidified by a conceptual network that specifies relationships between parts and wholes. Furthermore, perception has been introduced as an *interactive* process. Both the perceived object and the perceiving subject contribute forms which dovetail. If we need a medium where this interaction can occur, imagination would be a very plausible location. Imagination should be looked upon as the interface between perception (in the restrictive sense) and knowledge. [11] The percept of a common object is a pile of registered forms interspersed with imagined forms derived from conceptual knowledge. The imagined forms carry expectations when exploring an object, some of which are matched with input and some of which are attributed to the object without being checked.

Fig. 10.4. Rat-man. Depending on whether some partial configuration is seen as *snout-eye-ears* or as *eye-glasses-nose*, the figure is further developed as either rat or man. For both interpretations, some data driven processes have been indicated in terms of arrows pointing from figural parts to concept labels. Examples of concept driven processes are indicated by means of arrows pointing from concept labels to figural parts. Top down and bottom up processes relate to the vertical dimension of the conceptual networks. (Rat-man, after Bugelski, B. R. and Alampay, D. A., 'The Role of Frequency in Developing Perceptual Sets', *Canadian Journal of Psychology* 15 (1961), 206, Figure 1; *copyright © 1961 by Canadian Psychological Association*, reprinted with permission.)

Before investigating whether the proposed model can be fruitfully extended to the study of cognitive processes involved in science, it is worthwhile to notice that it goes beyond the classical doctrine of Gestalt psychology. The laws formulated by that school suggest an autonomous organizing activity operating upon pictorial elements independently of their conceptual interpretation. While it respects these findings as indicating some built in functional biases of our perceptual system,[12] the new model superimposes upon it a conceptual system that can interact with the perceptual system independently of configurational effects. Moreover, the new model (dis)solves the classical riddle of Gestalt psychology, i.e. that 'the whole is more than the sum of its parts'. This riddle can only arise in a sense-data oriented view where, as in classical empiricism, the elements are perceptual, contributed by the world, and the organization — the seeing of the whole — is contributed by the mind. The unity of the gestalt is then thought to result from some still undefined interaction between the elements. In the new approach, gestalt perception simply reflects knowledge of the structure of the object. The whole is not necessarily perceived through the perception of the parts. It has its own pictorial representation and is directly attainable. The formula 'the whole is more than the sum of its parts' could be changed into 'the whole is an additional part over and above its parts proper'.

Analysis of a Gestalt Switch in Science: Harvey's Discovery

The exemplary ambiguous figures allow for an interactive theory of perception which is a major requirement for overcoming the cognitive paradox. However, we are still far away from science. It might seem as if the tricky drawings with their reversal of interpretation are far too simple to function as genuine prototypes of scientific innovations. Is their simplicity misleading or illuminating? Instead of further toying with the idea, we should try a straightforward application to a genuine case of scientific discovery.

Harvey's discovery of the circulation of the blood is not a case of scientific innovation which figures prominently in the discussion of Kuhnian concepts. However, it has some characteristics which might make it seem highly suitable for an analysis along the lines of the Copernican Revolution. Like in the analysis of the movement of celestial bodies where Copernicus is opposed to Ptolemy, so can, in the study of the movement of the blood, Harvey be opposed to Galen. Also parallel to the astronomical innovation, it *can* be argued that the new paradigm of circulation stems from an attempt to solve major anomalies with the older tidal system. Therefore, in this chapter as well as in those which follow, we will focus applications of cognitive concepts to science on this illustrious and well-documented case which seems appropriate for testing a Kuhnian program. With AI rules of thumb in mind, we wager on an attempt to understand one 'simple' case thoroughly against a series of principles illustrated by means of acrobatic jumps from one case to another. In this chapter, we restrict it to the applicability of what we have discussed with respect to gestalt perception and gestalt switch.

Galen and Harvey as Angiological Duck and Rabbit [13]

The case of the discovery of the circulation of the blood is all the more interesting for our purposes because the major opponents, Galen and Harvey, are both strong proponents of observation as the method for obtaining adequate knowledge. At times it has been fashionable to present Galen as a highly speculative teleological reasoner and Harvey as a keen and cool observer of causal connections. However, consulting Galen's work, one feels confronted with an author who clearly is thoroughly experienced in dissection and anatomical research. To place Galen side by side with Harvey is not comparing an armchair thinker with a highly experienced practitioner. Both are skilled in exposing the interior parts of the body. Both have a conceptual system to account for the perplexing entanglement of tubes and structures that their method reveals.

Figure 10.5 indicates how, in a schematic representation of the vascular system, Galen's system as well as Harvey's yields an integrative picture of a great variety of observations and allows for the kind of interactive perception outlined above.

Galen makes a major distinction between a system for alimentation (the

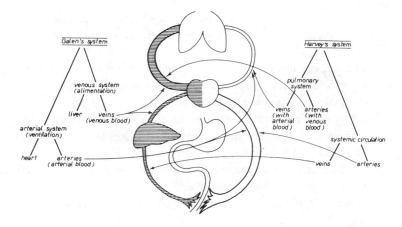

Fig. 10.5. Galen's and Harvey's model of the vascular system as angiological rabbit and duck. Galen's system emphasizes observational differences between an arterial system with as major organs the lungs and the heart and a venous system with as major organ the liver. It is congruent with observational differences between arterial blood and venous blood and between arteries and veins. It allows a two-way blood flow in conformity with a tidal model. Harvey's system does not emphasize the difference between arterial blood and venous blood and it interprets differences between arteries and veins in terms of differences in the direction of one-way blood flow in conformity with a circular model. In Harvey's system the heart is the major organ coupling the two circuits. In Galen's system there is a partial symmetry between arterial system and venous system, each having their own major organs (see Vesalius' diagrams of the two systems in Saunders, J. B. and O'Malley, 1973, p. 239 and p. 241).

veins) and a system for ventilation (the arteries). The notion of a ventilatory system based on arterial blood is a modification of pre-Galenic views which considered arteries to be comparable to the trachea and filled with air. On observational grounds, Galen rejects the notion of arteries containing only air but he retains some ventilatory function in terms of 'aerated' blood. In the venous system, the venous blood is considered alimentation produced in the liver from ingested food (via various stages in the stomach and intestine) and distributed through the body for consumption as a kind of fuel by means of the venous vessels. There are perceptual indicators, at various levels, to support the conceptual framework:

— there is a difference between venous blood and arterial blood in both color and viscosity: venous blood is bluish red and thick while arterial blood is highly red, thin and vaporous;

— veins differ from arteries: arteries have thicker walls while veins are thin and arteries pulsate, a feature that Galen considers a criterial attribute for distinguishing arteries from veins;

— respiration rhythm and pulse seem coupled: pulsation is what the respiratory system and arterial system have in common; it can be seen as a global characteristic of the system as a whole;

— the two systems relate to each other more or less symmetrically: the liver is to the venous system what the heart is to the arterial system and the portal vein is to the liver what the aorta is to the heart.

These and other observations link the Galenic concepts to the perceptual forms which dissection makes accessible. If one feature should be retained as predominant, it is the pulsating quality of arteries, heart and . . . lungs. Somehow, the rhythmic alternation of inhalation and evacuation seems a major theme for Galen. It is generalized into a *tidal* model allowing two-way movement in both the arterial and the venous system.

Harvey's picture is quite a different one. Along the line of observations leading to his discovery, we should notice that:

— scrutinizing the movement of the heart, one finds that the active phase is the contraction and not the expansion: the heart does not suck in blood but propels it away;

— the total quantity of blood is more or less fixed; the impressive amount pumped away at each contraction suggests that it is the same blood that is continually pumped around;

— ligatures show centripetal flow in veins while ligatures applied to arteries show centrifugal blood flow;

— the venous valves suggest centripetal blood flow.

Each of these points amplifies the importance of observational forms which connect to various levels of the conceptual system developed by Harvey. Some of the observational forms used by Galen are reduced in importance: the difference between venous and arterial blood and the difference between arteries and veins. It is the same blood on a one-way course through a double-looped circuit.

Both Galen and Harvey provide a conceptual system for combining perceptual forms ranging from global aspects to particular details of the vascular network. Both systems select various observable attributes retained at various levels and integrate them into object-like entities similar to the much-discussed duck and rabbit. They constitute, in terms of their conceptual and pictorial representation, the stored knowledge which the perceiver contributes in perceiving while dissecting or while looking at the result of a dissection. How they can provide detailed guidance in further research will become clear in our analysis of the gestalt switch.

Harvey's Gestalt Switch

Fourteen centuries separate Galen from Harvey. Gestalt switch suggests a sudden and apparently unprepared global reorganization. However, between Galen and Harvey, several important discoveries were made and, in some sense, Harvey's achievement can be seen as the completion of a development that started several generations before him. According to Harvey's own account of his discovery, as reported to Boyle, it were the valves in the veins that first made him think about the circulation of the blood.

It is quite plausible to attribute a special significance to these valves for Harvey because they were discovered by Fabricius, his professor of anatomy at Padua. Undoubtedly, they belonged to the research front findings he learned about in a period that was most formative for him. But obviously, valves in the veins added to previous knowledge do not automatically produce a gestalt switch. Harvey took courses in Padua from 1599 to 1602. His famous monograph on the circulation appeared in 1628. Yet, in the notes for his course as Lumleian lecturer from 1616, the venous valves are hardly mentioned. Apparently, he had other things on his mind during those preparatory years.

An orthodox Kuhnian analysis could locate the seed of the revolution in an anomaly. In the case of the cardio-vascular system, the problem of the pores in the septum separating the right and left ventricle of the heart was a major source of confusion. Though the Galenic doctrine considers the venous and arterial system as two distinct systems, it postulates a connection between them in order to account for the origin of arterial blood. Arterial blood is supposed to derive from venous blood brought into the arterial system through minuscule pores in the septum. Vesalius, however, reported in his epochal work (1543) that he could not find such pores, so that the connection between right and left ventricle became a serious problem for the Galenic paradigm. But at about the same time Realdus Columbus could 'save' that view by finding an alternative path from the right to the left ventricle: a route through the lungs which eliminated the problem of unfindable pores. Thus, a substantial part of the new paradigm, the pulmonary circuit, was first introduced as a way to save the old paradigm!

Harvey knew Columbus' discoveries and accepted them. Yet, his acquaintance with both the venous valves and the pulmonary circuit did not result in instant gestalt switch. Instead, impressed by Columbus analysis of cardiac activity he apparently embarked on a normal science type inspection of the movements of the heart. Though accounts of Harvey's discovery differ with respect to what his basic insight was and how he arrived at it, there is no doubt about the fundamental importance of his very careful study of the heart. Bylebyl (1973b) even argues that the first part of Harvey's (1628) revolutionary monograph (Chapters 1 to 7) was written years *before* the discovery and contains the accurate report of the basic findings in an highly significant preparatory phase. The critical issue is how and at what point this normal science type work turned over into a revolutionary view. In the case of Harvey, there are several well-documented and cogently argued positions. Again, few will deny Bylebyl's account according to which Harvey's observations made him realize that the heart induces fast and massive movement into the blood. But then opinions diverge.

According to Whitteridge (1971), Harvey saw clearly that the movement was due to the competence of the valves of the heart which by their functioning impose a one-way directional flow upon the blood. Transposing this newly verified knowledge to the venous valves, which are similar in structure, consequently led to the crucial suggestion of centri-petal flow in the veins, contrary to the accepted idea of centri-fugal flow. This is the decisive step: from the valves of the heart to the valves of the veins. What follows, the idea of a circle and various considerations and investigations, is elaboration and solidification. Whitteridge's version, as she herself emphasizes, integrates Harvey's account as reported by Boyle. The venous valves play a crucial role.

Pagel's (1967, 1976) reconstruction of Harvey's train of thought is different. He feels compelled to attribute more importance to another biographical element in Harvey: his Aristotelian education and sympathies. Puzzled by the massive movement inflicted upon the blood, Harvey ponders about how to avoid disequilibrium and distortion resulting from continuous vehement motion in one direction. Acquainted with Aristotle's concept of circular motion as perfect motion, he realizes that also in this case, circular motion might allow for both preservation and constant change. So, he wonders whether the blood does not move in a circle. The decisive step is from the notion of sustained vehement motion to the notion of circular motion. What follows, including the notion of venous return, is then the application of the notion of circle to the cardio-vascular system and its verification.

For our purposes, it is instructive to witness how the Whitteridge view and the Pagel view are played off against each other as respectively the *observation-view* and the *theory-view*. Whitheridge is somewhat allergic to theory, warning the reader from the very first sentence of her book that he will find "the history of a discovery, *not* an account of a *theory* of discovery" (Whitteridge, 1971, xi, italics ours). The recurrent theme is "observation", "simply . . . patient and repeated observation", "observation . . . the rock on which foundations can be built", "phenomena . . . manifest to the senses . . . totally indifferent to any man's opinion concerning them". The skepsis expressed in the introduction to her own translation of Harvey's *De motu cordis* casts doubt upon Pagel-type explanations: "it is most unlikely that the imaginative act that created the hypothesis of the circulation arose from reflecting upon any abstract philosophical ideas" (Whitteridge, 1976, xl). Pagel on the contrary doubts whether so brilliant an idea could be arrived at by simple observations, whatever their volume and accuracy. In his last book on the subject, he emphasizes that Harvey's discovery "was not the result of a 'mass of evidence' already assembled, but of a 'hunch' or idea that formed the challenge to produce such evidence *de novo*" (Pagel, 1976, 172, italics in the original).[14]

This strange polarization of positions between two outstanding specialists on Harvey indicates that more is at stake than proficiency in a history of science case. It expresses the category-mistake (Ryle, 1949) dealt with above, i.e. the conviction that elements of observation or data are details, stemming from the world, while ideas or theories are organizational frameworks stemming from the mind.

In our scheme of analysis, such an opposition does not arise. We decompose the dimension on which theory and observation are located into two axes orthogonal to each other. The vertical dimension harboring the conceptual hierarchy is independent of the horizontal dimension relating conceptual entities to tangible shapes. Apparently, current analyses of Harvey's discovery, including Whitteridge's and Pagel's, stress the importance of Harvey's preparatory study of the movements of the heart. This study we would label as mainly data driven, resulting in a refinement of Columbus model of the heart. Thus, on an intermediate conceptual level, a modified interpretation of the heart is entered into the conceptual pyramid of the Galenic system. Thereupon, Whitteridge and Pagel propose different roads. The Whitteridge interpretation suggests a top down path along the conceptual hierarchy to the venous valves and a concept driven observation of venous blood flow as centri-petal. The Pagel-interpretation suggests a bottom up path along the same conceptual hierarchy connecting at a higher level the venous and arterial blood systems by means of the notion of circle and the concept driven observation of a fixed quantity of the selfsame blood.

We do not have to make a choice between Whitteridge's and Pagel's version. They are not mutually exclusive. Our reconstruction can allow interactive perceptual processes at more levels either simultaneously or in close succession, up and down the conceptual ladder. The critical point is reached when conceptual reorganization affects the top level. That makes the old conceptual pyramid crumble. Before that level is reached, many findings which will later on solidify the new conceptual framework are first discovered and introduced as local corrections or extensions of the old.

The Infiltration of the New

The model of perception which we have outlined seems to be compatible with some of the major aspects of the paradigm view. At least it allows to conceive of perception, in science and in daily life, as controlled partly by knowledge embodied in conceptual frameworks and partly by registered perceptual forms. Acquired knowledge constitutes a source of expectations which fuse with available information to constitute a percept. However, while this allows to account for the shaping and suggestive function of paradigm-like entities, it is difficult to see how a perceiver can ever escape the grip of his knowledge on what he can see. The importance which the Kuhnian model attributes to anomalies seems to stem from this problem. Somehow, the governing paradigm should loose its hold upon the mind of the perceiver before he can see the new. Anomalies have a disheartening effect and undermine the sense of direction provided by the paradigm. Once the previous paradigm has deteriorated, the following one can be assembled from scratch. Is that what we discover in the case of Harvey?

Though several serious anomalies with the Galenic paradigm were known for decades, there is little evidence that they led to a gradual disintegration of that view. Harvey appears as an author who has remained faithful to the Galenic doctrine until forced to spectacular change by his own findings, much to his own surprise. How else can one explain the dramatic impact of the quantitative argument in the central chapter eight where the impossibility of the Galenic frame is derived from calculations of the amount of transmitted blood in relation to ingested food? However, the final blow to the old paradigm allows to the scene, almost instantly, a new one whose major components have apparently been assembled while the old one was still governing. The intriguing question remains: how could elements of the new emerge and develop within the old framework?

There is only one explanation for the penetration of new observations into established frameworks of knowledge: growth by accretion of extraneous frameworks. In line with the category-mistake of an organizing mind and a fact delivering world, we might be tempted to think of a paradigm as a mental device which uses up some integrative potential by elaborating and extending itself in accommodating an increasing number of facts. This would be growth by differentiation. The study of the discovery of the circulation of the blood does not support the idea that a paradigm as a major framework of knowledge for doing science develops according to some purely internal dynamics. Rather substantial developments result from hooking on knowledge structures available in the mind for other purposes. A major indication for this is the use of analogies.

While Knorr (1981b) is certainly right that analogies cannot be considered secure keys to scientific achievements, it cannot be denied that apparently they play an important role in a number of cases. Though Harvey uses a concise and sober style of writing, he does not refrain from using suggestive analogies and metaphors at crucial points in his argumentation. There is the analogy of the inflated glove illustrating the reception of the blood by the arteries, the analogy of the venous valves with flood gates, the Aristotelian rain-water-vapor-cycle. Each of these expresses in an undifferentiated way a new relationship between some levels of analysis, offering a global grasp of the relation between parts and whole. We need not to decide whether the analogy is instrumental in arriving at the insight or just expressing it. The point is that it reflects imported knowledge. Fragments of understanding, developed for no special purpose or stemming from acquaintance with daily routines or devices outside science might thus find their way into a scientific framework, brought in by analogies as extensions or corrections. It is in this way that a better understanding of the pulsation of the heart and the arteries could be hooked on to the Galenic doctrine as a local improvement that did not endanger the overall system although ultimately, like in the Bartlett experiment, it turned out to have been the detached detail that undermined the whole framework.

Drawings are deceptively simple compared to real objects of perception. Nevertheless, the gestalt switch in the simple case of duck and rabbit applies also to the complicated transition from Galen to Harvey. Though it seems to happen all of a sudden, it has not to be totally unprepared. As we have indicated in the simple case, local re-interpretations of sub-configurations are able to unbalance the conceptual system, forcing re-interpretation at levels of more global forms as well as at levels of finer detail. Harvey's detailed analysis of the movement of the heart clearly fits this category of local modifications which finally led to the overthrow of Galen's conceptual framework.

It should be clear that, although the new cannot enter the conceptual framework massively, it can in principle infiltrate everywhere, from the level of the finest detail to the level of the most global characteristics. In Harvey's case, the new creeps into the Galenic framework more or less like an infection, but one can imagine other cases where extraneous observation schemes attach directly to top-level concepts and are worked through downwards.[15]

There are two important consequences to this approach. First, despite the weight attributed to the integrative conceptual framework, it offers a view in which science can be considered as basically *open*. In principle, observation schemes, whether stemming from acquaintance with other specialties, routines from daily life, social interactions, hobbies, or whatever origin, are eventually attachable to integrative conceptual frameworks in science. Thus, the cognitive resources which a scientist has at his disposal for extra-scientific endeavors become occasionally important for his scientific work as well. Negligible as this form of openness might seem, it is of the utmost importance to overcome the cognitive paradox and the distinction between internal and external factors discussed in Chapter Five. Secondly, this view debilitates the notion of a monolithic scientific mind. In Chapter Six we invoked a cellular model for paradigm, introducing its components as distinct entities of various kinds, floating around more or less freely in some protoplasmic form. At this point it might be useful to extend that image to the entire mind, viewing it as a collection of free floating conceptual frameworks, some related to doing science, some related to down-to-earth daily routines, some very elaborated, some barely organized. Indeed, if the mental structures governing our scientific observations grow in terms of affixment of structures which are foreign to the leading framework, we have to conceive of the mind as a collection of such entities which we allow various forms of decomposition and recombination. This is the central position of an AI orientation which we discuss in the next chapter and which considers the individual mind as a community of experts. It should not necessarily be in conflict with our starting point according to which paradigms contribute to coherence within the individual mind. The paradigm can remain a pivotal conceptual framework in scientific activity without monopolizing the scientist's entire mind.

CHAPTER ELEVEN

PUZZLE-SOLVING AND REORGANIZATION OF WORLD VIEWS

> There is no doubt that even rather simple ideas sometimes are very elusive. – L. Pauling, 1974, p. 711.[1]

In 1960, Miller *et al.* expressed in their influential *Plans and the Structure of Behavior* the opinion that "the Image... must contain an amazing amount of knowledge all organized for fast access to attention. In some respects, apparently, our brains are still a great deal more complicated than the biggest computer ever built" (1960, p. 51).[1] Despite the enormous developments in computer technology the same statement could still be made today without losing any validity. It is not so much the absolute capacity of the knowledge store but rather the way the knowledge is organized which accounts for human superiority. That is the reason why AI engineers are now eager to find out how knowledge is organized in human brains and why AI is – maybe only temporarily – fusing with studies of human cognitive processes.[2] In the previous chapter we have explored an interactive model of perception that entitled stored knowledge to play an important role without reducing perception to hallucination. We now face the task of analyzing the internal organization of stored knowledge to find out how internal reorganization within that store contributes to discovering the new. As we will see, once cognitive processes and perception in particular have been made heavily dependent upon mediating cognitive structures in the knower, the complexity of the cocktail party is no longer so much out in the environment as it is within the individual attendant. Hence, selective attention no longer bears primarily upon the overwhelming multitude of stimuli impinging upon the senses, but on the multitude of cognitive units available in the mind and competing to govern behavior.[3]

The knowledge store of the mind is memory. Miller *et al.* (1960) make a distinction between *image* and *plan*, i.e. between a particular *world* view and plans for *action* in that world. While this segmentation is based on the distinction between programs and data-structures in computers, it is both confusing and leading into a basic difficulty. In the first place it is an arbitrary distinction with respect to computers. Programs are stored in memory as well as data-structures. We better view memory in a large sense as including all stored knowledge, i.e. programs as well as data. Secondly, adhering to a strict separation between knowledge (image) and action (plan), one ends up with a contemplative system that never comes to action. Miller *et al.* (1960, p. 9) refer to this anomaly as the gap between knowledge and action pointed out by Guthrie. Shepard (1968) considers it a major anomaly of the cognitive orientation in his review of Neisser's epochal *Cognitive Psychology*. To overcome this difficulty, we have to analyze how knowledge and action are connected and intertwined rather than handling them separately. Thirdly, recent history of AI indicates an increasing tendency to blur the distinction between image and plan or program and data-structure. This is apparent from the evolution of ideas on problem solving. As indicated in Table IV, problem solving programs have undergone a development similar to the changes in thinking about data-structures.

From an emphasis on simple principles of universal order and superpowerful search, units of analysis have shifted in the direction of heterogenuity and complex interaction. In the procedural representation of knowledge, the distinction between data structures (knowledge) and program (problem solving routine) tends to dissolve and one ends up with a notion of intelligence as stemming from interaction between hybrid memory structures. Thus, serious doubt has been cast upon the notion of intelligence as a set of rules detachable from specific knowledge. According to Winograd: "There is no magic principle which makes a mechanism intelligent. Rather, intelligence comes from having a broad flexible variety of mechanisms. . . " (1976, p. 8). Notice the similarity with Kuhn's reluctance to identify scientific research and creativity with the mastery of rules, whatever their cleverness or power. Indeed, to try to characterize intelligence in terms of a set of general principles is like trying to describe scientific thinking solely in terms of rules of methodology. There is no magic principle which makes a set of rules scientific and rules are no substitute for knowledge.

The most advanced ideas about the mind as a system of interacting analogous units are developed in the MIT-doctrine of the mind as a "community of interacting experts". 'Experts' refers to the "large, complex, symbolic

TABLE IV

The converging development in AI concepts of data structures and intelligent procedures illustrating an evolution from high systematicity and complementarity (between, e.g., algorithm and data structure) to rich local complexity and equality (i.e. equality between entities representing 'knowledge' and entities representing 'plans for action').

MEMORY (data structure or knowledge store)	INTELLIGENCE (program or search plan)
from	
file and matrice	algorithm
tree	heuristic rule
uniform net	domain specific heuristics
to heterogenious net	experts

The terms relate to structural models for memory versus search and retrieval programs. The dominating notion of the last decade has been the *semantic network* introduced in 1966 by Quillian. Since then, the graph-theoretical concept of *net* as a set of nodes connected by means of various types of labeled relations has been the skeleton for several proposals on the structure of memory. Obviously, when many types of relations are involved, such networks can become quite complicated. Earlier models which focussed on *lists* or *trees* or *files* and *matrices* might seem to have a more transparent structure. The development indicated above applies in particular to the MIT-approach in AI (Minsky, Papert). However, it is also compatible with a general trend away from earlier concentration on search management to the current focus on the construction and use of knowledge bases (Sridharan, 1978, p. 2).

structures" which function as units in such interaction process. At one time in the development of the MIT-ideas, such units were introduced in Minsky's (1975) rather controversial 'frame-paper' as *frames*. The characteristic which makes frames particularly important for our discussion is that they are inspired by the Kuhnian concept of paradigm which is extended in such a way as to cover all cognitive processes, i.e. those in ordinary life as well as those in science. One of the definitions of frame reads: "A frame is a collection of questions to be asked about a hypothetical situation; it specifies issues to be raised and methods to be used in dealing with them" (Minsky, 1975, p. 246). If one were to substitute the term 'paradigm' for 'frame', it could indeed be a definition taken from Kuhn's monograph. Minsky stresses this relation explicitly: "the frame idea is not particularly original — it is in the tradition of the 'schema' of Bartlett and the 'paradigms' of Kuhn" (p. 213). The extension is clearly pointed out in the statement that "while Kuhn prefers to apply his own very effective redescription paradigm at the level of major scientific revolutions, it seems to me that the same idea applies as well to the microcosm of everyday thinking" (p. 261). Thus, we might expect to gain insight into possible articulations and refinements of the concept of paradigm by studying frames.[4] Furthermore, the scope and importance of such an apparently mystifying notion as "a multitude of interacting cognitive agents" will be better explained and appreciated after familiarization with the frame-approach.

Frames

Anyone who has ever studied something by means of programmed instruction has met with a subject area segmented into what is also known as *'frames'*. Indeed, that technique makes use of micro-units of knowledge specified in terms of sentences which contain one or more blank spaces that have to be filled in by the student. The first lesson in a programmed text for computer programming might contain a frame like *'January 17' is a* ———— *data-item*. If the student has read the preceding lines attentively, he has no trouble in finding that in this line *alphanumeric* is the missing adjective. The idea of an entity which, consisting of some partial knowledge, indicates the type of additional knowledge which will be needed to complete it, is essential to Minskyan frames. The latter, however, are much more elaborate and more complicated than the simple and single-minded uncompleted sentences of programmed instruction. Better examples could be found among complex forms such as those used to reconstruct a person's medical history or to declare income for tax purposes.

The concept of frame is an attempt to describe data-structural units available for several procedures or actions. In his paper, Minsky introduces frame as "a data-structure for representing a stereotyped situation" (p. 212). A *room-frame* should contain the common knowledge of a most conventional room specifying that it consists of a floor, a ceiling and four walls with at least one doorway in one of the walls. Use of that frame for a particular purpose will involve further specifications and sometimes modifications, i.e. superimposed knowledge related to the specific purposes of the procedure. The frame represents some *tacit knowledge* of which we take no notice. Aspects of a situation which we do notice on the basis of perception or communication relate to additional knowledge-items which are 'filled in' on an underlying structure. This openness, suggesting completion with additional elements, is basic to this unit of analysis.

The frame-concept is proposed as an extension of Fillmore-type case grammars. Such grammars make an analysis of the meaning of a sentence in terms of cases or roles that different parts of the sentence fulfill in order to complete the information conveyed by the most crucial part: the verb. In the example quoted in Goldstein *et al.* (1976), the sentence *John tickled the girl with the feather* involves three cases which have to be filled in, in order to specify the main activity: the tickling. There is an *agent* which in this 'case' is *John*, a *patient* which is indicated here as the *girl* and an *instrument*: the *feather*. Frames extend this approach in the sense that for any major cognitive event, perception as well as language understanding, such a list of slots has to be filled in. Frames, however, differ from the case-grammar approach in that the selection of entities which fulfill the prescribed role is not based on superficial syntactic rules (e.g. 'instrument' can be indicated by means of a functional term 'with'), but on *examples*. It is specified in advance what kind of objects or entities one might expect to fulfill the functions or roles indicated by the frame. As such, the frame not only orients attention toward specific variables that have to be looked at, but it also contains *specific* and *detailed proposals* for determining their value! When confronted with the sentence *John tickled the girl with the funny nose*, a syntactically-oriented approach will not be able to disambiguate between an adjectival or adverbial use of either *with the funny nose* in this sentence or *with the feather* in the previous example. The frame approach, having incorporated *common sense knowledge* in terms of specific expectations for all cases, would favor an adverbial interpretation in the first and an adjectival interpretation in the second sentence. Indeed, a feather might well be the favorite instrument for tickling while a peculiar kind of nose might rather be used as a feature to identify a person. Logically of course, the feather as well as the nose can function in both ways. Incorporation of this common-sense knowledge is, however, thought of as part of the *strategy for 'bypassing logic'*, whenever the latter seems to impede understanding by producing a 'combinatorial explosion'.

It is in this use of preferred examples as constitutive elements that frames differ from more classical approaches which introduce variables that may take on any of the values of a broad and extensive set. Before dealing with these *default values* — the frame-approach label for preferred examples — we need to indicate the other frame-components (Minsky, 1975, p. 212–213). Frames are large complex symbolic structures which can be represented as a "network of nodes and relations". The "top level" nodes of a frame are fixed and stand for "things which are always true about the supposed situation" e.g. in a room-frame: that a room has "walls" or vertically supporting structures of some sort. At the lower levels, frames "have many *terminals* or *'slots'* that must be filled by specific instances or data'. These terminals are usually filled with typical instances which can be seen as prototypical data for the terminals and which are called *default assignments*, e.g. lower levels of a 'room frame' might have light-painted flat walls with a doorway somewhere and a window on the opposite wall as default assignments. At a particular level, a frame represents a specific view on a situation, e.g. how a room will look from a well-defined position. More general structures are called *frame-systems* and coordinate different frames by means of *transformations*. These transformations stand for actions which are responsible for systematic changes in the data that fill the terminals. When I walk around in a room, changes in its appearance corresponding to my shifting perspective are accounted for by those transformations which permit me to compute the values for the data in the terminals as they are going to be when I move. Since these transformations are comparable to Piaget's 'concrete operations' we shall deal with them in more detail in the next chapter. Here we should note, however, that the various frames of a frame system share the *same* terminals. Frame systems in their turn are combined into networks. If we furthermore note that terminals of frames may be filled with *subframes*, this combinability of frames into frame-systems and networks might suggest a clear hierarchical organization of knowledge. However, this might promote a too rigid and systematized picture of the knowledge store. As already has been mentioned, the orientation is toward a model which proposes a heterogeneous distribution of knowledge with many areas of highly specialized knowledge functioning as relatively independent and self-contained units, linked among one another by relatively weak bonds which permit high flexibility and ease of reorganization. The forces that hold frames together in frame-systems seem to be much stronger and of another sort than the weaker 'forces' that combine frames into networks. Nevertheless, both types of aggregates might, at some level, simply reappear in the description as *frames* and this might inspire a critical remark very similar to objections made with respect to Kuhn's concept of paradigm, i.e. the widely varying scope of the frame concept.

At one end frames become similar to bundles of features such as those used in pattern recognition, and a frame-unit might be involved in the identification of a single character. At that level they are dealt with in terms *of template filling structures*. At the other end one meets with *topical frames* or *thematical superframes* (p. 236) which have to account for the sequential structure of something like a novel or a detective story. To account in terms of frames for an activity like reading and understanding such a story, in between the levels mentioned, one needs to invoke frames for words, for sentences, for paragraphs, for chapters and possibly for even much more. As such, the frame-concept covers a wide span of very heterogeneous mental entities and one might wonder whether such a broad and diverse application does not make the term entirely hollow. Minsky himself, guarding against premature parsimony, is hesitant:

It is tempting to imagine varieties of frame systems that span from simple template-filling structures to implementations of the 'views' of Newell. . . . I feel uncomfortable about any superficially coherent synthesis in which one expects the same kind of theoretical framework to function well on many different levels of scale or concept. We should expect very different question-processing mechanisms to operate on our low-level stereotypes and on our most comprehensive strategic overviews (p. 247).

As is apparent from the example of reading, it is obvious that frames may be organized quite differently, e.g. from parallel character-frames toward sequential entities to account for stories. If the contents of frames can differ so widely, what then is the use of so broad a concept? Before attempting any global evaluation, we should also analyze the characteristics which govern the extensibility and combinability of frames.

Defaults and Exemplars

Frames represent units of knowledge one brings to bear upon a situation. As structures which have to be completed with data, they orient the information processing system toward specific aspects of that situation. A basic point of the frame theory is that frames are neither stored nor retrieved as empty or blank forms. The open slots, frame terminals, are filled in with 'weakly bound' default assignments, i.e. typical examples of the kind of concrete objects one expects to meet when the frame turns out to be applicable. Default assignments are to be seen as the products of imagination and constitute the bulk of the knowledge contributed by the knower but illusorily perceived as stemming from the known. In visual perception, they represent aspects of the situation which are thought to have been seen but which have not actually been checked with respect to their presence. In verbal communication they represent things which are thought to have been understood although actually they have not been explicitly expressed. A familiar experience which could yield an intuitive grasp of default assignments involves the comparison of a story which one might learn first from reading a novel and then once more from seeing a motion picture based on the novel. It is often said that motion pictures are disappointing when based on a novel that one has read before. Probably, the reason is that in reading the novel, many default assignments have been provided in one's own evoked frames which differ from the default assignments which constitute the interpretation of the novel by the motion picture director.

The main point is that, contrary to most traditions, the frame approach does not consider default assignments to be unimportant or even annoying byproducts of authentic knowledge, but core components which not only serve an important role in substantiating expectations but also in providing directions or suggestions for connections with other frames. In this respect, an extended use of Kuhn's concept of exemplar is proposed. In frame-theory, the notion of *exemplar* is introduced to denote a *typical instance* of a frame. In general one would expect the default assignments for the different terminals to be derived from or to be congruent with the values of the exemplar.

The basic idea behind the notion of exemplar is that, whenever we have a sufficiently rich body of knowledge of a certain object or a certain problem area, we apply it to similar objects or cases, even if there are apparent discrepancies. Discrepancies can be dealt with simply by noticing them and by remedying the problems caused by the overgeneralization. This is somewhat contrary to Minsky's reservations with respect to parsimony: different things are treated as being the same until it is necessary to differentiate them. However, it should be emphasized that differences are not ignored, but explicitly acknowledged and used to direct the patching up of the partly deficient stereotype which the default assignments or exemplar in fact is. Indeed, networks of local knowledge are thought of as *similarity networks*

(Winston's thesis 1970) in which related concepts are grouped and connected
to each other in terms of labeled links which indicate how the concepts differ
from the exemplar. In such a way, around the concept of *chair*, a network is
developed in which *bench* is entered as differing from 'chair' by being 'too
wide and having no back', *table* is 'too big and has no back', *stool* 'too high
and no back' etc.

We should keep in mind that a major impetus for the development of the
frame theory is to develop a model of cognitive functioning which avoids
the 'combinatorial explosion'. Minsky wants to take into account that "in a
complex problem one can never cope with many details at once. At each
moment one must work within a reasonably simple framework (p. 257)".
Therefore, one should never permit novelty or unknowns to enter massively
into the system. By thinking in terms of exemplars, a countless number of
combinations of attributes can be left out of consideration, and attention can
be limited to the filling in of terminals of familiar frames, with the assimilation of the unknown restricted to local modifications or additions under the
form of difference annotations. It is a strategy for keeping the unknown
under the control of the known all the time, permitting the known to be
extended (or changed) only gradually and in small steps. As such it goes back
to the rationale accounting for *focus gambling* described in Bruner's *A study
of Thinking* (Bruner *et al.*, 1956).

The frame approach emphasizes the importance of stereotypical but
also of specific knowledge. Stereotypic should not necessarily mean rigid
and impersonal. Frames seem to derive their specificity from these default
assignments and through them, the knowledge of individual systems even
acquires some idiosyncratic character, which according to Minsky might
account for personal styles of thinking (p. 228).

We have referred earlier to the widely varying scope of the concept of
frame and the difficulty of finding criterial attributes to distinguish superframes such as Newell's 'views' on a macro-level from subframes in template
filling structures on a micro-level. Applying the theory to our own attempt to
grasp the concept of frame, we should resist the tendency to search for an
exhaustive list of criteria and rather look for a frame whenever in a domain
of knowledge a network of concepts is connected to the rich subdomain of
default assignments provided by an exemplar. The detailed and idiosyncratic
knowledge embodied in the exemplar appears to be the most basic constituent.

We have described frames and defaults in a rather freewheeling way, as if knowledge can be dealt with on its own, independent of the purposes for which it is used. However, frames are expected to provide room for goals as well as for data: "when we go beyond vision, terminals and their default assignments can represent purposes and functions, not just colors, sizes and shapes" (p. 232). Goals are invoked to control the matching process which occurs to check whether or not an invoked frame is appropriate (p. 218). As such, frames and frame-networks are not free floating descriptive entities. They are always part of some procedure designed to fulfill some purpose or to solve some problem.

Problem-solving and Debugging

With respect to problem solving procedures, the strategy of the frame approach is the same on all levels. If there is a problem, try to arrive at a solution by applying a familiar (and simple) method, even if it does not seem entirely appropriate. When difficulties arise, try to overcome them by adapting the procedure in such a way as to make the difficulties disappear. This general procedure is called *debugging*, a label which suggests an analogy with the practice of computer programmers who, during the first runs of a program, eliminate all kind of *bugs*, errors due to partly inappropriate subroutines or to some unwanted interaction between subroutines. The frame orientation generally emphasizes domain specific knowledge with special skills providing an identity to each domain. However, for debugging, some generally applicable knowledge is invoked, consisting of a classification of bugs combined with a set of methods for eliminating them, whatever the procedure in which they occur. The distinction between general intelligence and specific intelligence(s) has been discussed in psychology for many years. The frame approach would clearly favor a theory which would distribute intelligence over a great number of specific skills, retaining as general factor only one type of skill: the skill to debug programs, i.e. to transform or rearrange procedures for solving specific and well-known problems into procedures for solving new problems. Gerald Sussman (1975) has worked on a program designed to develop skill with respect to manipulations in MIT's favorite microworld: the block world. A type of bug which can be illustrated in that domain, but which is also highly generalizable, stems from the widely used *linear approximation*: i.e. attempts to solve a problem by simple addition of partial solutions as independent modules. In the block world, the simple task of building a tower with three blocks has to be decomposed into two subtasks: putting block B on block A and putting block C on block B. At first, applying 'linear approximation', Sussman's manipulation program ignores any possible constraint with respect to the sequencing of subtasks and might put block C on B before attempting to put B on A. The special debugging program operating on that program is equipped to diagnose that linear approximation might not be entirely justified in this case, and that trouble probably arises because the second subtask undoes the result of the first subtask, or that the result of the first subtask makes the execution of the second subtask impossible. A reorganization of the sequence of subtasks is sufficient in this case to eliminate the bug.

Typically, problem solving within the frame approach avoids elaborate exploration and evaluation of alternative methods for the solution of a particular problem. Again, the risk of drowning in complexity when trying to take into account every possibility and eventuality is estimated to be very high. Therefore a strategy of 'ruthless generalization' (Sussman, 1975, p. 114) rather than careful investigation is followed. On the basis of vague similarities with a familiar problem, a new problem is tackled with the procedure associated with the familiar problem, relying on debugging knowledge for making the procedure fit the new case.

Puzzle-solving and Heterarchical Control

The concept of heterarchy has been invoked to characterize complex multiple goal-systems, such as man, by McCulloch (1945). In essence, the argument of his paper comes down to the position that purposive systems of higher flexibility and adaptability can be derived from systems which have their goals ordered in a circular fashion rather than in a linear hierarchy. In AI, heterarchy is used to arrive at computer programs which can use knowledge in a more flexible way than the more conventional hierarchical programs.

Most computer programs have a rigid hierarchical structure. As an example, we might consider a hypothetical program designed to 'read', i.e. to process written material in such a way as to understand its meaning. To describe that activity in some detail, a computer program can be developed in terms of segments which are parts of the higher order process. 'To read' can be decomposed into 'determining the meaning of each sentence of the series of sentences composing a text'. 'To determine the meaning of a sentence' can be broken down into the determination of the meaning of several phrases. Phrases can be decomposed into series of words, words into series of characters, characters into bundles of features. Each higher-order task is defined in terms of a number of more elementary lower-order tasks and 'the whole is the sum of the parts'.

It should be clear that this kind of rigidity makes the problem vulnerable to even minor disturbances. If reading implies the identification of the meaning of phrases, if this implies the identification of the meaning of words, if this implies the identification of characters ... what then would occur when one single character appears to be illegible? In a strictly hierarchical program, the success of the higher order task critically depends on sequential and successful execution of each of the lower order component tasks. In heterarchical control, however, an attempt is made to get rid of this unidirectional control of the whole over the parts. A program is envisioned in which the familiar strategies are embodied which people use fluently and easily. If, while reading, we encounter an illegible character, we do not endlessly fixate the indecipherable dashes of ink. Using context, almost without effort, we reduce the number of potential alternatives and construct a set of possibilities which guide our perceptive apparatus to disambiguate the problematic character. Programming this kind of procedure requires a withdrawal of control from the subroutine designed to recognize characters (because it is blocked by a problem it cannot solve) and a transfer of control to the subroutine designed to recognize words. When, however, an analysis on the level of words is introduced to disambiguate a character, the subroutine to recognize words becomes subordinated to the subroutine for the recognition of characters! Heterarchical organization of components is characterized by this kind of reversible control between higher order goals and subordinated goals. This is what makes it possible for routines to become a part of their own subroutine(s). According to Minsky and Papert

The very concept of 'part' as in a machine must be rebuilt when discussing programs and processes ... : So, the traditional view of mechanism as a hierarchy of parts, sub-assemblies and sub-sub-assemblies (. . .) must give way to a heterarchy of computational ingredients (Minsky and Papert, 1971, p. 20).

The reversal of control between levels in a hierarchical program provides a technique to account for context-sensitivity. But the concept of heterarchy can cope with a wider diversity of approaches than those represented by adjacent levels in a rigid hierarchy. In Chapter One, we referred to the different approaches in the study of pattern recognition, corresponding to stages in the development of the cognitive view: i.e. template matching, feature analysis, context analysis and analysis by synthesis. It has been indicated that successive stages have not replaced each other but have proven complementary to one another. A similar point has been made with respect to perception in the preceding chapter. In a heterarchical organization, these different levels of analysis should represent interchangeable components and subcomponents of a pattern recognition device in the sense that at one time one component may be part of the other, while the next time this relationship might be reversed. Winograd has developed a heterarchical coupling of a syntactic, a semantic and a world knowledge component in his famous program for understanding natural language (Winograd, 1972). His favorite example to illustrate heterarchical coupling in terms of a familiar experience is based on how we solve jigsaw puzzles, the same activity which Kuhn considers somehow as prototypical for scientific problem solving.

In the reconstruction of such a jigsaw puzzle, three kinds of subprograms are involved: *syntactic knowledge, visual semantics* and *pragmatics*. Puzzle syntax governs the combinability of pieces depending on their geometrical form, visual semantics is related to correspondences of colors, complementarity of figures on pairs of pieces, etc., while pragmatics is concerned with the knowledge of the theme of the puzzle, the personages depicted, pertinent objects, etc. Typically, in the reconstruction of a puzzle, subjects oscillate between these separate levels. Sometimes a piece is discovered on the basis of general knowledge, e.g. finding the sparkling shoe which we know has to be in a picture about Cinderella. Sometimes pieces are found to combine because the fingers on the one piece go with a hand which we see on the other piece. Sometimes pieces are combined because the complementarity of their geometrical forms is evident. In continuing alternation, one component produces a possible solution for a major subgoal and the other components are called in, in a servant role, to guide further search and to check the acceptability of the proposal. Thus, to eliminate the rigidity of a hierarchical organization, the components corresponding to the levels in such a hierarchy are freed and reintroduced as a 'community of experts'. Whenever such an expert can arrive at a proposal for a partial solution, others are called in to assist further elaboration and execution. Each expert corresponds to a different level of representation and a task like solving a jigsaw puzzle might be thought of as involving the flexible use of different levels of representation. Whereby 'flexible' means that each level can function temporarily, from the point of view of control, as top level while, at some other time, it is secondary to another level.

Procedural Aspects of Scientific Knowledge

Before assessing the merits of a pluralistic theory of the mind in a further exploration of Harvey's case, we need to consider one more basic attribute of the frame approach: procedural representation. Explaining it already allows us to indicate some relevant applications of the approach to science and scientific knowledge. The thesis of *procedural representation of knowledge* stresses the *functional organization* of knowledge. Knowledge is not stored in terms of a big heap of isolated knowledge-items expressed in single propositions, but in terms of procedures, i.e. plans and programs for the attainment of specific goals. As defined by one of the proponents of procedural representation of knowledge, C. Hewitt, the principle reads: "Knowledge is intrinsically tightly bound to the specialized procedures for its use" (Hewitt, 1974, p. 398). It is the *use* which gives coherence and directionality to a given set of knowledge items. Insofar as this orientation characterizes the frame approach, it could equally well be labeled the 'skill approach' or 'skill paradigm', as is apparent from Sussman (1975). After having quoted a Webster-definition of skill as "the ability to use one's knowledge effectively and readily in execution or performance", he elaborates on 'effectiveness' in alluding to skill in plumbing as an example:

It wouldn't be enough to memorize all of the facts in the plumber's handbook, even if that could be done. The knowledge would not then be in an effective, usable form. One becomes skilled at plumbing by practice. Without practice it doesn't all hang together. When faced with a problem, a novice attacks it slowly and awkwardly, painfully having to reason out each step, and often making expensive and time-consuming mistakes. Thus the skill, plumbing, is more than just the information in the plumber's handbook; it is *the unwritten knowledge derived from practice which ties the written knowledge together*, making it usable (Sussman, 1975, p. 1).

Typically, studying accounts of how science is learned by an aspirant scientist, one often finds emphasized that science is more than a set of knowledge items, that it is not something which one can learn from books alone but that it involves intensive practice and exercise. It is stressed that one becomes a scientist, not by memorizing texts but by practicing scientific research. Mastering of scientific knowledge, like plumbing, cannot be reduced to knowing a collection of facts or assertions in a given field. Reputed scientists who, after a successful career in science, take a self-reflexive turn in an attempt to analyze their field or science in general often emphasize the importance of tacit knowledge and find the kind of experience which cannot be obtained from books but from practice in a laboratory most important. In a formulation of Ravetz (1971, p. 103): " . . . in every one of its aspects, scientific inquiry is a craft activity, depending on a body of knowledge which is informal and partly tacit". And in Kuhn's emphasis on the necessity for exercise and practice in the use of an exemplar, there is an implicit suggestion that paradigms should be closely connected to knowing how, or procedural knowledge. But what is it that *use* adds to knowledge? What is the binding force that action or practice brings into collections of knowledge items?

For behavioristic psychology there is nothing magical about exercise. Exercise reinforces the proper habits. Progress achieved through practice is then accounted for by the gradual elimination of irrelevant parts or aspects of the response pattern, the latter achieving a degree of stability or solidity directly proportional to the number of successful uses. The frame approach follows an alternative route. Instead of having variety and diversity at the start which then gradually disappear in favor of a stereotypical stable habit, it puts the stereotypical structure at the start and has it modified and differentiated and complicated with use and exercise. The central idea is that exercise is not so much the drilling in by repetition of a basic response pattern or a powerful rule, but the subtle acknowledgement of specific points relevant to the application of the stereotypical pattern in specific situations. Exercise is not repetition but the acquisition and addition of the fine-grained knowledge that guarantees a fluent application of a skill. Exercise and practice superimpose large amounts of local and task-specific knowledge on the basic skeleton of knowledge provided by a frame. Here also, the fundamental orientation is not an attempt to circumvent knowledge but to recognize plainly that highly skilled activity presupposes a huge amount of highly specific and highly differentiated knowledge. Again, this corresponds closely to retrospective reports of scientists who emphasize that the acquisition of skill in scientific research is not so much the drilling in of the general rules of methodology as it is the mastery of very specific methods by learning to identify the many pitfalls (Ravetz's term) which the use of the method undoubtedly will contain. Ravetz's 'pitfalls' are very similar to the 'bugs' of the frame approach. His description is so close to this orientation that it is worth quoting extensively. He recognizes that

... most of the basic knowledge which a scientist will need to have available as tools can be organized into a systematic form in which the pitfalls are bypassed without any immediate need for their being identified (p. 99).

This corresponds to what Sussman calls the 'written knowledge' in the plumbers handbook. The 'unwritten knowledge', however, which ties the written knowledge together is derived from handling pitfalls:

For the pitfalls are encountered only when this knowledge is put to use, and they depend so much on the particular application, and are so various, that a comprehensive discussion of them would be quite impossible. In laboratory courses, students are given a gentle introduction to the pitfalls likely to be encountered in the use of physical tools for the production of data, and they learn the craft techniques for manipulating these tools reliably. However, this essential aspect of laboratory teaching is frequently ignored, and students usually believe that they are merely 'verifying' for themselves that certain standard effects can be reproduced (Ravetz, 1971, pp. 99–100).

Similar characteristics apply to higher aspects of the scientist's craft such as

... the formulation of problems, the adoption of correct strategies for the different stages of the evolution of a problem and the interpretation of general criteria of adequacy and value in particular situations.

Of these, Ravetz says:

... the body of methods governing this work is completely tacit, learned entirely by imitation and experience, perhaps *without any awareness that something is being learned rather than 'common sense' being applied* (Ravetz, 1971, p. 103, italics ours).

This interpretation of use and exercise in terms of the acquisition of subtle additional knowledge rather than solidification of acquired knowledge throws a new light upon several established facts about science and scientific practice. For one thing it suggests that exercises, emphasized by Kuhn as characteristic for paradigmatic disciplines, have not only a social function in establishing the dogmas (solidification) but also a cognitive function in the detailed elaboration of the paradigm. It is also an interpretation which is quite compatible with the frequently noticed fact that excellent scientists tend to have learned their trade from excellent scientists. Remember Ravetz's reference to the pattern of medieval crafts as a model for the transmission of skills in science (see Chapter Eight). If scientific practice were mainly a problem of adhering strictly and consistently to a body of generally recognized principles, it would not be so sensitive to subtleties of idiosyncratic interpretations conveyed in the teachings of a master.

Self-world Segmentation and Compatibility of World Views

Through concepts and mechanisms such as exemplars and defaults, heterarchical organization and procedural representation of knowledge, the frame approach can account for several features which careful description of scientific knowledge production have identified as important. The pivotal role of the exemplar is in line with Kuhn's (1970) later emphasis on the restricted interpretation of paradigm as a prototypical example. The flexibility offered by heterarchical control in problem solving appears to indicate a kind of shifting back and forth between levels of analysis which typifies scientific search and speculation as well as jigsaw puzzle solving. Both exemplar and heterarchy underline the significance of the vertical dimension dealt with in the previous chapter and might offer room for extended exploration of the function of metaphor, which is periodically brought up as crucial to creativity. However, because of their abundance, the possibilities offered by these mechanisms lead into a nightmarish 'combinatorial explosion' if we would not, at the same time, be able to account for the difficulty of discovery and the relative rarity of significant breakthroughs. If scientific discovery were so easily explained, why would it be so difficult to achieve? The frame-approach contains an intriguing suggestion with respect to the question why even simple solutions can be so elusive. It relates to what Papert (1980) calls the *syntonicity* of world views and it stems from following through the consequences of the distinction between *knowing that* and *knowing how* and going beyond it. The example of Harvey's discovery of circulation will again allow to illustrate this point.[5]

We are inclined to think of knowledge as concerned with the outside world only. When we introduce various alternative world models residing in the knower, we tend to conceive of them as a bundle of maps available to a driver. Based on some global characteristic of an area, e.g. the typical skyline of a city, he will select the relevant map to guide his actions in more detail. Generalizing this metaphor, we might consider the knower as a driver having 'maps' for 'restaurant eating', for 'skin diving', for 'lecture attendance', for 'chemical analysis' (if he happens to be a chemist), etc. When some globally identifiable conditions apply, he will select the appropriate world model and 'run' his behavior accordingly. The deeply misleading conception in this model is the notion of *one* driver with *several* maps. We should recall that the notion of knowledge or representation applies to the 'self' in the same way as it applies to the 'outside world'. As we have argued in Chapter Two, the notion of world model is prior to a segmentation into self and world. We cannot simply assume that that segment of our knowledge what we relate to 'self' remains identical throughout all the specific world models we have available. Each world model has its own 'self specification' and problems of decomposition and recombination of world models might stem from incompatibilities between the self-world segmentations in the models involved. Let us go back to Harvey's case to explain what we mean.

The previous chapter has allowed us to become acquainted with the major aspects of the discovery of the circulation of the blood. Once understood, circulation is quite simple and one wonders where the resistance resides that impedes easy discovery. It is not unusual to define creativity in terms of the ability to find new interpretations for familiar facts in the cross-section of incompatible world views. Koestler (1964) describes creativity as "the perceiving of a situation or idea in two self-consistent, but habitually incompatible, frames of reference" (p. 35). Bylebyl's model (1973b, 1979) of Harvey's discovery is particularly revealing in this respect because he indeed presents it as the outcome of a recombination of frames of reference which seem at first incompatible. Moreover, he formulates his analysis in terms of an interaction between *contexts* in the cognitive sense we want to elucidate (Bylebyl, 1979).

Bylebyl distinguishes between a *medical* context, an *anatomical* context and a *philosophical* context. The philosophical context is the Aristotelian conceptual frame-work with its preference for circles allowing for preservation and change at the same time. Its potential relevance has already been indicated in the previous chapter. The two other 'contexts' are crucial to the present discussion. The medical context is the conceptual system of traditional medical practice in which pathological events or processes are seen as connected with movement of bodily fluids according to a humoral pathology doctrine. The anatomical context is based on the Padua school of anatomy and relates to empirical research and new knowledge on the heart and the venous valves developed at the 'cutting edge' of Harvey's contemporary science. These are two different worlds which come into conflict for Harvey when he attempts to remain loyal to both.

Within the medical context of *humoral pathology*, movement of bodily fluids is perceived as an indication of distortion and illness. Medical treatment aims at preventing or counteracting such movement or at neutralizing its effects. Movement of the blood in particular is associated with pathological changes in bodily functions. A central notion is the concept of *flux*, humors moving to a particular part of the body, supposed to account for such phenomena as swelling, inflammation and in case of an external wound: bleeding. Emission of blood after an incision results from the flux of blood toward the location of the trauma. The incision is considered a kind of external traumatic event to which the body reacts with displacing blood toward that area. In general, therapy aims at controlling the flow by blocking it or by diverting it away through fermentation, massage, or phlebotomy. It is obvious that in this medical model, movement of bodily fluids, including blood, is synonymous with pathology while absence of movement indicates equilibrium and health. This doctrine constitutes the conceptual core of the medical tradition derived from Hippocrates, adapted by Galen and taught to medical apprentices for many centuries, up to Harvey's.

The anatomical context is different and seems to have no direct bearings upon the medical world view. *Anatomy*, revitalized by Vesalius almost a century before Harvey's publication, was still avant-garde to the medical practitioners of his time. A fair description is probably that physicians considered it interesting but irrelevant to their profession. What valid knowledge could be derived from *dissecting* dead bodies or from *vivisection*? It was evident that the procedure for investigation was so distorting to the object that it could hardly produce any result relevant to the art of healing, the sole and sacred purpose of medicine.

Bylebyl joins Pagel in considering Harvey 'a kind of split mind' and 'a dweller in two worlds' (Pagel, 1975, p. 1). In Bylebyl's interpretation, however, it is not the world of Aristotle which is juxtaposed to the world of modern anatomy but the world of traditional medical practice.

Through his education at the leading medical school of Padua, Harvey has been assimilated into the community of anatomy researchers. Through his medical practice in London, he is a member of the group of physicians who have the doctrine of humoral pathology as their major theoretical framework. His major interest is in the pulsation of the heart, and his careful investigations lead him to the idea of strong movement of the blood. Obviously, this constitutes a serious conceptual conflict. While anatomy leads into thinking of movement of blood as a normal and continuous process, medical thinking conceives of movement of blood as abnormal and a consequence of pathological events. Koestler's definition applies: Harvey's discovery involves the kind of perception that combines two incompatible frames. Vehement motion of blood, a major indication of illness should now be seen as characterizing the normal healthy process. While new, the idea is not intrinsically difficult. What is it that prevents its easy discovery?

Ultimately, the quantitative argument (see Chapter Twelve) will be crucial to Harvey's acceptance of the anatomical finding, but with respect to its conception we should notice that the incompatibility of the 'two worlds' relates to their incompatibility in self-world-segmentations.

What is basically involved is a different classification of events with respect to whether they are to be located in the outside world or to be related to the observing self. It is not that the movement of the blood is seen in one world model and not seen in the other. But while the anatomical analysis leads to seeing movement as a genuine attribute of the observed object, the medical approach considers it as a consequence of the action of the perceiving subject. Blood pouring out of the wound after you make an incision confirms the humoral model: the movement of the blood is a reaction upon the opening of a vein or artery. What you witness is a consequence of the act of making an incision, not something to be attributed to a healthy body left in peace. The effect you see reflects only your own action and reveals nothing new about the object.

Other famous discoveries in science allow for a similar approach. The difference between a Ptolemaic and a Copernican view can indeed be phrased in terms of attributing movement either to the observed object, the sun, or to the perceiving subject. One can argue for different perceptions in the cases where sunset is seen as the sun gradually sinking away behind the horizon compared to the earth as slowly turning away from a motionless sun. While the input is the same, sorting motion either on one side or on the other side of the boundary between subject and object yields a totally different picture. It would be an overly quick application of Winston's 'one two three infinity' rule [6] to generalize the scheme into a pattern for all scientific innovations. The major purpose of our example is to illustrate the relevance and the range of changes in this dual structure of self and world within world models.

We are not indulging in a kind of mystifying 'self-analysis'! The notion of 'self' and 'world' involved here are crisp and clear. The situation we describe is comparable to the problem faced by a moving observer who has to distinguish between perceived movement due to his own locomotion and perceived movement due to genuine changes in the environment. In his analysis of 'tacit knowing', Polanyi (1966) introduces the subject's action as the contribution of the knower in the acquisition of knowledge. Up to now, we have witnessed the elusive character of the tacit knowledge embedded in paradigms. According to Polanyi "meaning tends to be displaced *away from* ourselves" (1966, p. 13, italics in the original). However, in his description of tool use, he illustrates how this sometimes involves subtle shifting of the subject-object boundary. When one 'feels' with the tip of an instrument, the body is extended in such a way as to include the instrument. The key to tacit knowledge is self knowledge. The symmetric segmentation between a set of actions initiated by a self and a set of events located in an outside world constitutes the backbone of world models. Tacit knowledge becomes tangible when we look at the self-segment of world models.[7]

Paradigms and Perspectives

We are now in a better position to appreciate the scope of an approach which studies the mind as 'a multitude of interactive cognitive agents'. The same multiplicity of intellectual worlds represented by the many scientific specialties which recent sociology of science has discovered (see Part II) has reappeared within the confinements of the single mind. This proliferation of mental worlds will not be alarming if it leads indeed to a better grip upon the apparently elusive entity we have been chasing. Does it contribute to our understanding of paradigms?

The trade off model of paradigms whereby a set of programmatic principles is gradually transformed into an assessable body of knowledge (see Chapter Nine) is not highly compatible with the view of the mind that we have explored. While the latter emphasizes the interaction and combination of a multiplicity of world views, the former seems to support a monolithic entity which develops mainly according to internal laws. However, this does not necessarily lead to a total rejection of the notion of paradigm. As argued in the previous chapter, some cognitive structures can have a pivotal role without monopolizing the mind. Rather than thinking of the paradigm as generating its own 'secondary' discoveries, we should consider it as a conceptual frame to which, under certain conditions, other conceptual frameworks can be hooked on. While the paradigm, because of its peculiar configuration, definitely imposes restrictions upon the frames that can be attached to it, it is basically open. The frameworks which are hooked on it are externally supplied and do not stem from a purely internal development. Despite its being foreign to the Galenic framework, the pulmonary circuit could be hooked on to the Galenic paradigm. Partly because of the basic incompatibility described, Harvey's concept of systemic circuit could not. While preserving the cognitive function of paradigms, such an accretive model of paradigmatic growth relativizes the distinction between normal science and revolutionary science. Both types of activity involve the same type of combination and reorganization of world models. Revolutionary science differs herein that a pivotal segment of cognitive structure is pushed off while in normal science it sticks on. The accretive model also affects the discontinuity-thesis. The Galenic framework is at first instrumental in accommodating the pulmonary circuit. Later on, the pulmonary circuit notion combines with Harvey's newly discovered concept of systemic circuit upon which Galen's frame is eliminated, somewhat like in Newell's terms "scaffolding that is removed when the theory is complete" (Newell, 1973, p. 25).

When the paradigm is a core constituent but not the sole element in a configuration constructed out of world view segments, questions arise about the mechanism of composition of such larger units. How do these various cognitive structures cling together and what are the conditions of stability for such conglomerates? Again, in good AI tradition, it is instructive to look at the problem of coordination of world views in terms of a technique which seems to have solved it for the apparently simpler case of pictorial representation: linear perspective.

Kuhn's relativism in science has sometimes been interpreted as *perspectivism*. This label suggests a perceptual metaphor for reconciling the hardheaded realist with the idea of multiple world views. Like we can have different views of the same object when looking from different positions, we can have different paradigms which are equally valid in providing a cognitive map of the world. However, in this second sense, 'perspective' is much more frequently used than 'paradigm'. Any random selection of subtitles of recent books is likely to contain a few 'perspectives', an anthropological perspective, a psychohistorical perspective, a sociological perspective, etc. . . . on whatever subject one can study. What would make perspective such an apparently suitable notion?

Perspective is a remarkable device. As Goodman has indicated: "The adoption of *perspective* during the Renaissance is widely accepted as a long stride forward in realistic depiction" (Goodman, 1976, p. 10). The intriguing quality of perspective, however, is that while it is considered to be a major step forward in realism, it does so not by providing superior techniques for depicting the object but by providing techniques for specifying the position of the subject. Indeed, one should realize with Gibson that *all the information perspective gives is concerned with the location of the viewer*: "perspective . . . puts the viewer into the scene, . . . that is all. It does not enhance the reality of the scene". (Gibson, 1979, p. 283). What is experienced as progress in realism could be equally seen as a progress in subjectivism. Perspective adds information on the subject to a picture, even at the price of deforming the representation of the object. It illustrates the point made earlier, i.e. *that world models require specification of the subject's possible roles and actions as much as the object's* (the outside world's) *possible states*. A subject-object distinction is a solid component of any domain of knowledge, not just an epistemological presupposition to talk about knowledge. Improvement of knowledge might result as much from achieving a better representation of the subject-side as from obtaining a better representation of the object-side.

Since we prefer to think of scientific knowledge as 'objective' and accordingly try to remove any 'subjective' noise or distortion that might contaminate cognition, it seems all the more surprising that objectivity could be gained by using a method that amplifies 'subjectivity'. There is, however, one aspect of linear perspective that makes it fundamentally different from paradigms as we have encountered them in the multiple world view approach. When done according to the principles developed by Brunelleschi, formulated by Alberti,[8] perspective provides indeed a particular view on a scene, *but* it also allows, in principle, to reconstruct the view one could have from any other point of view. Perspective preserves basic information in such a way as to allow the derivation of other possible points of view. This hardly applies to the 'perspectives' that designate 'paradigms' in the subtitles of books mentioned above. There is no such clear-cut way to coordinate them. Or is it possible to compute a 'psychological' perspective when given a 'historical' one? Certain authors have been more impressed by the systematic coherence of all possible views, exemplified in the application of rules of perspective, rather than by the sheer multiplicity of subjective views these rules allow to produce. Not the multiplicity but the relationship between invariants and transformations is what perspective and science seem to have in common. In this respect, it is instructive to consider Russell's argument stating that

perspective, because of its logical recognition of internal invariances through all the transformations produced by changes in spatial location, may be regarded as the application to pictorial purposes of the two basic assumptions underlying all great scientific generalizations or laws of nature (Russell, B., *An Essay on the Foundations of Geometry*, 1897, quoted in Ivins, 1973, p. 10).

How does the subjective view relate to all possible views?

All along our analyses, we should have kept aware that a cognitive approach is only viable if it accepts the dynamic character of cognitive schemes. Otherwise, we lock the subject up into himself unable ever to see more than he knows. How to assure that cognitive schemes accommodate the possible and retain an open structure? Multiplicity of world views and recombination may be part of the answer but it does not provide a mechanism to generate *all* the possibilities in the way perspective allows their specification. Among the authors that have studied knowledge, Piaget is the one who has confronted the question of the relation between the subjective and the possible in a monumental genetic approach. His account of the growth of knowledge is an intriguing attempt to relate perspective in the weaker sense of paradigm or world view to perspective in the stronger sense of systematic preservation of all possible views.

CHAPTER TWELVE

CONSERVATION AND THE DYNAMICS OF CONCEPTUAL SYSTEMS

> The ambition of genetic epistemology has always been to link the problems that arise at the most elementary levels of knowledge to those raised by the theory of scientific thought itself. — Garcia, R. and Piaget, J., 1974, p. 139.

Confront an occasional companion with a phenomenon in nature. On a flight, ask your neighbor what it is that holds an airplane aloft. If one of you is more acquainted with aerodynamic theory than the other, your conversation will probably follow a typical pattern. Somewhat hesitatingly, the interrogated layman might put forward that the engines, by sucking in massive amounts of air, pull the airplane in a forward direction. The amateur aerodynamist will raise the question why one would expect the airplane to move towards the air that is sucked in rather than that the air would rush towards engines remaining static since, after all, the plane is much heavier than air. The educated layman will suddenly remember something about the applicability of Newton's principles and the law of action and reaction and modify his guess by proposing engines that push the plane in a forwards direction by throwing out the gases resulting from combustion in a backward direction. The interrogator will agree that this provides forward propulsion but he will then ask whether this can also account for flight. With a touch of pride, the partner will point to the slightly tilted position of the wings and invoke the pressure upon the underside of the wings as responsible for lift. But his expert companion will temper his enthusiasm by pointing to the problem of drag coupled to such an angle of incidence. If he perseveres, the conversation will continue and deal with the Magnus effect and Lanchester's insight, eventually touching upon the calculation techniques of Kutta and Joukowsky which are all, as pilot-philosopher Hanson (1970, pp. 257–273) convincingly demonstrates, essential to a genuine understanding of wing aerodynamics.

The conversation exemplifies a pattern of communication and interaction that figures as a celebrated technique for the acquisition of knowledge in some of science's most famous texts. It is the Socratic method illustrated in Plato's dialogues and it is the skeleton of Galileo's *Dialogue Concerning the True Chief World Systems*. It is also a major pattern of analysis for Jean Piaget (1896–1980), not as a rhetoric device for convincing occasional acquaintances, but as a systematic method for studying the construction of knowledge in children.

Scientific Knowledge and Children's Concepts

It is Gombrich's advice that "When a discussion has become tangled, it is always useful to trace one's steps back to its origins and see where the misunderstandings occurred" (Gombrich, 1960, p. 253). In Chapter Six, we have seen how such an attitude inspired Conant's orientation towards history of science. But it was Kuhn's experience that the study of the past does not lead automatically to the disentanglement of the complicated issues of the present. However, to achieve insight in the intricate structure of mature science, an alternative road is available: *psychogenesis*. For almost more than six decades (1920–1980), Piaget has studied the developmental cognitive psychology of children, not to provide parents with schemes for educational guidance or control, but to develop a grip upon the mechanism of knowledge construction. It is not that the views of children at any particular time are important or have special significance. The basic assumption is that the mechanism that allows children to move from inferior knowledge about some natural phenomena to superior knowledge is the same mechanism that is responsible for scientific progress, so that an understanding of the cognitive dynamics in children provides a key to understanding development in science. To study the psychogenesis of scientific knowledge, Piaget has collected thousands of Socratic- or Galilean-type dialogues with children. As the results of what he called his *clinical method*, they constitute a hallmark throughout the many volumes that report his theories and his findings.

The works of his first period of research in particular are almost solely based on that method. This earlier period of investigations coincides with his first academic appointment as a professor of philosophy of science and psychology at the University of Neuchâtel (1925–1929). His inaugural address, entitled *Psychologie et critique de la connaissance* (Piaget, 1925), clearly indicates how he conceives of history of science and cognitive developmental psychology as complementary methods in the study of scientific knowledge. But while some of his teachers such as Brunschvicg or Boutroux had emphasized history, his appreciation of trends (in 1925) is that psychology will now contribute its share: "la théorie de la connaissance semble en venir aujourd'hui `a se placer sur un terrain psychologique" (Piaget, 1925, p. 198). Piaget is a cognitivist from the very beginning stressing the Kantian point of view that "nous connaissons les choses `a travers seulement les formes et les *schémas* que notre esprit leur impose" (Piaget, 1925, p. 194, italics ours). The notion of scheme is a central concept in his system comparable to the notion of frame in the M.I.T.-school of AI. The mechanism by which we impose such schemes upon objects or collections of data is called *assimilation*. The mechanism that adapts these schemes to the specific objects to which they are applied is called *accommodation*. This is nothing but the cognitive view in Piagetian terminology. It is not seriously questioned. The problem is, according to Piaget, not whether or not such schemes exist but whether or not they are fixed or immutable. He considers it his task "à résoudre le plus important . . .des problèmes actuels de la théorie de la connaissance, celui de la fixité ou de la plasticité des 'catégories' de la raison" (Piaget, 1925, p. 208). In his view, the development of cognition involves more than the fine-grained local adaptation required in the accommodation of schemes. At times, radical reorganizations seem to take place which are to be considered genuine metamorphoses of the whole system of schemes and the way it operates. This leads to an approach which is in line with the kind of cognitive discontinuity model suggested by Kuhn (1962) while preserving a criterium for progress in this psychogenetic principle of development.

In *La causalité physique chez l'enfant* (1927) we find Piaget questioning children about natural phenomena in the style of the great classics of science. There is even a short section with conversations on the motion of airplanes in that 1927 book (pp. 257–262)! Children are confronted with questions like "Why does a boat stay on the surface of water?" "What are waves?" "Why does smoke go up in the air instead of falling down?" "Where does a shadow (of an object) come from?". From kindergarten age up to twelve, they provide sometimes amusing, sometimes perplexing answers. They say that the water can hold the ship because there is plenty of it in the lake. Or the ship can keep itself afloat because it is heavy and strong. The results of this volume and four others from the same period allow the construction of schemes for the dynamics of conceptual systems. The evolution is from animistic and teleological reasoning towards thinking in terms of inanimate objects and causal interaction. At times, parallels with history of science are indicated as, e.g., with respect to the physics of motion and Aristotelian and medieval concepts about it.[1] Piaget introduces the subject-object segmentation as the crucial dimension for characterizing the development as a whole. Children move from a purely subjectivistic or 'autistic' stage without genuine subject-object-segmentation to a mature stage of realism with relativized self-consciousness via a stage of *egocentrism* whereby the subject is aware of himself but unable to relativize his particular point of view or to relate it to other points of view.

The use of the term 'egocentrism' to describe the reasoning processes of 'innocent' children might have partly responsible for the heavy criticism that Piaget's earlier work elicited. Legitimate complaints were also based on his use of the clinical method and accused him of inducing many answers by luring children in some kind of conversational trap. But while some biographers are eager to point out that the basic concepts of Piaget's thinking are already expressed in his adolescent literary endeavor of 1918: the novel *Recherche*, Piaget himself deplores[2] the earlier phase of his popularity with scientific books because, in his own assessment, he failed to acknowledge the very basic relationship of knowledge with action. Conversational tracings with the clinical method restrict developmental patterns to what is reflected in verbal reports and consciousness. But the roots of knowledge go much deeper. Action or 'knowing how' is a form of knowledge too. To appreciate the dynamics of conceptual systems expressed in verbal reports, one has to understand their links with action. Even more generally, understanding the dynamics of cognitive processes requires insight into their relation with other mechanisms of adaptation and evolution.

Action and Adaptation

In autobiographic essays, Piaget praises himself for having been 'saved' from 'philosophy', his natural inclination, by an early education in a solid empirical science: biology. In his appreciation philosophy is perfect for posing incisive questions but only science is capable of providing acceptable answers. The pivotal role of biological concepts in his theoretical orientation shows that biology equiped him with more than some methodological sensitivity for empirical evidence and systematic research. In his foreword to Gruber and Vonèche's (1977) superb anthology, he declares

my efforts directed toward the psychogenesis of knowledge were for me only a link between two dominant preoccupations: the search for the mechanisms of biological adaptation and the analysis of that higher form of adaptation which is scientific thought, the epistemological interpretation of which has always been my central aim.

What is the relationship between scientific thought and biological adaptation?

White bears adapt to the polar cold by developing a thick fur for protection. Camels adapt to the drought of the desert by having special organs for storing extra amounts of water in their body. Striking anatomical equipment compensates for harsh environmental conditions. While much morphological features might function as exemplars of biological adaptation, we should keep in mind Waddington's (1961) remark that such features remain adaptive only when coupled to a specific pattern of behavior. The fur of the white bear is adaptive as long as he stays in his polar regions. It becomes more than a nuisance if he would move to tropical areas. Behavior and morphology go together and accomplish adaptation in a complementary pattern. Quite different behavioral patterns might evolve for coping with similar problems. To escape the cold of winter, some kinds of birds have developed periodic migration while other kinds of animals have developed hibernation. As an adaptive system, behavior and morphology are in a complementary relation to an ecological 'niche'. In a sense, an animal can be considered as a kind of mirror image of the environment in which it is fitted to live. Its structure and behavior reflect the kind of world to which it is sensitive and in which it can exist. In his neo-Kantian endeavor, Piaget lines up with authors who relate the Kantian a priori categories to the morphological and behavioral equipment of an organism. The roots of knowledge lie in the rules of operation of bodily functions and of organs that allow for interaction with the environment. The crucial entity is the pattern of action, in Piaget's terminology: the *scheme*.

The shape of the hand is complementary to a world containing graspable objects: branches of trees, tools, other hands. . . etc. As such, it is bearer of of some underspecified but nevertheless genuine tacit knowledge of an a priori nature. Its way of functioning: the scheme of grasping is a particular pattern of coordinated activity requiring the muscles of the fingers to flex around an object or to make a fist. The scheme is in the pattern of coordination rather than in the movement of the specific muscles. It could be specified in terms of a subroutine for a robot: close fingers until the sensed resistance reaches a certain value. Such a robot program is generalizable to any system having a comparable organization of parts. Some skilled movements done by means of the arms can be performed by means of the legs as well with a minimum of exercise.

The schemes of a newborn baby constitute his cognitive capital at birth. The developmental pattern of these schemes is such that at the outset they are underspecified while in the end they become detached from their locus of origin and are cultivated on their own. The 'underspecification' means that the scheme is, at the start, very 'schematic'. The grasp reflex of a baby is awkward and ineffective. Apparently, exercise is essential for finding out and filling in specific parameter values for specific applications. In this aspect of indefiniteness, the scheme expresses its potential as carrier of newly gained knowledge since in its adaptation to specific objects, newly discovered aspects of the external world are incorporated in the scheme. As it can be transposed to other configurations of organs or behavioral equipment, the scheme is detachable from the action in which it is embodied. Eventually it becomes reflected in consciousness where it can be cultivated on its own as a conceptual system independently of the action from which it originated. This is what Piaget calls 'reflexive abstraction'. Ultimately, such representations of action systems are organized and systematized in their own right and lead into very powerful systems for the organization and control of 'possible' actions. The formal reasoning exhibited in scientific thinking is an exemplification of the scope of abstract conceptual systems which are, however, the end product of a development that starts with simple sensori motor adaptations.

All this might seem rather abstract. But Piaget is able to apply these notions quite convincingly in the description of skills which children master easily and for which he analyzes both their *performance and* their *understanding* of their own performance as reflected in the descriptions they provide. His later works, in particular *La prise de conscience* (1974a), illustrate the importance and relevance of action as a generic source of ideas, while exhibiting the link with verbal thinking in classical applications of the clinical method.

Table Tennis Ball Expertise

A representative example of *La prise de conscience* confronts children, aged 4 to 12, with the following well-known trick. By means of the index-finger, a table tennis ball is propelled away over a flat surface while receiving an inverse rotation which makes it come back on an inbound trajectory along the same path that it followed on its outbound course. A similar situation is demonstrated with a hoop thrown in such a way that it seems to be trundled through its trajectory over the floor in both directions. When shown these boyish tricks, children are interested but not impressed. When invited to try it for themselves they prove in general to be quite dextrous and they reproduce the effect with the table tennis ball within a few trials. At the age of six, almost all subjects in Piaget's sample are able to perform the trick. Once the successful performance is established, conceptual understanding is checked. How do children explain the remarkable behavior of the table tennis ball?

While a boy might be perfectly able to accomplish the trick, it is quite obvious that he has at first theories about the physics of the situation which are substantially inferior to the level of his performance in practice. His very first opinions are tightly connected to the conviction that, when a spherical object moves over a flat surface, it has to rotate. Thus, most children will explain the situation by rotation in the direction of the movement for both up and down the trajectory. The annoying question is why the ball would stop at a certain location and in particular, why it would then start to roll again in the other direction without any apparent external interference. Some children arrive at ingenious ideas for saving the all rotational model suggesting a 180 degree turn of the ball so that no reversal of rotation is necessary. They claim to observe the rotation all along the course of the ball. This is a case of concept-driven top-down observation where theory prescribes what can be seen. Piaget emphasizes: "la lecture des observables est fonction de la compréhension et non pas de la perception lorsqu'il y a contradiction entre un fait et l'interprétation causale qui paraît s'imposer" (Piaget, 1974a, p. 54). Even when balls are used which have sections painted in bright colors such that the sense of rotation is more easy to be perceived, some children keep claiming that they *see* the rotation in the sense of the movement. The perception of the inverse rotation is obviously difficult to achieve. An observation which is apparently easier concerns their own movement. Children acknowledge that the finger which produces the propulsion of the ball makes a slight but significant backward movement, i.e. down and backward. But the kind of rotation this induces in the ball seems at first unnoticed. A crucial insight (or observation?) occurs when the hypothesis arises that the ball does not roll on the way up of the trajectory, but jumps over or trails along the surface. Suddenly, the composed nature of the phenomenon becomes transparent. The outward course is due to the propelling effect of the finger moving down, thereby displacing the ball with a certain force in a specific direction. The inbound course is due to the rotation inflicted upon the ball through the backward movement of the finger. It might then be seen that both processes interact, the forward movement resulting from propulsion overcoming inverse rotation while return is due to inverse rotation winning from straightforward propulsion.

The progression from some global 'misunderstanding', which is originally experienced as a sufficient description, toward a more differentiated conceptual system that is more adequate, is again along the lines of our earlier example: the properties accounting for the flight capacity of airplanes. In this simpler case, Piaget attributes progress in the construction of the conceptual system to an equilibrated elaboration of both the subject and object components involved. Being led to focus in an alternating pattern upon the behavior of the object and its own actions producing that behavior, the subject is forced to develop a detailed account of what specific effect goes with what specific action. In general, Piaget adheres to the principle that a process becomes conscious in some centripetal sense, i.e. we notice first the effects on the object (observables sur l'objet) before we take notice of our actions. In the case of the table tennis ball, the behavior of the ball is apparently so mysterious that scrutiny is first applied to the own action resulting in the 'discovery' of the dual composition of the finger movement. That does not alter the fact that the construction of an adequate conceptual system results from balancing two systems: one to account for the actions of the subject, one to account for the effects on the object. Their development is gradual and coupled: a discovery with respect to one's own action leads into a new analysis of the object and vice versa. This reinforces the position according to which self-knowledge is a basic constituent of genuine knowledge. Insight into the composition of its actions refers to the subject and is part of self-understanding. But this gradual increase in awareness about subject-object interaction is only one aspect of the development of skills.

The book *La prise de conscience* is part of a more elaborate endeavor, the other part can be found in *Réussir et comprendre* (1974b). Together these two volumes aim at clarifying the relationship between *knowing how* and *knowing that* by analyzing their developmental connections. *La prise de conscience* studies elementary skills which are easily learned. Piaget observes that elementary motor skills, such as the use of a sling or a catapult, are acquired on the level of performance without insight into the structure of the action. Oriented by a consciously selected goal, action apparently is self-organizing and regulates itself on the basis of unconscious sensorimotor feedback. On that level action is superior to conceptual thinking. There is 'success' (hit the intended goal) without 'understanding' (the ballistics, or more general, the physics of throwing or shooting), *réussir* without *comprendre*. Then, along the lines sketched above for the table tennis ball, a conceptual representation is gradually built up in consciousness mirroring in symbols what action accomplishes in fact. Conceptualization, inaccurate and irrelevant at first, makes up on action and comes to the same level. *Knowing how* is now rendered accessible in terms of *knowing that* as well. But conceptualization does not stop at the level of mere reflection of action. In his second book, *Réussir et comprendre*, Piaget tends to explore more complicated tasks. To give one example: a toy boat in a basin of water, exposed to heavy wind produced by a ventilator, has to be pushed through a specific course by positioning the boat, the rudder, the sail, etc. Although, in these and similar tasks, the solution is equally provided through some skillful sensorimotor act, success requires a detailed and complicated analysis in order to 'compute' the effects of the several factors involved and to predict the result of their interaction. Action can no longer take care of itself without conceptual guidance. Conscious analysis is needed to assemble a symbolic model of the situation that allows to derive, on theoretical grounds, the composed action required to reach the intended goal. Understanding has become a condition for success.

This development can be coupled to the more general stages of psycho-genetic development which Piaget has described in terms of *sensorimotor intelligence* (first years of life), *concrete operational thought* (achieved around seven and generalized up to age eleven or twelve) and *formal operational thought* (from age eleven or twelve on). The sequence is almost identical: first action on its own, then conceptual representation rejoining action, and finally, conception taking the lead and generating all possible action. We will not bother about these ages which pose problems that are not relevant here. As Piaget himself has repeatedly emphasized: it is the sequence that is crucial.

The functional significance of this development is the same at all stages. It has already been indicated how various forms of action are part of biological adaptation mechanisms. The general strategy yields an extension of the environment. Locomotion means a substantial enlargement of the environment by allowing many animals a more variable adaptation pattern than plants. They can run away from what is threatening and go searching for what is lacking. Conceptual representation prolongs this tendency allowing to generate in thought, possible environments and possible actions, without the necessity of actually going through them or performing them. In this respect, conceptual knowledge fits into the hierarchy of "vicarious knowledge processes" that D. T. Campbell [3] has explored so vigorously. In his concept of *reflexive abstraction*, however, Piaget sees a mechanism that allows to transcend sheer substitution by trial and error. By being decoupled from the specific actions and situations in which they are embodied, the regulative processes can be turned into an autonomous generative device that is not restricted to mimic action and actuality but to generate in a coherent and systematic way: possibilities. Formal thought is characterized by the predominance of the possible on the actual. This should become clearer when viewed in relation to Piaget's ambition to account for mathematics in psychological and biological terms.

While for the biologists action is only part of the adaptive process, for the psychologist it might seem some absolute beginning (*La prise de conscience*, p. 278). That is what he is given. His task is to identify the adaptive mechanisms that operate in action and to trace how they reappear metamorphosed into concepts which, in the end, become essential for extending the adaptive range of action. All this can be traced back in science which Piaget calls "la plus belle des adaptations de l'organisme humain au milieu extérieur" (Piaget, 1950, tome III, p. 112).

Piaget's Stages and the Finalization-model

In the preface to *Réussir et comprendre*, Piaget alludes to the relevance of that monograph for understanding the epistemological relation between technology and science. The developmental analysis of the link between *knowing how* and *knowing that* is, in his words: "de nature à éclairer la question épistémologique fondamentale des rapports entre la technique et la science" (Piaget, 1974b, p. 7). It is not uncommon to encounter in scientific circles the idea that application follows upon articulation of theory. It is considered almost as a truism: how could one speak of application before there is anything to apply? Hiding behind the pseudo-juvenile character of their discipline,[4] some social scientists in particular seem convinced of this principle. In his monograph on attention, Keele (1973) says "because the field is new, there have been few practical applications as yet" (p. 5). For the practical skills and techniques studied by Piaget, this sequence is reversed: *réussir* comes before *comprendre*, successful application before theoretical explanation and understanding. There are indications that science does not differ in this respect from practical skills. Kuhn (1971) illustrates how up to the nineteenth century, a pattern of technology *preceding* science has been the rule. Practical inventions were not the product of applied theoretical science. They resulted from alert handy men who hit upon new combinations of actions and procedures to yield effects that already had some value or that turned out to be valuable. Afterwards, scientists, partly intrigued by some of these results, developed conceptual systems to account for such effects in the same way Piaget's subjects learn to appreciate in conceptual terms what they earlier apprehended in action. Technological innovation purely based on theoretical understanding in the sense of practical means derived from theoretical knowledge is a relatively recent phenomenon. However, the most striking correspondence between Piaget's analysis of skills and scientific development lies in the close connection between his stages and the stages of the specialty life cycle according to the finalization-model (Chapter Nine).

The autonomous development of action in Piaget's first stage, whereby an effect is obtained and explored without the guidance of a conceptual model, corresponds to the explorative phase of the Starnberg approach. The second phase, during which a conceptual representation of the skill is developed and articulated, resembles the elaboration of a closed conceptual

network characteristic of theoretical science. The third phase in which conception has grown so powerful that it allows the computation of all possible
applications seems comparable to the finalization stage where conceptual
knowledge has attained such a degree of completion that it can be freely used
for many different and even incongruous ends. There are several intriguing
aspects in this parallel which suggest the same developmental process in both
science and skill. However, if Piaget's work has something to contribute to
the elucidation of this pattern in science, it should be his analysis of the
closure of conceptual systems which constitutes the central core. The finalization approach, while phenomenologically acknowledging this closure in its
second stage, does not unravel the underlying mechanism. One of the major
preoccupations of Piaget has been with the study of how conceptual networks become coherent and autonomous systems, only obeying internal rules
and immune to external influences. These studies relate to what Piaget has
studied as *conservation*.

Conservation and Closure of Conceptual Systems

In one of the most typical conservation studies (Piaget and Szeminska, 1941),
a child is confronted with a series of cylindrical glass containers. After two
bigger ones, identical in shape, have been filled with the same quantity of
water, the contents of one of the containers is poured into a beaker of different shape while the other remains untouched and is available for comparison.
The child is then asked whether there is still the same amount of water in
each of them. Before the age of seven, many children will say there is more
water in a narrower beaker because the water reaches a higher level and there
is less in a wider beaker because the level is lower. Similarly, when the liquid
of one of the initial containers is distributed over a series of smaller ones,
the children tend to say that there is more when there are more beakers,
regardless of their size. At some intermediate level, children will take into
account the compensation between thinner and taller or between wider and
lower and assert that the amount has not been changed. However, when asked
the same question with respect to one taller beaker distributed over several
smaller ones they fall back on their previous strategy and locate more water
where there are more beakers. A child reaches what Piaget calls 'conservation'
when it maintains throughout these tasks that the amount of liquid remains

unchanged, whatever the form or the number of vessels in which it is decanted. The child might support this position by arguing that all you do is pouring the liquid, sometimes called the identity-argument, it might point to the compensation-argument already encountered at the intermediate stage, or it might invoke reversibility: if the liquid is poured back into the original vessel, the initial situation will be completely restored. Conservation studies are done for several concepts and the various forms of conservation are acquired at different ages. Discrete quantities (e.g. marbles in the vessels) lead to an earlier conservation than liquids (continuous quantities) while physical quantities such as weight, mass and volume (Piaget and Inhelder, 1942) are 'conserved' between seven and eleven years of age. Both the theoretical interpretation and the observational procedure have made the Piaget-determined ages rather controversial.[5] But again, for our purposes it is more important to grasp the mechanism than to establish the age at which the mechanism is acquired.

The distinct character of genuine conservation, as distinguished from

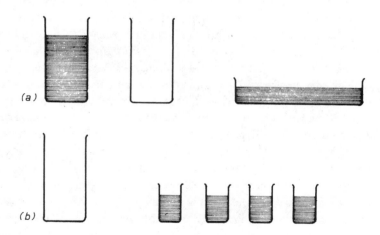

Fig. 12.1. The exemplar of Piagetian conservation studies: decanting liquids. In (a) the liquid of a tall and narrow beaker is poured into a wide receptacle. A beaker, identical to the one poured, is left for comparison. Children are asked whether the amount in the wide receptacle is still the same. In (b) the liquid is distributed over a number of smaller beakers. The general scheme is to introduce drastic changes in appearance by decanting and to check whether the amount of liquid is experienced as unchanged.

pseudo conservation, is the conviction of *logical necessity*. In the intermediate stage, children who argue for conservation on the basis of compensation of the dimensions can be brought into doubt by increasing shape differences. They will accept conservation for nearly similar beakers but fall back on pre-conservation responses when a very tall and thin receptacle is to be compared with a very flat and wide one. In a case of genuine conservation, the child even does not need to look. He knows for sure that the quantity is the same. It *could not* be different. From where this certainty and necessity?

According to Piaget, the fact that a position is achieved on the basis of logical necessity indicates that a cognitive structure becomes closed. As we have seen, although he obviously adheres to a cognitive orientation, his notion of cognitive structure is not primarily linked to symbols which represent perceptual elements. The models of the environment are embodied in the action schemes. Cognitive structures consist of systems of action schemes which go back on, e.g., manipulative actions with objects, such as composing a tower by means of blocks, taking an object to pieces, ordering a series of objects to achieve some specific configuration, etc. The characteristic of closure results from the coordination of such highly generalized actions into a self-contained system designated as 'groupement'. The arguments used in the justification of conservation only partially reflect this mental reorganization. When a child argues that nothing has been added or subtracted and that all that has been done is pouring the liquid, it indicates, in one out of several possible formulations, that a cognitive structure has been established in which operations such as addition and subtraction have been clearly fixated and defined in terms of each other and in relation to other operations such as partition and composition. Other arguments such as compensation or reversibility reflect the same structure viewed from another angle. What counts and what is essential is a system of operations that constitutes a coherent whole because of quite definite properties such as *additivity, reversibility, associativity* and *identity*. It is this kind of higher order systematization resulting from reflexive abstraction and mutual assimilation that assures conservation by embedding phenomena into fully equilibrated systems of operations.

Piaget's notion of 'groupement' is closely related to the mathematical concept of group.[6] This is no accident. Fascinated by the Pythagorean problem of the ontological status of mathematical entities, his work is in the line of Boole (1854) approaching mathematics as expressing 'laws of thought'. Mathematics is the codification of the rules of thought of which psychology studies the construction. Thus, it is no surprise to see references to mathematical entities as characterizations of cognitive structures. But one does not explain science by simply saying that mathematics is relevant and expresses scientific thought. Does conservation theory lead into new insights about scientific specialties?

Harvey: Conservation of the Blood?

In many areas of science, conservation principles play an important role. Conservation of momentum constitutes a basic rule of mechanics. In his book on the construction of number, Piaget (1941) points to the importance of the conservation of weight in the development of modern chemistry. Equally important is the principle of conservation of energy developed by several authors almost simultaneously in the nineteenth century (Kuhn, 1959 and other relevant papers in Clagett, 1959). As in the experiments with children, these cases involve complexes of phenomena which are considered to remain basically the same while changing drastically in appearance. Postulating immutable essences behind the appearances of things might seem suspiciously close to introducing invisible metaphysical entities. In particular Mach appears to be have been sensitive to this danger, insisting that principles such as the conservation of energy only serve as a vehicle for the "economy of thought" since "mere unrelated change, without fixed point of support or reference, is not comprehensible, not mentally reconstructible" (Mach, quoted in Cohen, 1959, p. 371). Expressed in the mathematical language of invariants, conservation principles in science can be related to a set of operations whose coherence seems to assure the stability of a whole system rather than the existence of an ephemeral entity. But is this true for simpler cases as well? Using the same mathematical concepts for his experiments with children, Piaget suggests it is. It is worthwhile to investigate how it functions in an intermediate case as Harvey's. There too, one can argue, is a sense of sudden insight and logical necessity indicating the closure of a cognitive structure.

Harvey's discovery of the circulation of the blood is indeed also the discovery of the conservation of the blood. Remember how in the Galenic

scheme ingested food was supposed to be transformed into blood that was to
be consumed by the body as some kind of fuel. Nothing impeded the idea of
a rather variable quantity of blood of even variable kinds. Now, one of the
most famous passages of Harvey's *De motu* expresses the discovery of the
fundamental identity of the blood as a consequence of his equally famous [7]
quantitative argument:

When I had for a great while turned over in my mind these questions, namely, how great
was the abundance of the blood that was passed through and in how short a time that
transmission was done, and when I had perceived that the juice of the food that had
been eaten could not suffice to supply the amount of the blood — nay, more, we would
have veins empty and altogether drained dry and arteries, on the other hand, burst open
with the too great inthrusting of blood, unless this blood should somehow flow back out
of the arteries once more into the veins and return to the right ventricle of the heart — I
began to bethink myself whether it might not have a kind of movement as it were in a
circle (Harvey, 1628, p. 75).

Aware of the fast and massive movement of blood expelled by the heart
Harvey sets out to measure the quantities transported. The result is incompatible with the Galenic view. There is no way to account for the needed
amounts of blood. Both the rate of production of blood from food and the
rate of consumption of blood in the body have to be unimaginably high.
Obviously, the difficulties disappear by introducing the notion of movement
in a circle. Once it is realized that the amount expelled through an artery
reenters at the other side of the heart through a vein, the perplexing effect of
the fast rate of transmission of blood dissolves. But notice how it is equally
plausible to fall back upon the accepted doctrine and to deny the relevance of
fast movement for a healthy organism. Given the observed rate of transmission
of blood, it is impossible that a normal organism functions that way. Nevertheless, because of some superior coherence in the dynamic conservation
model, Harvey overcomes the pull of the old schemes. Wherein resides the
superiority of conservation coherence?

It has been emphasized that with Harvey a quantitative orientation has
entered in the purely qualitative disciplines of physiology and medicine,
substituting precise measurement for vague speculation. But Harvey's measurements are far from precise. While impressed by the fast rate of transmission
of blood, his estimates are still far too low. Nevertheless, they are sufficient
to undermine seriously the old view and to sharpen the need for a new view.

Furthermore, the new view does not depend in any critical way upon the exactitude of the measurements. As the same mechanism of propulsion is involved, the quantity expelled always matches the quantity taken in, whatever the rate. Quantity enters here as an indication of coherence between different levels of analysis and not as the expression of precision at the lowest level of detail. This is also what Piaget tries to capture with his notion of conservation.

Conservation explains the coherence of a particular conceptualization of a given situation. In the tangle of drifting and fluctuating features that characterize a given display, whether pouring liquid in different vessels or watching the movement of the blood, conservation is finally arrived at by selecting a set of features which is in line with the kind of requirements that is defined in the 'groupement'. This means that certain features have to be actively ignored as much as other features have to be noticed or to be discovered.

In Chapter Ten, we have seen how such features are extracted at various levels so that when combined to designate an object they constitute an hierarchy. In Chapter Eleven, we have seen how recombination of features might be impeded by the fact that the relevant features belong to different and partly incompatible world views. In Piaget's criteria for conservation, we find an attempt to characterize the powerful principles of coherence which account for the superiority of the new combination.

But while this system of transformations might account for the compensations involved in conservation, it does not indicate the way in which this systematization is attained. In general, Piaget does not describe this process of gradually sorting out relevant features that fit into the cognitive structures. Neither is there much interest for where the features stem from: e.g. whether they are self-discovered or have been suggested by others. Apparently there are many different ways to arrive at acceptable results and again, the process has to be understood from the dynamics manifest through development in some global way rather than from detailed analysis of a single case.

Conservation expresses a solution of the ever recurring problem of the relation between parts and whole. It indicates that a mechanism has been developed for allowing substantial change in the parts while preserving the whole. In his autobiography, Piaget (1976) mentions the problem of the relation between parts and whole as one of the central preoccupations which he already attempted to solve while adolescent. He even admits: "si j'avais connu à cette époque (1913–1915) les travaux de Wertheimer et de Köhler, je serais devenu Gestaltiste ... " (Piaget, 1976, p. 7). Like he was "saved" from philosophy, this ignorance from the early Gestaltist work "saved" him from an orientation with a too prevalent position for perception.

Perspectives on an Object

As in the conservation studies with liquids, some of the major steps in the evolution of scientific concepts involve looking away from apparently salient features. To arrive at his conceptualization of falling objects, Galileo had to overcome the idea that the weight of an object is important, somewhat in the same way that children have to disregard the level of the liquid in the different vessels involved in a conservation experiment. According to Piaget, such "creative overlooking" does not result from any mysterious gestalt type vision or insight. It results from the increasing power of the operations of thought to overcome the pull of apparently salient perceptive features. The same alternating differentiation of action and perception invoked in the study with the table tennis ball trick applies in general. A differentiated and organized conceptualization of the actions of the subject allows to account for ever more changes in the appearances of the object.

Throughout cognitive development, the construction of invariants functions as a major mechanism in the progress of knowledge. Levels of invariants seem to constitute something like learning plateaus for an increasingly powerful representation system. The stabilities studied in the psychology of perception which deal with constancies of form, size, lightness, etc. indicate mechanisms oriented toward the extraction of the same features under varying perceptual conditions. The notion of the permanent object is already a higher order type of invariant which involves substantial development and elaboration. It is not the result of a simple figure-ground segmentation mechanism. It constitutes the product of a system of actions which allows to counteract the changes in various perceptual forms in which the object appears.

It would take us too far to go through the several stages in the construction of the notion of obejct which the child attains at about eighteen months of age.[8] Let us consider the perception and conception of a simple cube to sketch the dynamics Piaget illustrates and explains at large for many areas.

When we look at a cube from a certain position, the 'percept' is but a symbol for a complex of actions that can be performed on the object. The meaning of the symbol is not so much a visible entity in the world as it is the set of manipulations together with the visual changes they entail. But then, in what resides our knowledge of cubes? Is it perceptual, imaginal, conceptual, action, or all of them together? What is the cube really and what makes it a stable and invariant entity? Piaget's approach can be exemplified in the way he would handle visual perspective for this situation. At first, the visual images of a cube obtained from two different points of view are unrelated. In the action of moving from one position to the other, the coordination is built up that relates the two different images to the same object, i.e. a level of conservation. Later on, the object develops into a set of images connected through a system of coordinating actions. Now, this is a general pattern followed by Piaget. Conservation is achieved by identifying change in terms of the actions that produce it or undo it. The action of going from one viewpoint to the other or of going back to the original viewpoint maintains stability in a changing visual world. Superior knowledge is achieved when there is the possibility to generate the image obtained from viewpoints which actually have never been taken. This is accomplished through the organization of actions characterized by the 'groupement'. Coordination and systematization of actions into a coherent whole provides them with the kind of generic quality that can generate all possible points of view. What is our knowledge of the modest cube at that stage? The real constitutive knowledge is, according to Piaget, not any particular representation of the object in terms of images or concepts, but the system of transformations that allows us to move between any particular view or formulation. Applied to perspective: our knowledge of an object is not to be reduced to any particular perspective of it, nor to a combination of a few of them. These acquire their meaning from an underlying system of operations that, given any particular view of the object, can generate any other possible view.

The example of a particular view on a simple cube allows us to assess the importance of action in Piaget's theory of knowledge. While empiricism tries to account for knowledge in terms of perceptual residues left over from interaction with the object, Piaget sees knowledge basically as a construction out of action, in which perceptual features are embedded as regulating elements. Action schemes are the bearers of knowledge and provide the supporting frameworks on which perceptual knowledge is received and regenerated in terms of images.

Moreover, the example of perspective also allows to indicate in an illustrative manner, the bidirectional and symmetrical development of knowledge whereby gaining knowledge of the object means also specification of the position of the subject. In fact, Piaget considers 'objective' knowledge to result from the coordination of all subjective points of view. The coordination and systematization of operations not only yields a 'conservation' of the object but also an elaboration of the subject.

The construction of the notion of object is the prototypical cognitive process. As Piaget indicates, scientific concepts and world models are built along the same functional principles. But while objects are assembled in the course of real actions, e.g. with locomotion and changing visual viewpoints, concepts are worked out on symbolic representations. They do not involve overt action but some 'symbolic' action on representations. When keeping in mind the self-world segmentation dealt with in both the previous chapter and Chapter Two, this notion of 'interiorized action' does not need to be considered mysterious. Once we introduce the notion of representation, why would it be more difficult to accept representations of action than representations of external objects? However, in line with the centrifugal principle of consciousness (first the effects on the object, afterwards the action which produces or undoes it), the accessibility of these interiorized action is, so to speak, one round behind on representations of the object. Only in the stage of formal operations, operations are conceptually explicit and become in turn the object of higher order operations. With respect to visual perspective: different viewpoints can be coordinated and any single point of view can be assimilated on the level of concrete operations. However, it requires the level of formal operations to make the procedural aspect explicit. Once it is fully understood that linear perspective reflects a technique for linking changes in viewpoint to changes in appearance in a systematic way, it also becomes transparent that it is only one possibility of many. The representation technique itself becomes an object of exploration and various alternative techniques such as inverted perspective or even 'psychological' perspective,[9] become other ways of approaching this now 'formal' object.

Paradigms and Development

Arguing for the prototypical character of the construction of the object, which functions as a model for the elaboration of scientific worlds and viewpoints, Piaget makes what he himself calls a somewhat daring comparison:

> ... the completion of the objective practical universe resembles Newton's achievements as compared to the egocentrism of Aristotelian physics, but the absolute Newtonian time and space themselves remain egocentric from the point of view of Einstein's relativity because they envisage only one perspective on the universe among many other perspectives which are equally possible and real (Piaget, 1937, quoted from Gruber and Vonèche, 1977, p. 283).

From Aristotle to Einstein and from infant to Nobel prize level physicist — the span of this development reflects the scope of Piaget's ambitions. Is it possible to compare this monumental work on the whole of science and knowledge acquisition to the more modest analysis of scientific specialties in recent social studies of science?

The fascinating quality of the notion of paradigm stems from its link with the cognitive point of view that stresses the role of knowing subject in any act of cognition. Empiricism is just too simple to account for knowledge acquisition. The subject approaches the world and interacts with it armed with expectation-generating schemes that guide its search. The disturbing quality of the notion of paradigm stems from the same link with the cognitive orientation which seems to lock the subject up in a particular viewpoint from which it cannot escape. No way is provided to go beyond the suggestions of the scheme, to break out of the accepted categories for achieving genuine discovery instead of filling in rigidly fixed boxes. Piaget's schemes are the cognitivist's schemes. They shape specific ways to interact with the world and impose boundaries upon experience. But there is an escape out of the cognitive paradox: development. Action schemes, which for Piaget embody cognitive structures are vulnerable to change. They are not only modifiable in the way Minskyan frames are adaptable to specific local conditions. In the course of development, they undergo substantial reorganization and systematization which opens them up for an ever extending realm of new discoveries and experiences. In the idea that development overcomes the intrinsic limitations of the cognitive view lies one of Piaget's most important contributions. What with respect to paradigms in science?

With his seven-league boots approach to history of science, Piaget might well seem out of step with the small strides that characterize progress within specialties. The sequence Aristotle — Newton — Einstein is a sequence of the major revolutions, the kind that Kuhn originally had in mind when introducing the notion of paradigm (Kuhn, 1962). However, in such sequences, Piaget suggests, while accepting discontinuous plateaus, a basic underlying functional continuity and progressive development. This allows to maintain the notion of paradigms as cognitive schemes without accepting the non-cumulative and non-progressive character of scientific development. In this sense, Piaget's system encompasses Kuhn's. As Wolf (1966) indicates "Piaget a voulu. . . souligner la linéarité et la continuité du développement, laissant pour des recherches ultérieures les questions de différences individuelles et les formes multiples de différentiation" (p. 237). His methods reflect this preoccupation with the establishment of a chain of stages that links the most elementary adaptations to the superior products of mathematical thought. But to confront five-year-old children with some advanced theoretical concepts of physics is in a sense short-circuiting genuine development. Conversations of one 'who knows' with one 'who does not know' are not representative for the way knowledge is socially constructed. In the position of the members of a scientific specialty, there is no one to guide the group toward the superior stages. "Scientific activity comprises the construction and sustenance of fictional accounts which are sometimes transformed into stabilised objects" (Latour and Woolgar, 1979, p. 235). What preoccupies Piaget are these conditions of stability, not the way in which an individual or a group attains them. This is manifested in his attitude towards social factors.

Individual Discovery and Social Success

In response to questions about the degree of control attributed to the structures he locates in the mind, Piaget has indicated that he does not ignore the multifaceted character of our knowledge and the multiplicity of world views as stressed by the M.I.T. school. The highest level of cognitive performance — the stage of formal thinking — is only attained under privileged conditions and for a limited subset of domains, e.g. some scientific specialties. The multiplicity of world views is reflected in the multiplicity of the self. In a discussion with Osterrieth, Piaget admits "l'unité structurale de la personne,

. . .je ne l'ai vue nulle part, a aucun stade dans le développement de l'enfant. Je ne la vois pas non plus chez la plupart des adultes. Je suis moi-même une personnalité multiple, divisée et contradictoire" (Osterrieth and Piaget, 1956, p. 58). But some apparent contradictions in his attitudes raise questions about the range of his system.

While considered a pedagogical pioneer by some, Piaget has a rather peculiar orientation with respect to teaching and education. About creative scientists, he contends: "un physicien de génie est un homme qui a su conserver la créativité propre à son enfance au lieu de la perdre à l'école" (Piaget, 1979b, p. 64). At several occasions, he is quoted to have declared "Tout ce qu'on apprend à l'enfant, on l'empêche de l'inventer, ou de la découvrir" (quoted in Bringuier, 1977, opposite p. 97). For an author who has demonstrated as no other that psychological development is long and complex and that it takes children much time to reach the 'right' ideas, such skepsis in relation to teaching seems strange and puzzling. But this is only contradictory in appearance. Defending a theory of knowledge in which the knower *constructs* his knowledge rather than to receive it ready-made, Piaget distrusts educational practices which risk to neglect the importance of action. The risk is that, in the hope of speeding-up development, children are offered representational devices which they assimilate only very superficially. What they then achieve is a kind of mimicry rather than genuine reconstruction in their own terms. But while being consonant with his theory of knowledge acquisition this opinion on education reflects Piaget's limited appreciation for social factors which in turn indicates a fundamental problem of the whole system.

Piaget does not seem to underestimate the role of social factors in the functioning of science. He has declared repeatedly: "le critère de réussite d'une discipline scientifique est la coopération des esprits" (Piaget, 1965, p. 44). Some subtle sting is hidden in the word 'réussite'. It indicates that full cooperation and communication among minds is achieved when the task is done, i.e. when the knowledge is constructed. But in the laborious process of construction itself, cooperation is ignored. Piagetian development is a Robinson Crusoe type of endeavor where the knower elaborates his own world in solitude,[10] and meets his fellow men only to discover in the end that their efforts have converged and all of them have arrived at the same result. The coordination of perspectives is again the generic metaphor. At first, various individuals represent different and unrelated perspectives. As each of them elaborates his own cognitive system, their partial products gradually encompass more perspectives. In the end, they all reach the point where their own systems mutually incorporate the perspective of all the others.

Again, one could point out that Piaget is neither interested in the particular trajectory run through for achieving a certain plateau nor cares about the background of the materials that induce the development to that plateau. Whether ideas stem from social interaction or pressure or derive from private exploration is unimportant. What counts is that, once combinations of features obey certain rules of coherence, they become coercive and are experienced as such by both individual and group. Once the rules of cognitive coherence have been established with respect to a particular domain, the rules become social norms almost without effort. With mature conceptual systems, the distinction between cognitive and social dissolves.

Piaget will not deny the existence in science of communities which recruit their members on the basis of the intense pursuit of a particular idea or particular point of view. In fact, he complains about the importance of unstable fashions in the scientific research of both the United States and the Soviet Union: "Je suis inquiet du rôle des modes et des écoles là-bas (U.S.). Ils font tous la même chose 'a un moment donné, et puis tout 'a coup la mode change, et ils font tous ensemble autre chose, et de nouveau la même chose. En Russie, c'est pareil" (Piaget in Bringuier, 1977, p. 37). In Piaget's appreciation these schools are but social versions of 'centrations', attempts to come to grip with a problem by following through one particular feature or aspect. Centration effects recur at all levels of cognition as the result of a one-sided approach. According to Piaget: "on peut aller jusqu'à soutenir que toute connaissance (représentative comme perceptive) est déformante en ses débuts, 'a cause de 'centrations' de diverses natures et que seules des 'décentrations' conduisent 'a l'objectivité" (Piaget, 1963, p. 3). In the conservation of liquids situation, it is focussing upon the height of the column of liquid at the price of missing conservation. In perception, centration accounts for gestalt effects which equally introduce a deformation in the appearance of the object. Some selected particular feature of a situation is incongruous with a global characteristic. It is this incongruity which is experienced as a gestalt effect: the whole being irreducible to a transparent combination of parts.

Social factors, whether internal or external to science have similarly deforming effects when focussed explicitly. Group pressure may push forward certain approaches or solutions. External goals might pull tentatives in a certain direction. They might induce a pattern of conceptual development in the way the tennis table ball trick induced an exploration in the concepts of the physics of motion. As such, they are but peripheral causes that set the conceptual dynamics going but they do not control the process all the way. In the conceptual system that settles the issue, they are assimilated as particular views or uses counterbalanced by alternative views or uses. In the end, conceptual systems that reach maturity become independent of the 'context' in which they originated. Or when applied to science: finalized specialties are independent of their history, i.e. the particular sequence of centrations that led to their formation. All the time, the warning is: "c'est bien souvent en se désintéressant de tout but utilitaire que la science a fait ses conquêtes" (Piaget, 1950, II, p. 72).

While his views and findings about the coherence and closure of cognitive structures are highly appreciated, it is in particular this character of inevitability and the ahistorical nature of development that raises doubt about Piaget's system. Not only in the growth of children but even in the evolution of the universe, historical choices play a role (Prigogine, 1977). We can not exclude them from scientific development.

EPILOGUE

Notre science, longtemps définie par la recherche d'un point de vue de survol absolu, se
découvre finalement une science 'centrée' . . .
— Prigogine, Y. and Stengers, I., 1979, p. 291.

There is no doubt that the advance of the computer will deeply affect our ways of doing science as well as our daily life. Whether the firmly established formal communication system based on the publication of printed materials will be soon replaced by fast and fancy personal computers linked into huge communication networks is an open question. Even for the suspicious and the reluctant, it will probably be difficult to keep terminals out of their offices and laboratories. But whether computer networks will replace the traditional system or not, it is certain that the influence of their availability and use will be comparable to or will be even greater than the influence of the fifteenth-century introduction of the printing press which has been closely connected to the advent of science (Eisenstein, 1979).

It would be fortunate if science of science could come of age at a time when such a major event is affecting science. While it might not yet be capable of accompanying the large scale infiltration of the computer with specific suggestions and directives based on a body of finalized knowledge, it certainly could take profit from studying the dynamics of knowledge production during a turbulent period of innovation that will force the system of science to a higher level of self-analysis and self-understanding.

In line with Sarton, Eisenstein (1979, p. 454, note) indicates how the introduction of the printing press could enhance an apparent retrograde movement while, on a broader scope, it served the advancement of science. The first scientific books to be printed and distributed were the celebrated texts of ancient doctrines that would soon be surpassed by the discoveries of the scientific revolution. The critical orientation built in into the scientific method did not arise independently from the new means for rendering the

252

old and reputed doctrines more accessible. The availability of exact copies of a text was a condition for the sharper scrutiny and rigor of analysis characteristic of science. New means of expression entail new ways of analysis.

With the computer, we are passing the stage of simply using new technology for performing old and established routines. The sheer power of the technology forces upon us a reflexive orientation with a reconsideration of what we accept as established knowledge and a search for new standards of rigor. This cannot remain without consequences for the industrialized production of knowledge that modern science is. As one of the first intellectual fallout products of computer technology, cognitive science has employed the new powers of expression to render theories of knowledge development in psychology and philosophy more explicit, thereby exposing weak spots as well as new possibilities. Our exploration of the cognitive paradigm in the study of scientific knowledge should have revealed some of these both in terms of cherished pretensions that are challenged as in terms of new vistas that are opened up. Let us briefly indicate those that affect the disciplines considered closely connected to science of science, focussing upon the crucial notion of context. Context is the ubiquitous concept that recurs in all of them and that explodes the boundaries between *internal* and *external*, between *cognitive* and *social*, and even between *self* and *world*.

The Notion of Object As Prototypical Cognitive Structure

As indicated by Piaget, the perception of even the simplest objects is to be considered a prototypical case of knowledge construction similar in structure to the highest achievements of scientific reasoning. In Chapter Ten, we have described objects as combinations of forms retained at various levels of analysis held together by a conceptual framework that solidifies transversal connections. Usually, object perception is assumed to be a rather primitive process based on a peripheral mechanism such as the figure-ground segmentation. This figure-ground segmentation is in itself the prototype of the opposition between structure and context. Though it cannot be doubted that mechanisms of this kind operate at a pre-conceptual level, we have seen that it is in no way necessary to consider them as determining absolute boundaries for objects. Figure-ground segmentation accounts for the extraction of a major invariant of a perceptual object: the contour, but this is only one. The way intra-figural elements are handled involves very similar extractions of similar invariants at several levels. One could argue that the perception of an object is a combination of nested figure-ground distinctions with no outer boundary absolutely imposed. The figure-ground boundary is shifting almost continuously.

Our approach is in contrast with common concepts about perception which consider objects as more or less given entities and introduce context as an external source of distortion or deformation. Dember (1960, Chapter 6) deals with the classical gestalt effects as "the effect of context" (See Figure 13.1.)

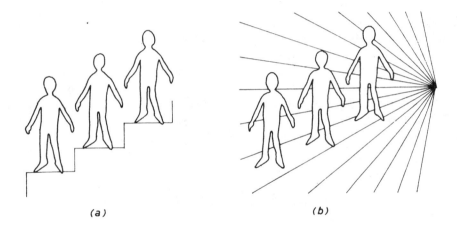

(a)   (b)

Fig. 13.1. The identical shapes of human figures in (a) and (b) are perceived as of the same size in (a) and of different size in (b). In (b) they are embedded in a classical gestalt effect: the Ponzo-illusion. Dember (1960) discusses such effects as 'effects of context'. This contrasts with a cognitive interpretation of context. For a comparison, see Figure 1.5, p. 10.

The perception of an object is modified according to the background against which it is perceived. We do not deny the genuine distortions introduced by gestalt effects which even resist correction through adequate conceptual knowledge. But to generalize them into *context effects* leads into a much more serious distortion of the concept of context. In the approach we have followed, the dualism of the figure-ground segmentation has no longer a privileged status. Perception involves the selection of various invariants retained at various scopes to be combined in an interpretation that construes both object *and* background. Context is not superimposed: it has a constitutive rather than an additive nature. If it is to be maintained as a meaningful notion, it should receive the more balanced interpretation provided by the cognitive orientation, i.e. a complete world model governing the organization of a selection of features retained from a complete array. In Chapter Ten we have seen how such an approach is compatible with an interactive theory of perception which is essential for preserving the cognitive view from a decline in solipsism. In Chapter Eleven, we have illustrated the decomposability of such world views and the possibility of reassembling new ones by means of recombining segments of other world views. In Chapter Twelve we have explored the mechanisms of coherence for such world views and we have discovered in Piaget's notion of conservation a principle that indicates how more elaborated conceptual systems achieve a similar kind of solidity that we attribute to perceptual objects.

We have repeatedly referred to the favorite AI-technique of looking for solutions for simplified versions of a problem first and then generalizing to more complicated cases. With object perception as an exemplar for scientific vision and thought, we have more than a heuristic device. The 'orthodox view' in philosophy of science takes object perception more or less for granted. When entangled with problems that arose from using a phenomenalistic approach to perception, early logical positivism saw no problem in shifting to a physicalistic 'thing-language' that assumed the perception of things sufficiently simple to be introduced as a primitive process. Focussing upon object perception, we transform what is assumed to be simple and peripheral into a challenge to be explained and make it a pivotal case in the argumentation for the cognitive view.

The Plurality of World Models

The impressive plurality of world views suggested by the cognitive view might encourage sociologists of science who have found a comparable plurality in the multitude of specialties which they have discovered to be the living cells of science. Philosophers of science are rather reluctant to accept such plurality, being allergic for what could easily develop into straightforward relativism. The unity of knowledge anchored in a single solid world is a serious stronghold. We have seen how the cognitive orientation, in particular the M.I.T. school of AI stresses the importance of multiple points of view. These are not necessarily related or situated with respect to each other through some super-coordination-frame. In one of the first pages of the *frame-paper*, Minsky quotes a passage from an essay of I. Berlin which depicts a duality of approaches one of which develops into a leitmotif of his own. It reads:

> For there exists a great chasm between those, on the one side, who relate everything to a single central vision, one system more or less coherent or articulate, in terms of which they understand, think and feel — a single, universal, organizing principle in terms of which alone all that they are and say has significance — and, on the other side, those who pursue many ends, often unrelated and even contradictory, connected, if at all, only in some de facto way, for some psychological or physiological cause, related by no moral or esthetic principle . . . (quoted by Minsky, 1975, p. 213 from I. Berlin *The Hedgehog and the Fox*).

Notice that with respect to science, the distinction is not between what Darwin called 'lumpers' and 'splitters': scientists who engage in the construction of all embracing systems versus scientists who, like a detective, pick out one particular and specific problem as their 'case' and forget about all the rest. It is not so much a question of wide scope versus limited scope but a question of *multiple* scopes: the possibility of using a finite body of knowledge in an infinite number of problem situations by combining or recombining subparts of the body of knowledge in an ad hoc way. Minsky affirms unambiguously: "We do not want to be committed to an inflexible, inclusion-oriented classification of knowledge" (Minsky, 1975, p. 251). The pluralism implied by this view might dishearten those who cherish the ideal of the unity of knowledge that carries with it feelings of certitude and security. It might attract others who like the idea of a world which can be known in a multitude of ways and which retains the potentiality for surprises and challenges all the time.

A major stumbling block with respect to a serious discussion of relativism is the rather crude conviction that relativism entails an 'anything goes'-attitude. It seems as if giving up the idea of one single solid world for a plurality of world views implies the acceptance of every possible view. But like not every number has to be even because there is an infinity of even numbers, not every proposed world view has to be accepted, even if there would be an infinity of acceptable world views. One can be relativist and nevertheless adhere to strict standards of acceptability. Whether the search for unity is nothing more than the expression of a need for security is doubtful. Though Minsky says "the desire for complete synthesis is probably a chimera, and should not be a theoretical requirement" (Minsky, 1975, p. 257), it can not be denied that sometimes substantial progress in science has resulted from projects of unification. But it is indeed questionable whether this mechanism of productive recombination of world views should be symbolized by a metaphysical entity or a methodological dictum.

The predominant feature arising from contemporary studies on science is one of multitude and diversity in views, techniques and approaches. It would be recommendable to explore the possible benefits of pluralism as well as its epistemological risks. In assessing the relative merits of realism and relativism in science, it might be instructive to look at the world of art.[1] The development of Renaissance art has been linked to the development of science (see Edgerton, 1980). Newly discovered pictorial techniques — such as linear perspective — have enhanced the idea of realism as an achievable ideal for representation that has dominated artistic preoccupations for several centuries. Modern art has gone beyond this monolithic fascination with realism to explore a multitude of other ways of representation and expression. While it might be puzzling and intriguing, it is difficult to see this development as a regression. Well-fed scholars and scientists should beware of looking down upon their more bohemian fellows in art. They might indeed be the avant-garde in representation techniques which scientists will discover, with some delay, to be relevant to their endeavors too. Modern science might still have to catch up with modern art.[2] Among the requirements could be: surpassing a monolithic notion of reality and a lop-sided preoccupation with the outside world that ignores the representation and multifaceted nature of the knowing subject.

Action, Purpose and Self-knowledge

Imminent confusion results from the ever recurrent dualism that is exemplified in the apparently obvious distinctions between internal and external,

cognitive and social, self and world. Most distorting, as a consequence of the
self-world segmentation, is the tendency to consider knowledge as mainly
concerned with the outside world. This puts the knower out of the scene
as an external and static observer what he manifestly is not. By recognizing
the representation of the self as genuine a problem of knowledge as the
representation of the outside world, the cognitive orientation restores the
balance between the contributions of subject and object in the process of
knowing. Piaget in particular has stressed the symmetry in the development
of knowledge whereby new discoveries about the outside world have to be
balanced by reorganization in the knowledge about the self, and conversely,
new discoveries about the subject entail reorganization of knowledge about
the external universe. In Chapter Eleven we have indicated the potential
importance of self-world realignments for theoretical knowledge and discoveries. It is of utmost importance for understanding the notion of practical
application of knowledge as well.

In the traditional approach where knowledge is related to 'external reality',
science is for finding out about the outside world while human goals and
purposes relate to subjective values that are to be determined by social
processes and policies. Here the dualism appears in the detrimental disguise
of the cognitive versus the social. The cognitive deals with the objective, the
world as it really is, the social is concerned with value, the world as the
community of subjects would like it to be. Such a segmentation ignores
the intricate relationship between action and knowledge explored in Piaget's
developmental approach.[3] According to that relationship, in the process of
finding out about the outside world, we also learn about ourselves. The table
tennis ball trick of Chapter Twelve is relevant here. Acquiring knowledge
about the object goes hand in hand with learning in more detail about the
action of the subject. Self-knowledge is but a more differentiated representation of the actions of the knowing subject linked to a more differentiated
representation of the object. Therefore, knowledge about the external world
can not be developed in function of unchangeable externally imposed goals.
The developmental approach makes it clear that the original goal for getting
involved is later on reduced to a peripheral legitimation for starting an inquiry.
By learning about himself, the subjects perspectives on possible goals change
and mature.

Ultimately, when dealing with the construction of applicable knowledge, there is no way to circumvent the Socratic dictum 'know thyself'. In the ever recurring opposition between the natural and the social sciences we currently experience the curious situation that natural sciences are made suspicious because of their success — reproaches for begetting disruptive technology — while social sciences are made suspicious because of their lacking that type of success (see Lindblom, C. E. and Cohen, D. K., 1979; and Weiss, C., 1981). This disequilibrium might well be related to the failure to recognize the constructive character of 'self-knowledge'. 'Selves' are made along with the external worlds we investigate and create. Problems to install or adapt to new worlds might stem from failure to recognize the self-image that goes along with it or the refusal to accept it. To be able to make choices these self-images should be made explicit.

As Prigogine and Stengers (1979) have pointed out, the understanding of scientific knowledge has been burdened with the belief that objectivity involves the removal of any reference to the observer: "l'objectivité scientifique avait longtemps été définie comme l'absence de référence à l'observateur" (p. 291). But as they indicate in their penetrating analysis of our knowledge of the physical world: the subject or observer has to be included in the picture! Indeed, "time has come for new alliances" (p. 296), new alliances between the various disciplines that study scientific knowledge, new alliances between natural sciences and social sciences which will reveal the intricate complementarity between knowledge of external worlds and models of ourselves.

NOTES

Notes Introduction

[1] *Cognitive Science* is the quarterly journal of the *Cognitive Science Society*, published by Ablex Corporation, Norwood, New Jersey. The Society organizes an annual meeting. The invited papers of the first meeting, held at San Diego in 1979, have been published in the fourth volume of *Cognitive Science* (1980). They are also available as a separate collection in Norman (1980). An editorial in *Cognitive Science* 1980 contains a few organizational changes in the journal and some general remarks on the status of the emerging discipline.

[2] Under the auspices of the *Fondation Archives Jean Piaget*, advanced summer courses in cognitive science are organized at the university of Geneva since 1979.

[3] James' 'classic stages of a theory's career' are in James (1907, p. 198). I found this reference in Merton (1967, p. 21) who deals with the topic under the heading of *adumbrationism* which he characterizes in terms of the following 'credo':

> "The discovery is not true
> If true, it is not new
> If both new and true, it is not significant."

[4] Among the few studies which apply a cognitive view to the study of science, or at least explicitly allude to it, Transgard (1972), Nowotny (1973) and Weimer (1975) should be noted. In the sociology of science there is an active interest in the use of cognitive concepts which is apparent in the work of Klima (1974), Weingart (1974), Whitley (1974), some papers and editorial comments in Knorr *et al*. (1975), and also in Mendelsohn *et al*. (1977). This is not to say that other studies of science have not been influenced by the cognitive approach. Papers such as Petrie's (1968) and Campbell's (1959) on methodology express a view that is clearly related to the cognitive approach but is not presented as such. The same remark applies even more strongly to Van Norren's (1976) monograph. However, as will be argued in this book, especially Chapter Twelve, the pioneering studies in the cognitive approach to science are Jean Piaget's, whose monumental work and efforts to establish 'genetic epistemology' – combining cognitive psychology with philosophy of science – will prove to be the foundation for cognitive science in general as well as for its application to science in particular.

[5] A similar approach in sociology of science which goes beyond the division is D. Bloor's 'strong program' (see Bloor, 1976). Friction and stress induced by the division were discussed in a memorable debate between L. Laudan, arguing for philosophy of science and D. Bloor at 'Science and Technology Studies – Toronto 80', the joint annual meetings of the History of Science Society, Philosophy of Science Association, Society for the History of Technology, and Society for the Social Studies of Science, held in Toronto, Canada, October 1980. See in this respect Restivo (1981). Transgressing the

same boundary is an unrestricted application of sociology of knowledge to sociology of science (see Mulkay, 1979).

6 In this respect, as an indication of the convergence of lines of thought, it is significant that T. S. Kuhn is among the invited speakers of the Third Annual Conference of the Cognitive Science Society, Berkeley, California, August 1981 with an address on 'From Revolution to Salient Features'.

7 The use of the concept of paradigm on two levels simultaneously might seem confusing. As used in this book, the notion of paradigm is both part of the cognitive view and applicable to it, i.e. the cognitive view is itself a paradigm for the study of knowledge in which the description of the structure and development of paradigms appears as a subtask. In Part III, it should become clear that this type of circularity is not necessarily detrimental to the knowledge developed within it.

8 LNR is an acronym for the research group directed by Lindsay, Norman and Rumelhart at the University of California, San Diego (see Preface of Norman, Rumelhart, 1975).

Notes Chapter One

1 AI or *artificial intelligence* is a subdiscipline of *computer science* originally developed "to build machines that perform tasks normally requiring human intelligence" (Nilsson, 1971, p. vii). Growing more mature, the field has gradually more emphasized *intelligent processes* or *procedures* rather than specific *devices* or *machinery*. Currently, the interest is in a *science of intelligence* which applies to both 'artificial' and 'natural' intelligence. See Boden (1977) and references of Chapter Eleven. A sociology of science analysis of the development of AI is provided by Fleck (1981).

2 We do not mean to oppose the *cognitive* approach to the *information processing approach*. The cognitive orientation can be considered a development within the information processing approach which stresses the importance of what is contributed by the subject when 'knowing' an object. The information processing approach tends to start from the object, analyzing in Neisser's terms "all the processes by which the sensory input is transformed, reduced, elaborated, stored, recovered, and used" (Neisser, 1967, p. 4). The shift from a one-way channel of information processing (i.e. from stimulus to concept) to a cognitive view is apparent in Norman's concept of 'conceptually driven information processing'. Information processing approaches which introduce a two-way channel of information processing (i.e. from stimulus to concept, Norman's 'data driven', and from concept to stimulus; 'concept driven') incorporate a cognitive view, such as Norman (1976) and Lindsay and Norman (1977). See Chapters Ten and Eleven in particular. See also Shepard's (1968) review of Neisser (1967).

3 Michie (1974, p. 113) introduces three levels, *monadic, structural* and *epistemic*, where I use four. The Michie-levels are well known to AI researchers. I am indebted to D. Michie and to B. Raphael who presented the same distinction in a lecture at an ASI on 'Artificial Intelligence and Heuristic Programming', Menaggio, Italy, August 1970. Raphael introduced the term 'context' to designate the third level. Support for making the distinction between 'contextual' and 'cognitive' (or 'epistemic') can be found in Winograd (1977).

4 For some audiences, the computer is a frightening product of a society (mis)guided by

a technological world view and as such a serious threat to humanistic values. It should be clear that in this book, the computer is looked upon as a technological innovation comparable to the printing press or even to the art of writing itself. Has the art of writing and of printing proven to be basically dehumanizing? Nobody would find any denigration to result from converging to paper noble feelings versified in a poem. Inasmuch as human feelings are not necessarily destroyed or dehumanized by being encoded in terms of strokes of ink on white paper, the human mind is not necessarily endangered or degraded by being encoded in terms of computer programs. The computer is only a powerful tool for expression. If we are to be afraid, it should be of the uses of the tool and not of the tool itself. That the metallic computer is not incompatible with human warmth and dignity is convincingly demonstrated in Boden (1977). See also Sloman (1978) and for a tempered view of an earlier computer enthusiast Weizenbaum (1976). An untempered enthusiastic view which does not ignore social and cultural aspects is Papert (1980). Also relevant are several papers in Dertouzos and Moses (1979).

5 'Gestalt' is only used here to designate an irreducible perceptual whole. I do not mean to suggest that the Gestalt psychology of perception with such principles as *proximity, similarity, closure*, etc. is embodied in template-matching.

6 With respect to perception as being mainly the product of the perceiver, David Waltz, in an oral communication at TINLAP-2 (see Waltz, 1978), referred to a description of perception as "controlled hallucination". He attributed the expression to Max Clowes. According to Leo Apostel (personal communication), the expression goes back on Freud.

7 Among the arguments that gave support to MT in the fifties, the opening up of foreign scientific literatures played an important role. In nuclear physics in particular, the cold war atmosphere made the idea prevail that the literature of adversaries should be accessible on a continuous and massive scale. Hence the receptivity for MT in institutions such as EURATOM.

8 There is an extensive literature on MT. A partial survey on both the development of the field and its literature can be found in Josselson (1971). With respect to paraphrasing or other types of language processing, they should not be considered intrinsically easier than translation. Bar-Hillel (1959, p. 791) remarks: "Abstracting and indexing are more complex intellectually than translating, and their complete mechanization therefore less likely than completely automatic translation".

9 The 'figure' and 'ground' phenomenon is also invoked here by Fillmore who emphasizes that

> ... whenever we pick a word or phrase, we automatically drag along with it the larger context or framework in terms of which the word or phrase we have chosen has an interpretation. It is as if descriptions of the meanings of elements must identify simultaneously 'figure' and 'ground' (Fillmore, C. J., 'The Case For Case Reopened'; quoted in Grosz, B. J. 'Focusing in Dialog', in Waltz, D. (ed.), *TINLAP 2*, New York, ACM, 1978, pp. 96– 103).

10 Some examples acquire a real exemplary status in the field. This one is based on Minsky (1968), p. 22.

11 For some workers in MT, the ubiquitous and evasive character of the notion of context is the deathblow to MT. The late Bar-Hillel left the role of enthusiast logician

pioneer of MT around 1960 to become "the leader of the destructive school against machine translation" (see Booth, 1967, p. vii). In an informal discussion at the university of Ghent (1978) Asa Kasher, Bar-Hillel's student and colleague for a long time, reported him saying: "It will take exactly thousand years to solve the problem of context".

[12] Obviously, the problem of context is not instantly solved by locating contexts in the minds of people in terms of the conceptual schemes they bring to bear upon a situation. If we think of contexts as triggered by external situations, we are again in need for the conceptual scheme to perceive the situation. Cognitive theory will have to explain how 'situational contexts' are mental constructions assembled on the spot for each particular occasion in interaction with that situation. This problem constitutes the background for a critical discussion of the cognitive interpretation of context by Petrie (1977).

Some psychologists seem as much or even more embarrassed with the notion of context than their colleagues in MT. Underwood is quoted as saying with respect to context that "never in the history of choice of theoretical mechanisms has one been chosen that has so little support in direct evidence" (Block, 1978, p. 11).

[13] It is not surprising that the cognitive view has been acknowledged and explored much earlier with respect to art than with respect to science. In art the contribution of the subject in his perception of the object is the core element. Scientific perception, so is often thought, requires the elimination of the subjective. A cognitive view, however, leads to a more 'artistic view' on science (see Wechsler, 1978). Major books in this respect are Gombrich (1960, 1979), Arnheim (1969), Goodman (1976) and Medawar and Shelley (1980).

Notes Chapter Two

[1] With respect to the 'knowledge' of a thermostat, see also McCarthy (1979) and, for a more general treatment of physical symbol systems, Newell (1980). Highly relevant is the discussion of Searle's (1980) paper on 'Minds, Brains and Programs', in *The Behavioral and Brain Sciences*, Vol. 3.

[2] The notion of *Umwelt* has been developed by von Uexküll (1909, 1920) to denote the 'subjective' world views of animals, i.e. those features or segments of the 'objective' world to which they are sensitive and to which they react as meaningful signals. The notion can be extended to include the world view of machines in terms of those features of the environment to which a machine is sensitive as an information processing device. In general, von Uexküll's concepts are very much in line with the cognitive viewpoint.

[3] According to the classical definition in epistemology, *knowledge* is *justified true belief*. Many philosophers have argued that this implies the so-called KK-thesis, i.e. 'knowing implies that one knows' (see *Synthese*, 1970, No. 2, which contains a symposium on 'Knowing That One Knows', with contributions by Hilpinen, Lehrer, Hintikka, Ginet and Castañeda). The prevailing idea is that for genuine knowledge a distinction should be made between the *belief* expressing the 'knowledge claim' and the *evidence* supporting it, so that any item of knowledge should be the outcome of a conscious process of weighting the evidence as either sufficient or insufficient to sustain the belief. Although this approach could be rephrased in terms of a cognitive framework, it is difficult to see what purpose could be served by keeping the notion of knowledge so restricted.

[4] The 'founding' paper of behaviorism, Watson (1913), is relevant here with its attack

on uncontrollable and unreliable introspective methods which are supposed to confound the information provided by behavioral data. From a cognitivist approach, the stimulus-response framework that guides the collecting of behavioral data is as legitimate a world view as the elusive world views of introspectionists. However, the difference is between world views — which are not even necessarily incompatible — not between objectivistic and subjectivistic methods.

[5] The basic segmentation of the world model into two parts has nothing mysterious and is almost inherent in any complex information processing system. Indeed a highly differentiated representation of the *outside world* makes sense only if it can be matched with a highly differentiated system for *actions*. It is of no use for a tree to be capable of recognizing the man with an ax as a lumberjack unless it is able to withdraw its roots from the soil to run away (I think this metaphor stems from G. A. Miller though I have not been able to find back any reference to it). A world model in a complex information processing system needs to accommodate the system's actions as well as the states of the outside world. The *self* can be considered the cognitive structure which harbors these actions in terms of values (underspecified goals), goals and intentions (unspecified action plans) and plans (more or less specific models for action).

[6] If we allow the system to have recursive properties, it can repeat the same segmentation (see Note 5) to arrive at a higher representational category such as the *I*. In a similar way, the *I* can be regarded as representing the system's action on a meta-level, reflecting the selection and combination of various specific world models by an executive agent. Thus, the basic segmentation at each level and in almost any world model is between actions (operations) and environment or outside world as object of actions.

[7] It should be realized that the categories of *self* and *I* can also be implemented for machines. Obviously, the control panels which inform users on the state or condition of a machine or some of its parts, already correspond to a kind of self-model for that machine. There is no basic incompatibility between the categories of self and I and a physical symbol system. Further elaborations on this issue can be found in Boden (1970) and Minsky (1968).

[8] An introduction to attribution theory can be found in Shaver (1975).

[9] Perception of the physical world as prior to communication is in line with the approach favored by Piaget. Communication guiding physical exploration is in line with Bruner's approach. See Note 9, Chapter Twelve.

[10] It is an intriguing fact of AI experience that different approaches combined yield a better performance than any single approach alone. In pattern recognition, the score of a monadic approach, template matching, is not overwhelming. Only half of a series of hand printed characters is identified correctly. The rate of achievement is hardly any better for feature analysis used as a single structural strategy: fifty percent correct. But a substantial gain in performance is achieved when template matching and feature analysis are combined in an eclectic way, rising as high as ninety percent correct identifications. These percentages are based on the lecture of B. Raphael referred to in Note 3 of Chapter One.

Notes Chapter Three

[1] For synopses of the work of Comte, Spencer and other positivists within the context of nineteenth-century thought, see Mandelbaum (1971).

[2] An excerpt of Comte's suggestion with respect to a science of science can be found in Note 9 of Chapter Five.

[3] Besides Comte's sequence of theology, metaphysics and science, other sequential models of the cultural evolution of thought have been developed to explain the emergence of scientific thinking. One very famous model is Frazer's who argued extensively for the sequence *magic, religion* and *science*. When religion is discussed as a prefatory stage for science, the shift from polytheism toward monotheism is often emphasized as crucial.

[4] This is in line with the ancient doctrine of *simulacra* or *eidola* according to which objects emit small copies or images of themselves. When caught by the senses, these copies allow for the identification of the object that is present. Notice that we should not too easily ridicule this older view since for the sense of smell it might turn out to be a quite valuable model.

[5] Helmholtz is often quoted as the author of the unconscious-inference-theory of perception. In so far as it relates to the combination of previous knowledge with current data, it should be obvious that this is the classical empiricist position that goes back to Locke, Berkeley and Hume. Typical for Helmholtz is his emphasis on the sign character of sensations and of relational patterns between sensations rather than sensations per se.

[6] Some behaviorist psychologists who are interested in cognition and who presume that they more or less know how the world really is, tend to consider their human subject as a black box whose interior is inaccessible to the observer. They feed inputs (stimuli) to the black box which, in return, provides them with outputs (responses). From an analysis of the complex patterns of input-output-relationships, they hope to derive the unobservable wiring design in the box. The Helmholtzian model is this approach inside out. The *real* real world is inaccessible for us. *We* are *in* the black box, trying to figure out, from what happens to the box, how it is in the world outside.

[7] Hume's remark reads "Philosophers begin to be reconcil'd to the principle, *that we have no idea of external substance, distinct from the ideas of particular* qualities. This must pave the way for a like principle with regard to the mind, *that we have no notion of it, distinct from the particular perceptions*" (Hume, 1739, p. 677).

[8] Occam's razor refers to the so-called *law of economy* proposed by the fourteenth century monk Occam. The notion of razor is used metaphorically to indicate an instrument for cutting down on theoretical entities. In animal psychology it has been reintroduced as *Lloyd Morgan's canon* to indicate that no higher faculties such as consciousness should be invoked when lower faculties such as reflexes prove sufficient to explain a given behavior. Besides these labels, the law of economy is also known as the *law of parsimony*, i.e. being parsimonious with theoretical entities.

[9] The choice of sensations as basic units should be understood in the light of nineteenth-century developments in psychophysics. Weber's research on differential thresholds led to the establishment of 'units of sensation' in terms of what are now called JND's. One JND (just noticeable difference) is the minimal increase in intensity required for making a stimulus perceptibly different from its initial presentation. As an example: presented with a weight of 200 grams, an increase with about 4 grams is required to make the weight perceptibly different, i.e. heavier. This is not to say that either Mach or Pearson adhered to any particular system of psychophysics. With its measurement of physical magnitudes on one side and subjective magnitudes on the other, psychophysics might

have been considered too much of a dualist endeavor. All we want to say is that the notion of 'unit of sensation' had been around for some decennia.

[10] On the relation between Pearson's philosophy of science and his statistical work, see Norton (1978).

Notes Chapter Four

[1] See, e.g., von Mises (1951, p. 51): "the aim is always to attain a kind of unambiguity of meaning ... ".

[2] Russell's method is often referred to as his 'rule of construction' formulated in his *Mysticism and Logic* (1917) in terms of this maxim: *Wherever possible, logical constructions are to be substituted for inferred entities* (p. 115, italics in the original).

[3] Structures are abstract systems of relations to be described in terms of that kind. Consider the favorite example of kinship systems. The notion of family consists of a collection of particular relations: father, mother, brother .... 'Being father of' and 'being mother of' are non-symmetrical relations which are also intransitive. 'Being brother of' can be symmetrical and is transitive but non-reflexive. The idea is that these properties capture aspects of structure that are independent of the nature of the elements involved. They might allow a characterization of the notions of family as they are embodied in biological or legal kinship systems as well as in zoological classifications.

[4] A rationalist component is acknowledged by Carnap who adds to his statement that "the designation 'empiricism' is more justified" (for logical positivism): "That it is not a *raw* empiricism needs hardly to be emphasized in view of the importance which construction theory attaches to the form components of cognition" (Carnap, 1928, p. 296). The abstractness of notions like 'experience' and 'empirical' in logical positivist methodology has been pointed out by Wright Mills (1959) who designates the movement as *abstract empiricism*.

[5] It is interesting to notice how models have entered further developments of logical positivist philosophy of science as some kind of third level entities. In Nagel (1961), in between the levels of the abstract calculus and the concrete materials of observation and experimentation, some space is allocated to an interpretative model which "supplies some flesh for the skeletal structure" (p. 90). As is apparent from the discussion in Suppe (1974, pp. 95–105), this notion of model assimilates both the roles of model in a logico-technical sense, i.e. as an interpretation for a formal system, and in the metaphorical sense, i.e. as a heuristic device for further exploration and explanation. This third level entity makes the logical positivist model more translucent but also exposes its major weakness. A three level description which distinguishes between a *formal* component (the abstract calculus), a *model* and *data* (the empirical system) provides a neat separation of empirical and structural components: structure is embodied by the formal system, experience is in the data. However, in as far as logical positivism set out to analyze and explain the interlocking of theory (language) and data (experience), the introduction of models as interfaces is to locate the problem in entities which are rather alien to logical positivism. An understanding of models as interfaces between theory and data is the central question for the cognitive orientation. Up to now, the logical theory of models (formal semantics) has not been sufficiently developed to become suggestive in this respect. As Bunge (1972, p. 239) indicates: "the semantics of factual theories is hardly under way".

[6] The classical distinction between 'context of discovery' and 'context of justification' goes back to Reichenbach (1938).

[7] The similarity between some basic Chomskyan concepts and Carnapian ones cannot be accidental. As pointed out by Yorick Wilks (1976, p. 214), Chomsky's basic rules, *phrase-structure* and *transformations*, are very similar to Carnap's *formation-rules* and *transformation-rules* introduced in *The Logical Syntax of Language* (1935).

For our further purposes it is also interesting to note how prominent spokesmen of opposing schools can share one very similar conceptual system and differ widely on others. While Carnap developed his concepts to work out an empiricist doctrine, Chomsky uses his to defend a rationalist position. Of empiricism, Chomsky says

> The grip of empiricist doctrine in the modern period, outside of the natural sciences, is to be explained on sociological or historical grounds. The position itself has little to recommend it on grounds of empirical evidence or inherent plausibility or explanatory power. I do not think that this doctrine would attract a scientist who is able to discard traditional myth and to approach the problems afresh. Rather, it serves as an impediment, an insurmountable barrier to fruitful inquiry, much as the religious dogmas of an earlier period stood in the way of the natural sciences (Chomsky, 1975, p. 12).

Notes Chapter Five

[1] A substantial part of the sociology of science literature is related to various aspects of the norms of science. In the collection of Merton's papers on sociology of science (Merton, 1973), twelve of the twenty-two papers selected by editor N. S. Storer can be considered to be directly related to the 'norms'. They are grouped into three sections: the *normative structure* of science, the *reward system* of science and the *processes of evaluation* in science (the other sections are on the sociology of knowledge and the sociology of scientific knowledge). Ben David and Sullivan's (1975) review of sociology of science provides an indication of the importance of the norms and their related areas in the whole field.

[2] The idea that cultural and societal factors (or internal and external factors) are both *contextual* aspects of science, and not only the external providing the context for the internal, is also in Merton's original 1938 monograph. In the final pages he touches upon the problem of the combination of contexts emphasizing "the necessity for explicitly defining the context within which generalizations apply" (Merton, 1938, p. 224) and introducing "the cultural *context*" as the "environment of these social relationships" (discussed previously) (p. 225, italics ours). However, his treatment of the concept of context is not followed through so that it is difficult to find out his position with respect to a number of alternative models on the relations and interactions between contexts.

Sometimes, one has the impression that the conceptual framework could be depicted by means of a number of concentric circles with the central circle representing the core problem (in structural terms) and each larger circle indicating a larger context. In this model, larger contexts 'support' more specific ones, though without influencing them in any specific way. One has to have a specific kind of culture in order to have chances for science to develop but given that kind of culture, science will evolve and grow according to its own 'internal' principles. However, in order to allow for 'external' factors such as specific technological questions to have specific influence on scientific development, one has to allow for specific contributions of the larger (external) contexts to the more restricted (internal) ones. But, once we introduce the possibility of symmetrical or reversible relations between contexts, any reference to a 'larger' context or a distinction between internal and external context is merely confusing. A better framework to depict relations between contexts does not require concentricity of the circles representing the different contexts but only some overlap between them. Our major criticism in this chapter concerns the attempts to introduce contexts as strictly ordered and hierarchically embedded. As is argued in Chapter Six: contexts should be placed on an equal footing and events and developments in science should be understood in terms of interactions between them. Other highly relevant critical discussions of the internalist-externalist-distinction can be found in Barnes (1974), Johnston (1976) and MacLeod (1977).

[3] The classical reference with respect to age and scientific productivity is H. C. Lehman's *Age and Achievement* (Princeton University Press, Princeton, 1953). The whole area is thoroughly discussed in Zuckerman and Merton: 'Age, Aging and Age Structure in Science', in Riley *et al.* (eds.), *A Sociology of Age Stratification* (Russell Sage, New York, 1972); reprinted in Merton, R. K., 1973, pp. 497–559. See also Stern (1978).

[4] Though personality profiles and personality related factors constitute a substantial part of the literature of psychology of science, as is apparent from Fisch (1977) and Eiduson and Beckman (1973), they by no means exhaust the topics and problems in the psychology of science. See Note 12 for Watson and Campbell's (1963) listing. Besides Fisch (1977), Brandstädter and Reinert (1973) provide a short overview and bibliography of the field. An evaluative review is provided by Mahoney (1979), complementary to his book (Mahoney, 1976). A topic which is commonly regarded as belonging to psychology of science and which is highly relevant to the cognitive paradigm is the so-called *Rosenthal effect*, also known as *experimenter effect*, or *experimenter bias*. This by now well-documented phenomenon goes back to a series of publications by Robert Rosenthal and a number of collaborators reporting research and results on the social psychology of experimental research in psychology (for a recent sample, see Rosenthal, 1976, and Rosenthal, R. and Rubin, D. B., 1978). The Rosenthal effect refers to the results of experiments which turn out to be largely unreliable when it can be shown that the experimenter has contaminated the subject, without being aware, by his expectations. It appears that in many experimental situations subjects are quite ingenious in using minimal cues to infer from the behavior of the experimenter (his tone of voice, the glance in his eyes, maybe pupil size . . . , etc) what response he is expecting from them. It even applies to animals. As a matter of fact, Rosenthal considers his own investigations as an extension of the famous Clever Hans case handled by Pfungst. It is interesting to notice that the Rosenthal effect is dealt with in terms of methodological precautions. A good experimentalist should avoid experimenter bias. However, from the point of view of communication, the Rosenthal effect provides convincing evidence that the subject in a psychological experiment is not a passive information processor, mechanically obeying the instructions of the experimenter. The effect shows that, in trying to infer (probably without being aware of it) from the behavior of the experimenter what kind of response is expected from him, the subject is attempting to assemble

his own conceptual model of the situation. In order to understand the experimenter in his deeds as well as in his words, the subject uses all available information to specify the world model that governs the behavior of the experimenter. In this, depending on how strong the Rosenthal effect happens to be, the subject might prove to be a better psychologist than the experimenter.

McGuire (1969) has distinguished three stages in an experimenter's dealings with artifacts: ignorance, coping and exploitation. In this area, the exploitation stage could well develop into a real 'paradigmatic' shift towards a cognitive orientation. Experimenters in psychology, instead of attempting to get rid of the Rosenthal effect, should make it the central topic of their investigations. How does the subject in a psychological experiment arrive at a model of the situation which permits him to make sense out of the behavior of the experimenter? It is only a specific case of the problem how participants in an encounter arrive at attributing knowledge (or intentions) to each other in order to make communication and interaction possible (see in this respect also Farr, 1976).

Merton has treated the Rosenthal effect in terms of the "self-fulfilling prophecy" (Merton, 1957).

[5] See on this difference between Copernicus and Kepler, T. S. Kuhn's *The Copernican Revolution* (Harvard Univ. Press, Cambridge, 1957), Chapters 5 and 6.

[6] The term 'monster of rational perfection' as a description of scientists has, according to R. K. Merton (1966), been used by Augustus de Morgan; see Merton (1969), reprinted in Eiduson and Beckman (1973), p. 602.

[7] In Fairbank, J. K., 'Our Vietnam Tragedy', *Newsletter* (Harvard Graduate Society for Advanced Study and Research, June 1975).

[8] The various sources available to a historian of science are described in Knight (1975).

[9] Comte's description of a science of science reads as follows:

> Le véritable moyen d'arrêter l'influence délétère dont l'avenir intellectuel semble menacé, par suite d'une trop grande spécialisation des recherches individuelles, ... , consiste dans le perfectionnement de la division du travail elle-même. Il suffit, en effet, de faire de l'étude des généralités scientifiques une grande spécialité de plus. Qu'une classe nouvelle de savants, préparés par une éducation convenable, sans se livrer à la culture spéciale d'aucune branche particulière de la philosophie naturelle, s'occupe uniquement, en considérant les diverses sciences positives dans leur état actuel, à déterminer exactement l'esprit de chacune d'elles, à découvrir leurs relations et leur enchaînement, à résumer, s'il est possible, tous leurs principes propres en un moindre nombre de principes communs, en se conformant sans cesse aux maximes fondamentales de la méthode positive. (Comte, A., *Cours de philosophie positive*, tome I, 1830, p. 17).

[10] Some philosophers have indeed claimed the label 'science of science' as the best designation of their discipline, e.g. Kotarbinsky.

[11] Walentynowicz' (1975) impressive list contains the following subdisciplines of science of science:

> philosophy of science, theory of science; ethics of science; sociology of science; classification of fields and disciplines of science; problems

of creative abilities and psychology of research workers; history of the organization of science and scientific societies; organization, administration and management of research and development activities; economics, productivity, efficiency, financing of research and development; statistics of science and technology; research and development planning, governmental policy in science and technology; technological forecasting, futurology; comparative-science policy; and legislation in regard to science and technology.

Irrespective of the division adopted as the most convenient in a given case, we should be aware of two particular sections of the science of science. The problems of the first group are of a fundamental character. They belong to those sciences which are sub-fields of certain fundamental disciplines and include the problems of the philosophy of science, the logic of science, the history of science, the sociology of science, the psychology of science, the economics of science (if we take science in its broad sense, as including the technology based upon it). The second group is more complex and touches upon many fields; it is the one most frequently encountered in practice. This group consists of policy-related problems, studied by research teams composed of specialists from different fields of science, or by researchers engaged professionally in the study of the problems of the science of science.

12 According to current standards and sensitivities, Watson and Campbell's (1963) short programmatic description of science of science could still serve as a valuable program. As an indication of its up-to-dateness, consider the listings of subfields for sociology of science and psychology of science:

The province of the sociology of science extends broadly. Illustrative problems include: (a) the social determination of discovery and invention, as illustrated through the occurrence of multiple independent discoveries of the same thing; (b) the relation of ideology to scientific belief, both at the individual level and at the level of social class and society; (c) the effect of social systems in furthering or hindering science; (d) the impact of science upon society; (e) the social validation of scientific truth; and (f) the institution of science, considered as a social system *sui generis* (p. vi).

At its present development, the psychology of science seems to have these problems areas: (a) the psychology of cognitive achievement as applied to the achievements in science – the psychological explanation of scientific creativity, discovery, problem-solving, trial and error learning, etc., (b) the psychology of cognitive bias applied to the biases and blind spots of scientists (Francis Bacon gave this area a good start in his list of the biases or 'idols' he found among his fellow philosophers); (c) the motivational psychology of scientists – the role of curiosity, aggressiveness, self-esteem, vanity power, and other needs in shaping the final scientific product; (d) personality and science – the tendency of certain personality types to be attracted to science, and within science, the tendency for personality differences between those who take various roles and positions; and (e) psychological epistemology – the role of psychological experience in establishing the inductive base of all sciences, the psychological description of the criteria of evidence and proof used by scientists, psychological aspects to the mind-body problem, and the innumerable other points where psychological problems border epistemological issues (p. vii, lines 15–30; reprinted by permission).

[13] A suggestion in line with the orientation of this book is Lenat's (1977) proposal to consider 'science of science' a major goal for AI (p. 283). Sneed (1976) introduces the notion of an empirical science of science to indicate what he calls partial rational reconstructions of scientific theories. If the formalism he is using (set-theory) can be worked out such as to represent the specific knowledge and prerequisite knowledge involved in particular specialties, his approach will be entirely in line with a cognitive science approach to science of science. But anyhow, Sneed's work is to be considered post-Kuhnian science of science.

[14] Copernicus, N., *De Revolutionibus Orbium Caelestum*, 1543, Preface, quoted in Kuhn, T. S. (1957), p. 139.

Notes Chapter Six

[1] The origin of the concept of scientific revolution and its links with astronomy are discussed in Cohen (1976).

[2] Neither in Kuhn (1957) nor Kuhn (1962) is the distinction between internal and external factors explicitly rejected by Kuhn. In fact, he uses the distinction and seems to approve of it. In a note to the 1962 preface, referring to previous work on "external intellectual and economic conditions upon substantive scientific development", he indicates that "with respect to the problems discussed in this essay . . . I take the role of external factors to be minor" (p. xii). In Kuhn (1957) he had already emphasized that "we need more than an understanding of the internal development of science" (p. 4) and the book can indeed only be understood as an integrated approach combining both internal and external factors. In the first pages it is indicated in advance that

we shall gradually discover how difficult it is to restrict the scope of an established scientific concept to a single science or even to the sciences as a group. Therefore, in Chapter 3 and 4, we shall be less concerned with astronomy itself than with the *intellectual* and, more briefly, the *social* and *economic milieu* within which astronomy was practiced (1957, pp. 2–3, italics ours).

In Kuhn (1971) it is quite obvious that the distinction between internal and external factors poses problems for him. He complains that "the internal-external debate is one which historians have lived with rather than studied" (p. 27). The sequential interpretation which I stress here is most clearly expressed in Kuhn (1968). See in this respect Johnston (1976) who brought my attention to some of these passages in Kuhn and whose discussion of the internal/external dichotomy is highly relevant. Another relevant paper of Kuhn is his critical review of Ben David (1971) where the distinction is again a basic issue (Kuhn, 1972).

[3] Kuhn's adherence to the continuity-thesis with respect to common sense and science is clearly expressed in the introduction of the *Structure of Scientific Revolutions*: "If

these out-of-date beliefs are to be called myths, then myths can be produced by the same sort of methods and held for the same sorts of reasons that now lead to scientific knowledge" (1962, p. 2).

4 Obviously, the capacity to recognize an anomaly and to recognize it *as* an *anomaly* is a major problem for the cognitive view. If perception is guided by our expectations, how do we perceive the unexpected? To Kuhn, the perception of anomalies is also a major research problem for his own approach: "the manner in which anomalies, or violations of expectation attract the increasing attention of a scientific community needs detailed study" (1962, p. xi). With respect to related questions dealing with the concept of the crisis-state, Kuhn remarks "the questions to which it leads demand the competence of the psychologist even more than that of the historian" (1962, p. 85).

5 For a treatment of both Feyerabend and Kuhn's account of scientific development in terms of meaning change, see Shapere (1966).

6 Torricelli's idea that "the earth in which we live is surrounded by a sea of air which by virtue of its weight exerts pressure" is one of the examples Conant uses to illustrate his notion of conceptual scheme (Conant, 1951, p. 35).

7 Kuhn's (1957) distinction between *economical* and *psychological* functions of conceptual schemes is also reminiscent of the logical positivist distinction between *context of justification* and the *context of discovery* (see Chapter Four).

8 An analysis of the psychological function of metaphysics has been developed by John Wisdom. While they might be wrong, metaphysical notions are very valuable. By being provocative, they both function as attention attracting devices and offer a global grasp of an orientation. See Wisdom (1957).

9 The prototypical problem in this respect that is cited by Kuhn (1962, p. 45) is Wittgenstein's analysis of 'family resemblances'. The central point is that the application of a general concept to a variety of objects which differ markedly in appearance is not necessarily based on the presence of one critical attribute. Wittgenstein argues that with respect to many concepts "these phenomena have no one thing in common which makes us use the same word for all, − but that they are *related* to one another in many different ways . . . we see a complicated network of similarities overlapping and criss-crossing: sometimes overall similarities, sometimes similarities of detail" (Wittgenstein, 1953, pp. 31−32). The set of problems to which a given exemplar conceptualization applies forms in a similar sense a 'family'. Wittgenstein's term 'family resemblances' is designated to indicate this intricate set of relations: "for the various resemblances between members of a family: build, features, colour of eyes, gait, temperament, etc. etc. overlap and criss-cross in the same way" (p. 32).

10 Textbooks are available for almost any field. The degree of consensus of paradigms should therefore not be based on the availability of textbooks but on the degree of overlap in content and structure of textbooks in a given field. In a pilot study, we explored the possibility of using the subject index for measuring the relationship between books. We explored a reordering of terms according to their frequency of use, i.e. the number of references made to that particular term in the book. We are then able to express the similarity of content between pairs of books in terms of the degree of overlap among the most characteristic index terms. Preliminary results indicate a high similarity between textbooks in psychology and economics what would indicate that these fields are comparable with respect to their 'degree of paradigmaticity'.

11 In particular Popper and his followers have criticized the notion of 'normal science'

arguing that it promotes an irrational model of the scientist, depicting him as a servile partisan of authorities or fashions rather than as an autonomous and mature investigator. With Feyerabend, the Popperians seem to agree that revolution in science should be permanent with iconoclasm all the time. See in particular Lakatos and Musgrave, 1970.

[12] A balanced review of some critics on Kuhn can be found in Heimdahl (1974).

[13] An English translation of Fleck's book has been prepared and is now available, including a preface by Kuhn. See Fleck (1979).

[14] I borrow the term *paradigm-hunters* from Belver Griffith who introduced it at the first International Research Forum in Information Science, London, 1975.

[15] With Milton's famous *Paradise Lost, Paradise Regained* in mind, the combination of both Morin's and Boneau's title suggests paradisiacal conditions in paradigmatic areas. However, only on the basis of its suggestive title, should Boneau's (1974) paper be associated with the group of paradigm-hunters. With respect to its content, it belongs to the third group of paradigm-dissectors, clearly groping to assess the cognitive import of Kuhn (1962) from a behavioristic point of view.

[16] The similarity between Whorfian and Kuhnian theories is not restricted to the ambivalent criticism which both have encountered. Also with respect to content, Whorf's ideas are quite comparable to Kuhn's. What Kuhn claims paradigms to do for particular sciences is in Whorf's theory attributed to world views in their relation to specific languages. If one would substitute the term 'paradigm' for 'linguistic background' in a typical Whorfian statement, one arrives at a perfectly Kuhnian expression. See e.g. Whorf's favorite expression "All observers are not led by the same physical evidence to the same picture of the universe, unless their linguistic backgrounds are similar, or can in some way be calibrated" (quoted in Carroll, 1956, p. 214).

Notes Chapter Seven

[1] Not all critics have welcomed Kuhn's 1969 option with respect to the circularity in defining the cognitive content of a paradigm by referring to group structure, while defining the social scope of a paradigm by referring to shared beliefs. See e.g. Musgrave (1971) who argued that

> ... philosophers of science, if they paid any attention at all to the socio-logical dimension of science, would presumably have a ready solution to this Kuhnian puzzle. They would insist that the content of science is primary, so that if scientists do organise themselves into different groups this must be a sociological reflection of their different problems, theories and techniques. Such philosophers of science would, therefore, define the group by the common scientific content of its activities, and not *vice versa* (p. 287).

[2] In the U.S., the major society is 4S (*Society for Social Studies of Science*), and in Europe PAREX (contraction of *Paris* and Sus*sex*) constitutes a major coordinating group. The central journal is *Social Studies of Science* published by Sage, originally started at the University of Edinburgh in 1971 under the title *Science Studies*. A volume which is generally considered as representative for the area is Spiegel-Rösing and Price (1977). The European tradition is mainly to be found in the series *Sociology of the*

Sciences, a Yearbook, the first volume of which appeared in 1977: Mendelsohn *et al.*
(1977). Other representative collections of papers are Stehr and König (1975), Knorr
et al. (1975), Lemaine *et al.* (1976), Blume (1977), Merton and Gaston (1977), and
Gaston (1978). However, the establishment of identity in terms of a professional society
has not led to the cultivation of intellectual isolation. The 1980 annual meeting of 4S
was part of Science and Technology Studies – Toronto 1980. It combined the annual
meetings of the History of Science Society, The Philosophy of Science Association, the
Society for the History of Technology and 4S. By several participants, the prevailing
impression of genuine complementarity between the disciplines involved has been
experienced as an indication of the viability of a combined effort in science of science.
As Sloan reports in *Isis* (1981): "the metascientific disciplines have reached a new stage
in their evolution" (p. 259).

3 See in this respect also Weiss (1960) on the growth of knowledge.

4 Finer grain analyses of the way in which scientific work is assimilated into the formal
literature of science have been performed by Garvey and Griffith. The mean times
provided for the field of psychology are 30 months between initiation of the work and a
journal publication. It takes another 30 months to appear in *Annual Review* and more
than five years to appear in *Psychological Bulletin* (integrative papers covering a particular issue or area). See Garvey (1979, in particular the figure opposite p. 134) and Lin
et al., 1970).

5 Since the early 1960's, E. Rogers has published a bibliography on the diffusion of
innovations over several years. Our comparison is based on three of them: 1962, 1964
and 1971. An up-dated version, which has subsequently been made available (Rogers,
1977), is not included. The items of the bibliographies are classified according to several
criteria, e.g. empirical versus non-empirical studies. Some further use of them is made in
Chapter Nine which also contains a brief discussion of the diffusion studies.

6 The distinction between 'paradigmatic' and 'non-paradigmatic' is reminiscent of
similar dichotomies which apparently, at some time, aimed at providing the ultimate
demarcation criteria. Snow's (1959) distinction between the *two cultures* can be compared with it and so can James' distinction between the *tough-minded* and the *tender-
minded*. James captures a subtle but significant difference between the two groups by
indicating that "each type believes the other to be inferior to itself, but disdain in the
one case is mingled with amusement, in the other it has a dash of fear" (James, 1907,
p. 23).Considering the amusement with which some bibliometricians diagnose the lack
of inertia in the unsure social sciences, they line up with the tough. A more refined
analysis which attempts to go beyond this polarization is Whitley's (1977) exploration
of Pantin's distinction between 'restricted' and 'unrestricted' sciences.

7 Sociometry is a method for studying group structure developed by Moreno (1934).
It is based on the members' responses and choices with respect to questions as with
whom they would like to go on vacation, do a particular job together, etc.

8 Citation indexes of ISI list all documents, mostly publications and occasionally
unpublished items, that appear in the reference lists of all papers in a few thousands
of leading scientific journals. They come in telephone-directory-like volumes listed
alphabetically according to cited author and they are compiled and published every three
or four months and recombined into yearly volumes. The *Citation Index* is accompanied
by a *Source Index* which contains the bibliographic information, including the list of
references that go with the papers on which the citation index is based. More information

on its conception and organization can be found in Garfield (1979), a book by the founder of SCI and SSCI.

[9] *Science indicators* is another label for designating quantifiable aspects of scientific activity as dealt with in *scientometrics*. See Elkana *et al.* (eds.), 1978. A special double issue of *Scientometrics*, edited by Zuckerman and Miller (1980), is devoted to *Science Indicators* – 1976. Notice that the editors, in their own contribution, express the expectation that "the present preoccupation of science indicators research with the quantity of scientific activity will soon be countered by new and vigorous efforts to get at cognitive and qualitative aspects of science" (p. 352).

[10] Citation studies are periodically at issue in *Social Studies of Science* and the *4S Newsletter*. *Social Studies of Science* (1977, Vol. 2, No. 3) consists of a theme issue on 'Citation Studies of Scientific Specialties'. The *4S Newsletter* (1977, Vol. 2, No. 3) contains a compact supportive note by Griffith, Drott and Small and a critical note by Edge. A more extensive critical review is Edge (1979). Also Wersig (1973) has some critical notes. In general, citation studies have rather narrowly been aiming at spectacular results by handling massive amounts of data without an in-depth analysis of what those data stand for. Citations have more been counted than analyzed and we need to know more about them qualitatively in order to appreciate what citation maps depict.

[11] Unobtrusive measures use by-products or incidental results to investigate some principal process (see Webb *et al.*, 1966). Because they focus on incidental and unintended aspects, they are considered to be immune to the experimenter's bias or Rosenthal effect (see Note 4, Chapter Five). Widespread suspicion against citation studies is based on the conviction that the use of citations for, e.g., evaluation of individual scientists, has made them very obtrusive and conspicious. If some scientists cite others in function of the expected use that will be made of their citations in citation counts or studies, citations are no longer unobtrusive data.

Notes Chapter Eight

[1] An extensive study of communication patterns in psychology was organized by the American Psychological Association (APA), published by APA as the *Reports of the American Psychological Association's Project on Scientific Information Exchange in Psychology* 1963–1969, 3 volumes, 21 reports.

[2] The 'Matthew effect' is based on a verse of the Gospel according to St. Matthew: "For unto every one that hath shall be given, and he shall have abundance: but from him that hath not shall be taken away even that which he hath" quoted and applied by Merton (1968).

[3] *Choices* here mean sociometric responses provided as answers to a questionnaire regarding communication in the specialty. In citation patterns, a segmentation between center and periphery is substantiated by the fact that core authors tend to cite other core authors more than they cite authors from the peripheral group, while the peripheral group highly cites core group members. In sociometric choices, the segmentation is supported by the fact that most choices refer to a small number of participants which receives a number of choices that is above the mean number of choices received per member. As with citations, in addition, the choices of the small selected group tend to go to members of their own group.

[4] Cole (1949, p. 331) refers to the label 'Invisible College' as stemming from Boyle's letters of 1646–1647. He also describes other groups, i.e. other groups than the Royal Society of London, which have an invisible college structure, in particular the Private College of Amsterdam. The perennial nature of this type of social unit is apparent from a remark by Schrödinger (1935) about workers "of *one* branch of science and of *one* epoch". About these scientists, Schrödinger says "These men practically form a unit. It is a relatively small community, though widely scattered, and modern methods of communication have knit it into one. The members read the same periodicals. They exchange ideas with one another . . . " (quoted in Hagstrom, 1965, p. 18).

[5] Chubin (1976) makes a distinction between *invisible colleges* and *social circles* mentioning that the membership of circles is indeterminate while invisible colleges are restricted in number. He also quotes some relevant literature with respect to the notion of 'social circle'.

[6] As Roger Krohn (1971) has indicated, scientists are, to varying degrees, involved in more than a single specialty. While actively participating in one area, they monitor developments in several other areas as well. Discoveries might trigger sudden shifts in priorities which could account for the onset of rapid growth in the earlier phases of a specialty.

[7] To study the pressures to uniformity on groups, Asch (1956) induces *cognitive conflict* in a subject by making him a member of a group which unanimously provides an incorrect response to simple perceptual tasks. Many subjects yield to such majority pressures which apparently are very strong.

[8] The first model of information transfer through mass media became popular at the time of the adoption of the radio. One piece of impressive evidence for this model, the *injection*-model, had been Orson Welles's dramatization of Well's *War of the Worlds*, the story of an invasion of Mars reported over the radio in 1938. It seemed as if a piece of information could be injected into an audience like a dose of drug into a body (see Cantril, 1940). The two-step flow model stems from Katz and Lazarsfeld (1955).

[9] Vlachy (1979) provides a bibliography on mobility in science. Though, according to the subtitle, it deals with "scientific career migration, field mobility, international academic circulation and brain drain", brain drain is the dominant theme.

[10] Though there are indications that the psychology community 'recognizes' the existence of AI, it is still not clear whether immigrants from AI will be treated as barbarian invaders, members of a liberation army, or just another branch to be added to APA. The influence is certainly considered as important. Boyd (1979) notices: " . . . the prevalence of computer metaphors shows an important feature of contemporary theoretical psychology: a concern with exploring analogies, or similarities, between men and computational devices has been the most important single factor influencing postbehaviorist psychology" (p. 360). When making the comparison with astronomy, we should realize that the developments in computer technology provide AI with an instrument that is unparalleled in the study of products of the mind (except maybe for the invention of the art of writing and to a lesser degree the art of printing). When AI is compared to radio astronomy, there is no equivalent of the optical telescope in psychology. Edge and Mulkay (1976) quote as epigraph a statement of S. Milton that describes radio astronomy as "a revolution in our knowledge of the Universe that is paralleled only by the historic contributions of Galileo and Copernicus". Considering Kant the Copernicus of cognitive science, the computer might well become the Galilean telescope.

[11] To the brilliant students applies Snow's (1959) statement that "provided the schools and universities are there, it doesn't matter all that much what you teach them. They will look after themselves" (p. 37). In that sense, weaker students take relatively more profit from carefully designed teaching programs than the very best students who are less dependent on the quality of teaching.

Notes Chapter Nine

[1] I came upon this quotation of Sarton through Eisenstein (1979, p. 124).

[2] The reported negative reaction toward a critical discussion of the distinction between competence and performance goes back to personal experience at Harvard's *Center for Cognitive Studies* 1966–67 and also to a seminar on the psychology of language directed by Roger Brown (Research Problems in the Psychology of Language, Social Relations 290).

[3] The use of the serious statistical terminology of errors of type I and II to characterize risks in stage two and three respectively can be extended into semi-serious error of type III to characterize feelings in stage four. A popular version of a type III error reads: "We have solved the wrong problem".

[4] Other relevant analyses of organizational life cycles, including some of universities, are to be found in Kimberly, Miles *et al.* (1980).

[5] The dichotomy between the 'cognitive' and the 'social' is reminiscent of the distinction between 'I' and 'world' discussed in Chapter Two. Like there is no privileged access to the 'I' in the I-world-segmented world-model, there is no privileged access to the needs of society either. We do not know ourselves any better than we know the environment, nor does society knows itself any better than what science provides. The determination of needs and values is as much a problem of construction of knowledge – knowledge about ourselves – as is the construction of knowledge about the world – misleadingly thought of as outside. As such, a representation of specific needs is not basically different from other conceptual systems. Finalization then is nothing more than combining a conceptual system representing needs with another conceptual system describing a particular part or aspect of the world. It is 'cognitive' all the time. Obviously, there remains a basic problem of relating action, value and knowledge. That, however, is a general problem of the relation between knowledge and purpose. How purpose and knowledge connect is as much a problem of individual psychology as well as of societal use of science and, in line with some AI traditions, we estimate that the psychology of the simplest cases, e.g. purposeful sensori-motor actions, is apt to provide the clearest picture. It is characteristic of Piaget's approach with which we deal in Chapter Twelve.

[6] With respect to the basic conceptual tools of diffusion of innovations research, see Rogers and Shoemaker (1971). For later refinements, see Zaltman *et al.* (1973) and Roessner, J. D. (1980). A general assessment is to be found in Radnor *et al.* (1978).

[7] Similar results are suggested for a much larger group of studies on 'small groups' by McGrath and Altman (1966).

[8] See also Note 7 of Chapter Four on Carnap's empiricism and Chomsky's rationalism.

[9] An approach which integrates science into an economic system of knowledge production and utilization is Machlup (1981), extending a pioneering view already indicated in Machlup (1962). I owe this reference to R. F. Rich.

Notes Chapter Ten

[1] As will become apparent further in this chapter, we should be careful not to identify visual perception of objects with the perception of pictures of objects. As J. Gibson (1979, p. 274) has indicated, a picture is a highly selective representation of invariants noticed by its creator. Nevertheless, we will assume that some mimetic qualities of pictures are representative for some aspects of the perceptual process.

[2] The availability of new kinds of apparatus and techniques is emphasized by N. Moray (1970, p. 5).

[3] The constitution of the *International Association for the Study of Attention and Performance* can be found in Dornic (1977, pp. 742–750).

[4] The Rorschach-inkblot-test is based on the idea that stable personality factors are revealed in the interpretation of basically ambiguous stimuli: inkblots.

[5] Obviously, E. Gibson would object to an interpretation of her view as 'cognitivist' since she considers herself 'outside of the network'. Her rejection of cognitivism would be justified in as far as it would imply a view according to which the input is 'only bits and pieces'. However, as this chapter should further illustrate, cognitivism does not have to be incompatible with a Gibsonian view if 'wholes' can be input as well. Another highly relevant author who would normally not be classified as a cognitivist is P. J. Galperin. His theory of attention as 'internalized' action is also highly compatible with a view that stresses task demands (see Galperin, 1967).

[6] For a discussion of perception as a multi-level process, see Sloman (1978), in particular Chapter 9: "Perception as a computational process".

[7] Simple geometrical elements as pictorial labels (Goodman, 1976, pp. 30–31) are used in both AI studies of perception and the art of drawing. With respect to this use in art see Arnheim (1969) and Gombrich (1960). An AI application is exemplified in Uhr (1973), in particular Chapter 9.

[8] See the volume edited by Lipkin and Rosenfeld (1970). For more recent developments: Hanson and Riseman (1978), Rosenfeld (1979).

[9] Here, we tend to differ from Palmer and other members of the LNR group. In what appears a regression to some kind of sense-data doctrine, they use as synonyms concept-pairs like *top-down/bottom-up* and *concept-driven/data-driven* while these distinctions are largely theirs. Palmer (1975) says: "It is clear that sensory data must play a bottom-up role" (p. 297). Such a connection seems unnecessary and reintroduces a confusion which the distinction between the two dimensions (top-bottom and concept-data) set out to overcome. Similarly confusing terminology is used by Arnheim (1977) when he is referring to perception as "the ground floor of mental structure" (p. 5).

[10] If as Gombrich (1979, p. 132) suggests, we tend "to read any shape from outside in", it should not be difficult to find a combination of data-driven top-down processes. A global shape giving rise to a conceptual identification at top-level might be conceptually analyzed top-down, leading eventually to a concept driven data-identification at some lower level.

[11] This is not incompatible with current research on imagination. See Kosslyn (1978).

[12] A bias toward the perception of objects (concatenations of forms) seems to be one of them. See Sutherland (1973) and Miller and Johnson-Laird (1976).

[13] As major sources with respect to Harvey's discovery of the circulation of the blood

we have used Pagel (1967), Pagel (1976), Bylebyl (1979) and Whitteridge (1971). For Harvey (1628), we use Whitteridge's translation of 1976.

[14] In his 1967 book, Pagel still wrestles with the cumbersome choice between theory and observation:

> ... we cannot separate the single 'spark', in his case the venous valves, from the tangle of ideas which seem indissolubly bound up with it and form the complex background of his discovery. Nor would it be easy to say which came first: idea or observation, or even philosophy or observation, ... there is support for the view that indeed the idea came first. Harvey had been a staunch Aristotelian all the time by up-bringing as well as by inclination. Perhaps this and all that it had taught him about the heart and the blood *influenced the way in which he looked upon the venous valves* (Pagel, 1967, p. 210, italics ours).

This is indeed the challenge and this is what a model of perception should achieve: to determine how ideas influence the way one looks at objects.

[15] A famous case to quote in this respect is Forman's (1971) analysis of the influence of the Weimar Culture on Heisenberg's contribution to quantum mechanics. A global designator of the political climate is directly imported into theoretical physics and worked out as the 'uncertainty principle'.

Notes Chapter Eleven

[1] I came upon this quotation through Portugal and Cohen (1977, p. 266).

[2] An indication of the differences between computer performance and human intelligent activity is provided in Norman (1979). Recognizing that "many of our most frustrating attributes may turn out to be essential for a creative, intelligent system. Many of our apparently intelligent virtues may turn out to be deficits" (p. 37), Norman explores the importance of self-awareness as a critical feature of human superiority. This is in line with the emphasis on 'knowledge about one's own knowledge' which we will further encounter as essential for the kind of flexibility exhibited in highly intelligent behavior.

[3] This suggests an approach which compares 'attention' to the control unit in computers, a line of research which has been recognized in both psychology (Kahneman, 1970) and AI (Winston, 1977).

[4] There exists an already extensive literature on specific AI applications in science. A special issue of the journal *Artificial Intelligence*, edited by Sridharan (1978) discusses various applications in medicine, chemistry, biology and other sciences. Since they aim at the representation and automated utilization of expert knowledge, they are obviously highly relevant for understanding paradigms and their routine application in 'normal science'. The bulk of pioneering efforts has been accomplished at Stanford University which has established an impressive project in the area. The brochure *Heuristic Programming Project 1980* (Stanford, 1980) contains a description of relevant work on expert systems, including a bibliography for further readings. Relevant work is also covered in the *The Handbook of Artificial Intelligence* (Barr and Feigenbaum, 1981–1982), Volume II. The Stanford approach is not directly concerned with psychological

mechanisms. Other leading AI-centers which are more concerned with psychological relevance have also done some psychological analyses of the representation and use of scientific knowledge: see Bhaskar and Simon (1977), Larkin *et al.* (1980a, b), and Langley (1981) for some Carnegie-Mellon work. We focus upon the M.I.T.-approach because of its obvious link with Kuhn's notion of paradigm and also because of its general challenge to established concepts in the theory of knowledge and the catalyzing effect this might have upon ideas in science studies. The *frame*-version is only one tentative formulation of a theory which is still developing. Papert (1980) provides some background information. See also Minsky (1977, 1979 and 1980).

5 This classical distinction of Ryle (1949) which opposes *knowing how* to *knowing that* corresponds to a distinction between 'declarativist' and 'proceduralist' positions in AI. While for some time, discussion seemed to indicate that these were alternative approaches to the representation of knowledge, there is, as Winograd (1975) pointed out, no opposition between the two positions. The distinction has been overcome in the recognition that, in the description of knowledge, both action and descriptive knowledge are to be characterized and are complementary to each other. Action relates to the subject-side while descriptive (declarative) knowledge relates to the object-side. See further Chapter Twelve on Piaget and the role of action in the acquisition of knowledge.

6 The 'one-two-three-infinity'-rule is Winston's label (Winston, 1977, p. 236) for a generalization-principle that has a pictorial counterpart in what Gombrich (1979, p. 99) designates as the 'etcetera-principle'.

7 While we argue that the articulation of the subject part in any world model contributes to the applicability and combinability of the knowledge involved, this is in itself not a sufficient procedure for problem solving and the recombination of world models. In addition, we need an agent or entity which can operate upon these self-world models. We have touched upon this problem in Chapter Two. Current emphasis in cognitive psychology on *metacognition* is in line with Minsky's emphasis upon the importance of 'knowledge about the structure of one's own knowledge'. But in both AI and psychology, this metacognitive subject is still rather undefined. Further search and analysis should bear out whether or not this notion opens the back door through which general intelligence reenters as a capacity for recursive representation.

8 For an apprehensive view on the discovery of linear perspective by Brunelleschi and Alberti in fifteenth-century Florence, see Edgerton (1975).

Notes Chapter Twelve

1 Some of Piaget's favorite applications to science relate to the earlier stages of mechanics. He notices that the stages that characterize the development of the notions of motion and force in history of science are the same that characterize the development of these notions in children. In the evolution of ideas about the transmission of motion between two objects in the situation where one hits the other, he distinguishes between four stages:

    — the Aristotelian theory of the two motors located each in an object and the connected notion of antiperistasis;

    — the external motor theory which no longer locates a source of motion in the object but lacks a differentiation between the notions of force, motion and moment;

– the impetus theory of Buridan introducing an intermediary transmitted from the moving to the passive object;

– the notion of acceleration and conservation of momentum of Galileo and Newton.

[2] With respect to Piaget's autobiography, we refer to Piaget (1976), an updated version of Piaget (1952) which first appeared in English. There exist several summaries and discussions of his work. Well known and widely used is Flavell (1963). Short and synoptic is Boden (1979). A most representative selection of Piagetian papers and segments of his books is Gruber and Vonèche (1977).

[3] In spirit, D. T. Campbell's interest and analysis of the relationship between evolutionary mechanisms and knowledge processes is very similar to the central preoccupations of Piaget. Campbell (1959) provides an impressive hierarchy of adaptive mechanisms in which each higher level process fulfills some representative function (in Campbell's terms: 'vicarious' role) with respect to lower adaptive functions and thereby extends the range of adaptation. Campbell (1977) expands his 'evolutionary epistemology' (Campbell, 1974) into a 'descriptive epistemology' aiming, like Piaget, at linking the normative aspect of epistemology with descriptive findings on knowledge acquisition. The basic difference between Campbell and Piaget is the strong neo-Darwinian orientation of the former while the latter adheres to some subtle version of neo-Lamarckism (see Piaget, 1976).

[4] The 'myth of eternal youth of the social sciences' has been indicated by Wax (1969) who also exposes some consequences of its cult.

[5] An introduction to Piaget which includes references to more recent replications and critical studies of his works is Brainerd (1978). With respect to conservation, see also Scandura and Scandura (1980).

[6] Piaget finds support for his mathematical characterizations of cognitive structures in Bourbaki's 'mother structures' which he considers a foundation of mathematics consonant with his psychological claims. Rothman (1977) criticizes this position by relativizing the importance of Bourbaki for the foundation of mathematics, arguing for an approach that focusses on the role of communication in the construction of mathematics. According to him, Piaget

> misunderstands the nature and status of proof, seeing it as a relatively unimportant part of mathematical thought subsidiary to the invention of discovery of structure; and it (Piaget's characterization of mathematics) relies on a completely individualistic view of mathematical creativity which denies any serious role to language or to the social context of thought (Rothman, 1977, p. 131).

See also Note 9.

[7] Harvey's use of numerical measures is often considered the characteristic of his method that qualifies it as truly scientific. To quote a recent example, Cohen (1980) argues: "Here we may see how numerical calculation provided an argument in support of theory: an excellent example of how numbers entered theoretical discussions in the new science" (p. 18). Cohen (1980) explores a concept of 'transformation' to account for revolutionary achievements in science in a way that appears highly relevant to our purposes. By the time I could obtain a copy of the book, it was too late to include a discussion of it in the present monograph.

[8] The construction of the notion of object is dealt with at large in *La construction du*

réel chez l'enfant (1937). It describes how a stable and invariant object arises as the result of a long process of linking and combining various experiences with objects. Important exerpts are available in Gruber and Vonèche (1977, pp. 250–294).

[9] Inverted perspective presents objects farther away as larger than objects close to the viewer. Psychological perspective depicts the sizes of objects related to the 'psychological' importance of the depicted subject.

[10] Bruner, while being a great admirer and promotor of the Piagetian approach in American psychology, has illustrated in his own studies of communication Piaget's one-sided preoccupation with subject-object-interaction. In Piaget's classical approach, the 'other' enters the scene only after the notion of object has been completed and after the child has discovered that he is but a physical object among the other objects that constitute the world. But studying interaction between mother and baby, Bruner discovered that long before the notion of object in Piaget's sense is established, communication patterns arise which indicate intellectual coordination and cooperation. A five-month-old baby can 'understand' the pointing finger of his mother, i.e. extrapolate the line indicated by the finger and look in the direction that is pointed in (Bruner, 1973). This is a remarkable achievement suggesting that primitive subject-subject-interactions are as impressive as the child's manipulative skills manifested in subject-object interactions. Given the importance of social interaction in specialty development, it would be worthwhile to follow the growth of cognition in communication patterns as well as in action patterns. It would be particularly interesting to see whether this could lead to a 'social' alternative or complement for the construction of entities which Piaget would consider as most crucial: mathematical structures. Rothman's (1977) criticism seems to point in this direction. Bloor (1976) is arguing along a similar line when claiming a primary and constructive role for social interaction in the establishment of mathematical entities.

Notes Epilogue

[1] Occasionally, the comparison between structure and development in art and in science has been a suggestive topic for analysis and discussion. Kuhn (1969) has dealt with it and Medawar and Shelley (1980) provide an intriguing report of a conference devoted to it. Equally relevant is Wechsler (1978). However, a fascinating approach which analyzes the issue of representation in a way relevant to art as well as to science and which focusses upon a plurality of world views is Goodman (1978). Instructive criticism by Hempel (1980) and Scheffler (1980) and revealing comments by Goodman (1980) can be found in *Synthese* **45** (1980), No. 2.

[2] Along these lines, Waddington (1970) discusses modern physics in relation to modern art, in particular painting. I owe this reference to Roger Krohn.

[3] A Piaget-based interpretation of action differs from action theory as developed in sociology where action is analyzed in terms of the (implicit) motives of the actor. This implies a rather static motivation structure, i.e. a representation of the self in terms of goals and purposes. It ignores the dynamic and opportunistic character of motivational structures which contemporary sociology of science has found to be operative in scientific work (Latour, 1980, Knorr, 1981b). Goals and motives develop along with the construction of knowledge and are in fact constructed in the same way as knowledge about the outside world.

BIBLIOGRAPHY

American Psychological Association, *Reports of the American Psychological Association's Project on Scientific Information Exchange in Psychology*, 3 vols., APA, Washington DC, 1963–1969.

Amosov, N. M., *Modeling of Thinking and the Mind*, Spartan, New York, 1967.

Apostel, L., *Matière et forme. Introduction à une épistémologie réaliste*, 2 vols., Communication and Cognition, Ghent, 1974.

Arnheim, R., *Visual Thinking*, Univ. of California Press, Berkeley, 1969.

Arnheim, R., *The Dynamics of Architectural Form*, University of California Press, Berkeley, 1977.

Asch, S. E., 'Studies of Independence and Conformity. A Minority of One Against a Unanimous Majority', *Psychological Monographs* 70 (1956), No. 9.

Austin, J. L., *Sense and Sensibilia* (1962), Oxford University Press, London, 1976.

Barber, B., *Science and the Social Order*, Collier, New York, 1967.

Bar-Hillel, Y., 'The Mechanization of Literature Searching', in *Proceedings of the Symposium on the Mechanisation of Thought Processes, Teddington 1958*, H.M.S.O., London, 1959, pp. 789–802.

Bar-Hillel, Y., *Aspects of Language*, The Magness Press, Jerusalem, 1970.

Barnes, B., *Scientific Knowledge and Sociological Theory*, Routledge & Kegan, London, 1974.

Barr, A. and Feigenbaum, E. A. (eds.), *The Handbook of Artificial Intelligence*, William Kaufmann, Los Altos, Ca., 1981–1982 (forthcoming).

Barron, F., 'The Needs for Order and for Disorder as Motives in Creative Activity', in Taylor and Barron (eds.), *Scientific Creativity, Its Recognition and Development*', Wiley, New York, 1963, pp. 153–160.

Barron, F., *Creative Person and Creative Process*, Holt, Rinehart & Winston, New York, 1969.

Bartlett, F. C., *Remembering. A Study in Experimental and Social Psychology*, Cambridge University Press, Cambridge, 1932.

Bateson, G., *Mind and Nature*, Wildwood House, London, 1979.

Beck, A. T., 'Cognitive Therapy: Nature and Relation to Behavior Therapy', *Behavior Therapy* 1 (1970), 184–200.

Ben-David, J. and Collins, R., 'Social Factors in the Origins of a New Science: The Case of Psychology', *American Sociological Review* 31 (1966), 451–465; reprinted in Sexton and Musiak (eds.), *Historical Perspectives in Psychology: Readings*, Brooks and Coole, Belmont, 1971, pp. 98–122.

Ben-David, J. and Sullivan, T. A., 'Sociology of Science', *Annual Review of Sociology*, 1975, pp. 203–222.

Bernal, J., *Science in History*, Watts, London, 1957.

Bertelson, P., 'Central Intermittency Twenty Years Later', *Quarterly Journal Experimental Psychology* 18 (1966), 153–163.

Beverdige, W. I. B., *The Art of Scientific Investigation*, (1950), Heinemann, London, 1961.

Bhaskar, R. and Simon, H. A., 'Problem Solving in Semantically Rich Domains: An Example from Engineering Thermodynamics', *Cognitive Science* 1 (1977), 193–215.

Bitterman, M. E., 'The Comparative Analysis of Learning', *Science* 188 (1975), 699–709.

Block, R. A., Review of B. J. Underwood, *Temporal Codes for Memory: Issues and Problems* (1977), *Contemporary Psychology* 23 (1978), 11–12.

Bloor, D., *Knowledge and Social Imagery*, Routledge & Kegan, London, 1976.

Blume, S. S. (ed.), *Perspectives in the Sociology of Science*, Wiley, Chichester, 1977.

Bobrow, D. and Collins, A. (eds.), *Representation and Understanding: Studies in Cognitive Science*, Academic, New York, 1975.

Boden, M. A., 'Intentionality and Physical Systems', *Philosophy of Science* 37 (1970), 200–214.

Boden, M. A., *Purposive Explanation in Psychology*, Harvard Univ. Press, Cambridge, Mass., 1972.

Boden, M. A., *Artificial Intelligence and Natural Man*, Harvester Press, Hassocks, 1977.

Boden, M. A., *Piaget*, Fontana, London, 1979.

Bohm, D., 'Science As Perception-Communication', in Suppe, F. (ed.), *The Structure of Scientific Theories*, Univ. of Illinois Press, Urbana, 1971, pp. 374–391.

Böhme, G., Van den Daele, W., and Krohn, W., 'Alternatieven in der Wissenschaft', *Zeitschrift für Soziologie* 1 (1972), 302–316.

Böhme, G., Van den Daele, W., and Krohn, W., 'Die Finalisierung der Wissenschaft', *Zeitschrift für Soziologie* 2 (1973), 128–144.

Boneau, C. A., 'Paradigm Regained?', *American Psychologist* 29 (1974), 297–309.

Boole, G., *The Laws of Thought* (1854), Dover, New York.

Booth, A. D. (ed.), *Machine Translation*, North-Holland, Amsterdam, 1967.

Boring, E. G., 'A New Ambiguous Figure', *American Journal of Psychology* 42 (1930), 444–445.

Boring, E. G., *History, Psychology and Science: Selected Papers*, Wiley, New York, 1963.

Boring, E. G., 'Cognitive Dissonance: Its Use in Science', *Science* 145 (1964), 680–685.

Boulding, K. E., *The Image* (1956), University of Michigan Press, Ann Arbor, 1977.

Boyd, R., 'Metaphor and Theory Change: What Is "Metaphor" a Metaphor For?', in Ortony (ed.), *Metaphor and Thought*, Cambridge University Press, Cambridge, 1979, pp. 356–408.

Brainerd, C. J., *Piaget's Theory of Intelligence*, Prentice Hall, Englewood Cliffs, 1978.

Brandstädter, J. and Reinert, G., 'Wissenschaft als Gegenstand der Wissenschaft vom menschlichen Erleben und Verhalten: Uberlegungen zur Konzeption einer Wissenschaftspsychologie', *Zeitschrift für Allgemeine Wissenschaftstheorie* 4 (1973), 368–379.

Bringuier, J. C., *Conversations libres avec Jean Piaget*, Lafont, Paris, 1977.

Brittain, M., *Information and Its Users*, Bath University Press, Claverton Down, 1970.

Broadbent, D. E., *Perception and Communication*, Pergamon Press, London, 1958.

Brookes, B. C., 'Review of J. C. Donohue, *Understanding Scientific Literature: a Bibliometric Approach*', *Nature* 249 (1974), 496–497.

Brown, R., 'In Reply to P. Schönbach', *Cognition* 4 (1976), 186–187.

Bruner, J. S., 'From Communication to Language – A Psychological Perspective', *Cognition* 3 (1973), 255–287.

Bruner, J. and Goodman, C. C., 'Value And Need As Organizing Factors in Perception', *Journal of Abnormal and Social Psychology* 42 (1947), 33–44.

Bruner, J. S., Goodnow, J. J., and Austin, G. A., *A Study of Thinking* (1956), Wiley, New York, 1962.

Bunge, M., *The Myth of Simplicity*, Prentice-Hall, Englewood Cliffs, N. J., 1963.

Bunge, M., 'Metatheory', in UNESCO (ed.), *Scientific Thought: Some Underlying Concepts, Methods and Procedures*, Mouton, The Hague, 1972, pp. 227–251.

Bush, M., 'Psychoanalysis and Scientific Creativity with Special Reference to Regression in the Service of the Ego', *Journal of the American Psychoanalytic Association* 17 (1969), 136–190; reprinted in Eiduson and Beckman, 1973, pp. 243–257.

Bylebyl, J. J., 'Harvey William', in Gillispie, C. (ed.), *Dictionary of Scientific Biography*, Vol. VI, 1973, pp. 150–162 (a).

Bylebyl, J. J., 'The Growth of Harvey's "De motu cordis"', *Bulletin Hist. Med.* 43 (1973), 434–438 (b).

Bylebyl, J. J., 'The Medical Side of Harvey's Discovery: The Normal and the Abnormal', in Bylebyl (ed.), *William Harvey and His Age. The Professional and Social Context of the Discovery of the Circulation*, John Hopkins University Press, Baltimore, 1979, pp. 28–102.

Campbell, D. T., 'Methodological Suggestions from a Comparative Psychology of Knowledge Processes', *Inquiry* 2 (1959), 152–182.

Campbell, D. T., 'Evolutionary Epistemology', in Schilpp, P. A. (ed.), *The Philosophy of Karl Popper*, Open Court Publishing, La Salle, Ill., 1974, pp. 413–463.

Campbell, D. T., *Descriptive Epistemology: Psychological, Sociological and Evolutionary*, Preliminary draft of the William James Lectures, Harvard University, 1977.

Cantril, H., *Invasion from Mars: A Study in the Psychology of Panic*, Princeton University Press, Princeton, 1940.

Carmichael, L., Hogan, H. P., and Walter, A., 'An Experimental Study of the Effect of Language on the Reproduction of Visually Perceived Form', *Journal of Experimental Psychology* 15 (1932), 73–86.

Carnap, R., *Der Logische Aufbau der Welt* (1928), translated by R. A. George as *The Logical Structure of the World*, Univ. of California Press, Berkeley, 1969.

Carnap, R., *Philosophy and Logical Syntax* (1935). Reprinted in Alston, and Nakhnikian, G. (eds.), *Readings in Twentieth-Century Philosophy*, Macmillan, New York, 1963, pp. 424–460.

Carroll, J. B. (ed.), *Language, Thought and Reality, Selected Writings of Benjamin Lee Whorf*, M.I.T. Press, Cambridge, Mass., 1956.

Cartwright, D., 'Determinants of Scientific Progress, The Case of Research on the Risky Shift', *American Psychologist* 28 (1973), 222–231.

Cattell, J. M., 'The Time It Takes to See and Name Objects', *Mind* 2 (1886), 63–65.

Chandrasekaran, B., 'Artificial Intelligence – The Past Decade', in Rubinoff, M., and Zovits, M. C. (eds.), *Advances in Computers*, Vol. 13, Academic Press, New York, 1975.

Charniak, E., 'Organization and Inference in a Frame-like System of Common Sense Knowledge', in Schank, R., and Nash-Webber, B. (eds.), *Theoretical Issues in Natural Language Processing*, ACL, Cambridge, Mass., 1975, pp. 42–51.

Cherry, E. C., 'Some Experiments on the Recognition of Speech with One and Two Ears', *Journal of the Acoustic Society of America* 25 (1953), 975–979.

Chomsky, N., *Reflections on Language*, Pantheon, New York, 1975.

Chomsky, N., 'Rules and representations', *The Behavioral and Brain Sciences* 3 (1980), 1–61.

Chubin, D. E., 'The Conceptualization of Scientific Specialties', *The Sociological Quarterly* 17 (1976), 448–476.

Cicourel, A. V., *The Social Organization of Juvenile Justice*, Wiley, New York, 1968.

Cicourel, A. V., *Cognitive Sociology, Language and Meaning in Social Interaction*, Penguin, Harmondsworth, 1973.

Clark, T. N., 'Institutionalization of Innovations in Higher Education: Four Models', *Administrative Science Quarterly* 13 (1968), 1–25.

Clowes, M. B., 'Transformation Grammar and Organization of Pictures', in Grasseli, A. (ed.), *Automatic Interpretation and Classification of Images*, Academic, New York, 1969.

Cochran, T. C., *The Inner Revolution*, Peter Smith, Gloucester, 1970.

Cohen, I. B., 'Conservation and the Concept of Electric Charge: An Aspect of Philosophy in Relation to Physics in the Nineteenth Century', in Clagett, M. (ed.), *Critical Problems in the History of Science* (*1959*), University of Wisconsin Press, Madison, 1969, pp. 357–383.

Cohen, I. B., 'Review of R. K. Merton, *Science, Technology and Society in Seventeenth-Century England*', *Scientific American* 228 (1973), 117–120.

Cohen, B. I., 'The Eighteenth-Century Origins of the Concept of Scientific Revolution', *Journal of the History of Ideas* 37 (1976), 257–288.

Cohen, I. B., *The Newtonian Revolution*, Cambridge University Press, Cambridge, 1980.

Colby, B. N., 'Culture Grammars', *Science* 187 (1975), 913–919.

Cole, F. J., *A History of Comparative Anatomy* (1949), Dover, New York, 1975.

Cole, J. R. and Cole, S., 'The Ortega Hypothesis', *Science* 178 (1972), 368–375.

Cole, J. R. and Cole, S., *Social Stratification in Science*, University of Chicago Press, Chicago, 1973.

Cole, J. R. and Zuckerman, H., 'The Emergence of a Scientific Specialty: The Self-exemplifying Case of the Sociology of Science', in Coser (ed.), *The Idea of Social Structure*, Harcourt Brace Jovanovich, New York, 1975, pp. 139–174.

Comte, A., *Cours de philosophie positive*, tome I (1830); republished: Schleicher, Paris, 1907.

Conant, J. B., *Science and Common Sense*, Yale Univ. Press, New Haven, 1951.

Coward, R. M., 'Tracking Scientific Specialties: Indicator Applications of Time Series Co-citation Clusters', Paper presented at O.E.C.D. Science & Technology Conference, Paris, September 1980.

Craik, K. J. W., *The Nature of Explanation*, Cambridge University Press, Cambridge, 1943.

Crane, D., *The Environment of Discovery*, Doctoral Dissertation, Columbia University, 1964.

Crane, D., 'The Nature of Scientific Communication and Influence', *International Social Science Journal* 22 (1970), 28–41.

Crane, D., *Invisible Colleges. Diffusion of Knowledge in Scientific Communties*, University of Chicago Press, Chicago, 1972.

Crawford, S., 'Informal Communication Among Scientists in Sleep Research', *Journal of the American Society for Information Science* 22 (1971), 301–310.

Dember, W. N., *The Psychology of Perception*, Holt, Rinehart & Winston, New York, 1960.

Dertouzos, M. L. and Mozes, J. (eds.), *The Computer Age: A Twenty-Year View*, M.I.T. Press, Cambridge, Mass, 1979.

Deutch, J. A. and Deutch, D., 'Attention: Some Theoretical Considerations', *Psychological Review* 70 (1963), 80–90.

DISISS, *The Use of Citation Linkages and Networks for Information Retrieval in the Social Sciences*, Working paper No. 6, Bath University Library, Bath, 1973.

Dornic, S. (ed.), *Attention and Performance VI*, Erlbaum, Hillsdale, 1977.

Downey, K. J., 'The Scientific Community: Organic or Mechanical?', *Sociological Quarterly* 10 (1969), 438–454.

Edge, D., 'Quantitative Measures of Communication in Science: A Critical Review', *History of Science* 17 (1979), 102–134.

Edge, D. O. and Mulkay, M. J., *Astronomy Transformed. The Emergence of Radio Astronomy in Britain*, Wiley, New York, 1976.

Edgerton, S. Y., *The Renaissance Rediscovery of Linear Perspective*, Harper & Row, New York, 1975.

Edgerton, S., 'The Renaissance Artist As a Quantifier', in Hagen, M. A. (ed.), *The Perception of Pictures*, Vol. 1; *Alberti's Window: The Projective Model of Pictorical Information*, Academic Press, New York, 1980.

Eiduson, B. T., *Scientists, Their Psychological World*, Basic Books, New York, 1962.

Eiduson, B. T., 'Psychological Aspects of Career Choice and Development in the Research Scientist', in Eiduson and Beckman, 1973, pp. 3–33.

Eiduson, B. T. and Beckman, L. (eds.), *Science as a Career Choice, Theoretical and Empirical Studies*, Russell Sage, New York, 1973.

Einstein, A. and Infeld, L., *The Evolution of Physics* (1938), Simon & Schuster, New York, 1966.

Eisenstein, E. L., *The Printing Press as an Agent of Change* (1979), Cambridge University Press, Cambridge, 1980.

Elkana, Y., Lederberg, J. et al. (eds.), *Toward a Metric of Science: The Advent of Science Indicators*, Wiley, New York, 1978.

Farr, R. M., 'The Man-in-the-street As "Scientist" and Scientist As "Man": Naive and Scientific Causal Attributions', Contribution to Symposium on Models of Man in Social Psychology, XXIst International Congress of Psychology, Paris, July, 1976.

Feigl, H., 'The "Orthodox" View of Theories: Remarks in Defense As Well As Critique', in Radner and Winokur (eds.), *Minnesota Studies in the Philosophy of Science Vol. IV*, Univ. of Minnesota Press, Minneapolis, 1970, pp. 3–16.

Feldman, J., 'Bad-mouthing Frames', in Schank, R. and Nash-Webber, B. (eds.), *Theoretical Issues in Natural Language Processing*, ACL, Cambridge, Mass., 1975, pp. 92–93.

Fisch, R., 'Psychology of Science', in Spiegel-Rösing and de Solla Price (eds.), *Science, Technology and Society*, Sage, London, 1977, pp. 277–318.

Fleck, J., 'Development and Establishment in Artificial Intelligence', *Sociology of the Sciences Yearbook*, 1981 (in print).

Fleck, L., *Entstehung und Entwicklung einer Wissenschaftlichen Tatsache. Einführung in die Lehre vom Denkstil und Denkkollektiv*, Benno Schwabe, Basel, 1935.

Fleck, L., *Genesis and Development of a Scientific Fact* (Translation of Fleck (1935) by F. Bradley and T. Trenn), University of Chicago Press, Chicago, 1979.

Fodor, J. A., 'Methodological Solipsism Considered As a Research Strategy in Cognitive Psychology', *The Behavioral and Brain Sciences* 3 (1980), 63–109.

Fores, M., 'Science of Science — Substantial Fraud', *Higher Educational Review* 9(3) (1977), 21–34.

Forman, P., 'Weimar Culture, Causality and Quantum Theory, 1918–1928', *Historical Studies in the Physical Sciences* III (1971), 1–115.

Galilei, G., *Dialogue Concerning the Two Chief World Systems — Ptolemaic & Copernican* (1632), transl. by Stilleman Drake, University of California Press, Berkeley, 1967 (second revised edition).

Galilei, G., *Dialogues Concerning the Two New Sciences* (1638), transl. by H. Crew and A. de Salvio, Encyclopedia Britannica, Chicago, 1952.

Gallant, J. A. and Prothero, J. W., 'Weight-watching at the University: The Consequences of Growth', *Science* 175 (1972), 381–388.

Galperin, P. J., 'Zum Problem der Aufmerksamkeit', in Lompscher (ed.), *Problem der Ausbildung geistiger Handlungen*, Volk und Wissen, Berlin, 1972, pp. 15–23.

Garcia, R. and Piaget, J., 'Physico-geometric Explanations and Analysis', in Piaget, J. (ed.), *Understanding Causality*, Norton, New York, 1977, pp. 139–185.

Garfield, E., *Citation Indexing*, Wiley, New York, 1979.

Garfinkel, H. and Sacks, H., 'On Formal Structures of Practical Actions', in McKinney, J. C. and Tiryakian, E. A. (eds.), *Theoretical Sociology Perspectives and Developments*, Appleton-Century-Crofts, New York, 1970, pp. 337–366.

Garvey, W. D., *Communication: the Essence of Science*, Pergamon, Oxford, 1979.

Gibson, E., 'How Perception Really Develops: A View from Outside the Network', in Laberge and Samuels (eds.), *Basic Processes in Reading: Perception and Comprehension*, Erlbaum, Hillsdale, 1977, pp. 155–173.

Gibson, J., *The Ecological Approach to Visual Perception*, Houghton Mifflin, Boston, 1979.

Gilbert, R. M. and Sutherland, N. S. (eds.), *Animal Discrimination Learning*, Academic Press, London, 1969.

Goffman, W., 'Mathematical Approach to the Spread of Scientific Ideas: The History of Mast Cell Research', *Nature* 212 (1966), 449–451.

Goffman, W., 'A Mathematical Model for Analyzing the Growth of a Scientific Discipline', *Journal Association Computing Machinery* 18 (1971), 173–185.

Goffman, W. and Harmon, G., 'Mathematical Approach to the Prediction of Scientific Discovery', *Nature* 229 (1971), 103–104.

Goffman, W. and Warren, K. S., *Scientific Information Systems and the Principle of Selectivity*, Praeger, New York, 1980.

Goldmann Eisler, F., *Psycholinguistics, Experiments in Spontaneous Speech*, Academic, London, 1968.

Goldsmith, M. and Mackay, A., *The Science of Science, Society in the Technological Age*, Souvenir Press, London, 1964.

Goldstein, I. and Papert, S., 'Artificial Intelligence, Language and the Study of Knowledge', *Cognitive Science* 1 (1977), 84–123.

Gombrich, E. H., *Art and Illusion*, Pantheon, New York, 1960.

Gombrich, E. H., *The Sense of Order*, Phaidon, Oxford, 1979.

Goodman, N., *Languages of Art*, second edition, Hackett, Indianapolis, 1976.

Goodman, N., *Ways of Worldmaking*, Harvester Press, Hassocks, 1978.

Goodman, N., 'On Starmaking', *Synthese* 45 (1980), 211–215.

Goodwin, B. C., 'On Some Relationships Between Embryogenesis and Cognition', *Theoria to Theory* 10 (1976), 33–44.

Goodwin, B. C., 'Cognitive Biology', in M. De Mey *et al.* (eds.), *CC77 International Workshop on the Cognitive Viewpoint*, Communication & Cognition, Ghent, 1977, pp. 396–400.

Goodwin, B. C., 'A Cognitive View of Biological Process', *Journal Social Biol. Struct.* 1 (1978), 117–125.

Greiner, L. E., 'Evolution and Revolution as Organization Grows', *Harvard Business Review*, July–August (1972), 37–46.

Griffith, B. C. and Mullins, N. C., 'Coherent Social Groups in Scientific Change', *Science* 177 (1972), 959–964.

Gruber, H. E. and Vonèche, J. J. (eds.), *The Essential Piaget*, Basic Books, New York, 1977.

Hagstrom, W. O., *The Scientific Community* (1965), Southern Illinois University Press, Carbondale, 1975.

Halle, M. and Stevens, K. 'Speech Recognition: A Model and Program for Research', *IRE Transactions on Information Theory* 8 (1962), 155–159.

Hanson, A. R. and Riseman, E. M. (eds.), *Computer Vision Systems*, Academic Press, New York, 1978.

Hanson, N., *Patterns of Discovery*, Cambridge Univ. Press, Cambridge, 1958.

Hanson, N. R., 'A Picture Theory of Meaning', in Colodny, R. G. (ed.), *The Nature and Function of Scientific Theories*, University of Pittsburgh Press, Pittsburgh, 1970, pp. 233–273.

Hanson, N. R., *Observation and Explanation. A Guide to Philosophy of Science*, Allen & Unwin, London, 1972.

Harvey, W., *An Anatomical Disputation Concerning the Movement of the Heart and Blood in Living Creatures* (1628), transl. by G. Whitteridge, Blackwell, Oxford, 1976.

Haugeland, J., 'The Nature and Plausibility of Cognitivism', *The Behavioral and Brain Sciences* 1 (1978), 215–260.

Hayes, P. J., 'Computer Programming As a Cognitive Paradigm', *Nature* 254 (1975), 563–566.

Hayes-Roth, B., 'Implications of Human Pattern Processing for the Design of Artificial Knowledge Systems', in Waterman, D. A. and Hayes-Roth, F. (eds.), *Pattern-directed Inference Systems*, Academic Press, New York, 1978, pp. 333–346.

Heidbreder, E., 'Functionalism', in Henle, Jaynes *et al.* (eds.), *Historical Conceptions of Psychology*, Springer, New York, 1973, pp. 276–285.

Heimdahl, C., 'The Structure of Scientific Revolutions: Kuhn and His Critics', *Communication and Cognition* 7 (1974), 33–60.

Helmholtz, H., *Popular Lectures on Scientific Subjects*, Longmans and Green, London, 1884.

Helmholtz, H., *Epistemological Writings* (1921), Reidel, Dordrecht, 1977.

Hempel, C. G., *Fundamentals of Concept Formation in Empirical Science* (1952), reprinted in Neurath *et al.* (eds.), *Foundations of the Unity of Science, Toward an International Encyclopedia of Unified Science*, Vol. II, Univ. Of Chicago Press, Chicago, 1970, pp. 651–745.

Hempel, C. G., 'Comments on Goodman's Ways of Worldmaking', *Synthese* 45 (1980), 193–199.

Hewitt, C. A., 'Larger Context', Review of Meltzer, B. and Michie, D., 'Machine Intelligence', *Information and Control* 26 (1974), 292–400.

Higham, J., 'Intellectual History and Its Neighbors', in Wiener, P. P. and Noland, A. (eds.), *Ideas in Cultural Perspective*, Rutgers, New Brunswick, 1962, pp. 81–89.

Hirst, R. J., *The Problems of Perception*, Allen and Unwin, London, 1959.

Holton, G., 'Models for Understanding the Growth and Excellence of Scientific Research', in Graubard and Holton (eds.), *Excellence and Leadership in a Democracy*, Columbia University Press, New York, 1962, pp. 94–131.

Holton, G., *Thematic Origins of Scientific Thought. Kepler to Einstein*, Harvard University Press, Cambridge, Mass., 1973.

Huey, E. B., *The Psychology and Pedagogy of Reading*, Macmillan, New York, 1908; republished by MIT Press, Cambridge, Mass., 1968.

Hume, D., *A Treatise of Human Nature* (1739), Penguin, Harmondsworth, 1969.

Ivins, W., *On the Rationalization of Sight*, Plenum Press, New York, 1973.

James, W., *The Principles of Psychology* (1890), Dover, New York, 1950.

James, W., *Pragmatism* (1907), New American Library, New York, 1974.

Johnston, R., 'Contextual Knowledge: A Model for the Overthrow of the Internal/External Dichotomy in Science', *Australian and New Zealand Journal of Sociology* 12 (1976), 193–203.

Josselson, H. H., 'Automatic Translation of Languages Since 1960: A Linguist's View', in Yovits, M. C. (ed.), *Advances in Computers*, Vol. 11, Academic, New York, 1971, pp. 1–58.

Kahneman, D., 'Remarks on Attention Control', *Acta Psychologica* 33 (1970), 118–131.

Katz, E. and Lazarsfeld, P., *Personal Influence: The Part Played by People in the Flow of Mass Communications*, Free Press, New York, 1955.

Keele, S. W., *Attention and Human Performance*, Goodyear Publishing, Pacific Palisades, 1973.

Kemble, E. C., *Physical Science, Its Structure and Development*, Vol. 1: *From Geometric Astronomy to the Mechancial Theory of Heat*, M.I.T. Press, Cambridge, Mass., 1966.

Kessler, M. M., 'Comparison of the Results of Bibliographic Coupling and Analytic Subject Indexing', *American Documentation* 16 (1965), 223–233.

Kimberly, J. R., Miles, R. H. *et al.* (eds.), *The Organizational Life Cycle*, Jossey-Bass, London, 1980.

King, M. D., 'Reason, Tradition and the Progressiveness of Science', *History and Theory* 10 (1971), 3–32.

Klima, R., 'Scientific Knowledge and Social Control in Science: The Application of a Cognitive Theory of Behaviour to the Study of Scientific Behaviour', in Whitley (ed.), *Social Processes of Scientific Development*, Routledge & Kegan Paul, London, 1974.

Knight, D., *Sources of the History of Science 1660–1914*, Cornell University Press, Ithaca, 1975.

Knorr, K. D., Strasser, H., and Zilian, M. G. (eds.), *Determinants and Controls of Scientific Development*, Reidel, Dordrecht, 1975.

Knorr, K., 'The Research Process: Method Reconsidered', in Knorr, Krohn and Whitley (eds.), *Yearbook of the Sociology of the Sciences*, Vol. 5, Reidel, Dordrecht, 1981(a).

Knorr, K., *The Manufacture of Knowledge. Toward a Constructivist and Contextual Theory of Science*, Pergamon, Oxford, 1981 (b).

Koestler, A., *The Act of Creation*, Macmillan, New York, 1964.

Kosslyn, S. M., 'Imagery and Internal Representation', in Rosch and Lloyd (eds.), *Cognition and Categorization*, Erlbaum, Hillsdale, 1978, pp. 217–257.

Kotarbinski, T., *Praxiology: An Introduction to the Sciences of Efficient Action*, Oxford, 1965.

Krantz, D. L., 'Schools and Systems: The Mutual Isolation of Operant and Non-operant Psychology As a Case Study', *Journal History of Behavioral Sciences* **8** (1972), 86–102.

Krohn, R., *The Social Shaping of Science* Greenwood, Westport, Conn., 1971.

Kuhn, T. S., *The Copernican Revolution*, Harvard Univ. Press, Cambridge, 1957.

Kuhn, T. S., 'Energy Conservation As an Example of Simultaneous Discovery', in Clagett, M. (ed.), *Critical Problems in the History of Science* (1959), University of Wisconsin Press, Madison, 1969, pp. 321–356.

Kuhn, T. S., *The Structure of Scientific Revolutions*, Univ. of Chicago Press, Chicago, 1962.

Kuhn, T. S., 'The Function of Dogma in Scientific Research', in Crombie (ed.), *Scientific Change*, Heinemann, London, 1963, pp. 347–369.

Kuhn, T. S., 'The History of Science', in Sills (ed.), *International Encyclopedia of the Social Sciences*, Macmillan, New York, 1968, pp. 74–83.

Kuhn, T. S., 'Comment', *Comparative Studies in Society and History* **11** (1969), 503–412.

Kuhn, T. S., 'Reflections on My Critics', in Lakatos, I. and Musgrave, A. (eds.), *Criticism and the Growth of Knowledge*, Cambridge University Press, London, 1970, pp. 231–278.

Kuhn, T. S., 'The Relations Between History and History of Science', *Daedalus* **100** (1971), 271–304.

Kuhn, T. S., 'Scientific Growth: Reflections on Ben David's "Scientific Role" ', Review of Ben-David, J., *The Scientist's Role in Society: A Comparative Study, Minerva* **10** (1972), 166–278.

Kuhn, T. S., 'Historical Structure of Scientific Discovery', in Henle, M., Jaynes, J. *et al.* (eds.), *Historical Conceptions of Psychology*, Springer, New York, 1973, pp. 3–12.

Kuhn, T. S., 'Second Thoughts on Paradigms', in Suppe, F. (ed.), *The Structure of Scientific Theories*, Univ. of Illinois Press, Urbana, 1974, pp. 459–499.

Kuhn, T. S., *The Essential Tension. Selected Studies in Scientific Tradition and Change*, Univ. of Chicago Press, Chicago, 1977.

Kuhn, T. S., Interview with Frans Boenders. B. R. T. (Belgian Radio and Television), 1978.

Kuhn, T. S., 'The Halt and the Blind: Philosophy and History of Science', Review of Howson, C. (ed.), *Method and Appraisal in the Physical Sciences: The Critical Background to Modern Science, 1800–1905, British Journal for Philosophy of Science* **31** (1980), 181–192.

Lachman, R., Lachman, J., and Butterfield, E., *Cognitive Psychology and Information Processing: An Introduction*, Erlbaum, Hillsdale, 1979.

Lachman, R. and J., Review of F. Suppe (ed.), *The Structure of Scientific Theories, Contemporary Psychology* **20** (1975), 388–390.

Lakatos, I. and Musgrave, A. (eds.), *Criticism and the Growth of Knowledge*, Cambridge Univ. Press, Cambridge, 1970.

Lakoff, G. and Thompson, H., 'Introducing Cognitive Grammar', in C. Cogen *et al.* (eds.), *Proceedings of the First Annual Meeting of the Berkeley Linguistic Society*, Berkeley Linguistics Society, Berkeley, 1975, pp. 295–313.

Langley, P., 'Data-driven Discovery of Physical Laws', *Cognitive Science* 5 (1981), 31–54.

Larkin, J., McDermott, J., Simon, D. P., and Simon, H. A., 'Expert and Novice Performance in Solving Physics Problems', *Science* 208 (1980), 1335–1342 (a).

Larkin, J., McDermott, J., Simon, D. P., and Simon, H. A., 'Models of Competence in Solving Physics Problems', *Cognitive Science* 4 (1980), 317–345 (b).

Latour, B. and Woolgar, S., *Laboratory Life. The Social Construction of Scientific Facts*, Sage, Beverly Hills, 1979.

Latour, B., 'Is It Possible to Reconstruct the Research Process?: Sociology of a Brain Peptide', in Knorr, Krohn, and Whitley (eds.), *The Social Process of Scientific Investigation*, Reidel, Dordrecht, 1980, pp. 53–73.

Law, J., 'The Development of Specialties in Science: The Case of X-Ray Protein Crystallography', *Science Studies* 3 (1973), 275–303.

Leeper, R. W., 'A Study of a Neglected Portion of the Field of Learning – The Development of Sensory Organization', *Journal Genetic Psychology* 46 (1935), 41–75.

Lemaine, C., Macleod, R., Mulkay, M., and Weingart, P., *Perspectives on the Emergence of Scientific Specialties*, Mouton, The Hague, 1976.

Lenat, D. B., 'The Ubiquity of Discovery', *Artificial Intelligence* 9 (1977), 257–285.

Levelt, W. J. M., 'Recente Ontwikkelingen in de Taalpsychologie', *Forum der Letteren* 14 (1973), 235–254.

Lin, N., *The Study of Human Communication*, Bobbs-Merrill, Indianapolis, 1972.

Lin, N., Garvey, W. D., and Nelson, C. E., 'A Study of Communication Structure in Science', in Nelson, C. E. and Pollock, D. K. (eds.), *Communication Among Scientists and Engineers*, Heath, Lexington, Mass., 1970, pp. 23–60.

Lindblom, C. and Cohen, D., *Usable Knowledge*, Yale University Press, New Haven, 1979.

Lindsay, P. H. and Norman, D. A., *Human Information Processing*, second edition, Academic, New York, 1977.

Lipkin, B. S. and Rosenfeld A. (eds.), *Picture Processing and Psychopictorics*, Academic, New York, 1970.

Lotka, A., *Elements of Mathematical Biology*, Dover, New York, 1950.

Mach, E., *Die Analyse der Empfindungen und das Verhältnis des Psychischen zum Physischen* (1886), transl. by C. M. Williams, revised and supplemented from the fifth German edition as *The Analysis of Sensations*, Dover, New York, 1959.

Machlup, F., *The Production and Distribution of Knowledge in the United States* (1962), Princeton University Press, Princeton, N. J., 1972.

Machlup, F., *Knowlege: Its Creation, Distribution, and Economic Significance*, Princeton University Press, Princeton, N. J., 1981.

Macleod, R., 'Changing Perspectives in the Social History of Science', in Spiegel-Rösing and de Solla Price (1977), pp. 149–159.

Mahoney, M. J., *Scientist as Subject: The Psychological Imperative*, Ballinger, Cambridge, Mass., 1976.

Mahoney, M. J., 'Reflections on the Cognitive-Learning Trend in Psychotherapy', *American Psychologist* **32** (1977), 5–13.

Mahoney, M. J., 'Psychology of the Scientist: An Evaluative Review', *Social Studies of Science* **9** (1979), 349–375.

Mandelbaum, M., *History, Man and Reason. A Study in Nineteenth-Century Thought*, John Hopkins Press, Baltimore, 1971.

Mandler, G. and Kessen, W., *The Language of Psychology*, Wiley, New York, 1959.

Marshakova, I. V., 'Citation Networks in Information Science', *Scientometrics* **3** (1981), 13–25.

Masterman, M., 'The Nature of a Paradigm', in Lakatos, I. and Musgrave, A. (eds.), *Criticism and the Growth of Knowledge*, Cambridge University Press, Cambridge, 1970, pp. 59–89.

Martins, H., 'The Kuhnian "Revolution" and Its Implications for Sociology', in Nossiter, T. J. *et al.* (eds.), *Imagination and Precision in the Social Sciences*, Faber and Faber, London, 1972, pp. 13–57.

Maruyama, M., 'Paradigmatology and Its Application to Cross-Disciplinary, Cross-Professional and Cross-Cultural Communication', *Cybernetica* **17** (1974), 136–156, 237–281.

Maslow, A., *The Psychology of Science: A Reconnaissance*, Harper & Row, New York, 1966.

McCarthy, J., 'Ascribing Mental Qualities to Machines', in Ringle, M. (ed.), *Philosophical Perspectives in Artificial Intelligence*, Harvester, Brighton, 1979, pp. 161–195.

McClelland, D., 'The Psychodynamics of Creative Physical Scientists', in Grüber, Terrell *et al.* (eds.), *Contemporary Approaches to Creative Thinking*, Atherton, New York, 1962, pp. 141–174.

McClelland, J. L., 'Letter and Configuration Information in Word Identification', *Journal of Verbal Learning and Verbal Behaviour* **16** (1977), 137–150.

McCulloch, W. and Pitts, W., 'A Logical Calculus of Ideas Immanent in Nervous Activity', *Bulletin of Mathematical Biophysics* **5** (1943), 115–137.

McCulloch, W., 'A Heterarchy of Values Determined by the Topology of Nervous Nets', *Bulletin of Mathematical Biophysics* **7** (1945), 89–93.

McGrath, J. E. and Altman, I., *Small Group Research. A Synthesis and Critique of the Field*, Holt, Rinehart & Winston, New York, 1966.

McGuire, W. J., 'Suspiciousness of Experimenter's Intent', in Rosenthal and Rosnow (eds.), *Artifact in Behavioral Research*, Academic, New York, 1969.

Meadows, A. J., *Communication in Science*, Butterworths, London, 1974.

Medawar, P., Shelley, J. (eds.), *Structure in Science and Art*, Excerpta Medica, Amsterdam, 1980.

Menard, H., *Science: Growth and Change*, Harvard Univ. Press, Cambridge, Mass., 1971.

Mendelsohn, E., Weingart, P., and Whitley, R. (eds.), *The Social Production of Scientific Knowledge*, Vol. 1, Sociology of the Sciences, A Yearbook, Reidel, Dordrecht, 1977.

Menzel, H., 'Scientific Communication: Five Themes from Social Science Research', *American Psychologist* **21** (1966), 999–1004.

Merton, R. K., *Science, Technology and Society in Seventeenth-Century England* (1938), Harper & Row, New York, 1970.

Merton, R. K., *Social Theory and Social Structure*, Free Press, New York, 1957.

Merton, R. K., 'Priorities in Scientific Discovery', *American Sociological Review* **22** (1957), 635–659.

Merton, R. K., *On Theoretical Sociology, Five essays, Old and New*, Free Press, New York, 1967.

Merton, R. K., 'Behavior Patterns of Scientists', *American Scholar* 38 (1969), 197–225; reprinted in Eiduson and Beckman, 1973, pp. 601–611.

Merton, R. K., *The Sociology of Science; Theoretical and Empirical Investigations*, N. W. Storer (ed.), University of Chicago Press, Chicago, 1973.

Merton, R. K., 'The Sociology of Science: An Episodic Memoir', in Merton, R. K. and Gaston, J. (eds.), *The Sociology of Science in Europe*, Southern Illinois Univ. Press. Carbondale, 1977, pp. 3–141.

Merton, R. K. and Gaston J. (eds.), *The Sociology of Science in Europe*, Southern Illinois University Press, Carbondale, 1977.

Michie, D., *On Machine Intelligence*, Edinburgh Univ. Press, Edinburgh, 1974.

Miller, G. A., Galanter, E., and Pribram, K. H., *Plans and the Structure of Behavior*, Holt, New York, 1960.

Miller, G. A. and Johnson-Laird, P. N., *Language and Perception*, Harvard Univ. Press, Cambridge, Mass., 1976.

Mills C. Wright, *The Sociological Imagination*, Oxford Univ. Press, New York, 1959.

Minsky, M., 'Matter, Mind and Models', in Minsky, M. (ed.), *Semantic Information Processing*, MIT Press, Cambridge, Mass., 1968.

Minsky, M. (ed.), *Semantic Information Processing*, MIT Press, Cambridge, Mass., 1968.

Minsky, M., 'A Framework for Representing Knowledge', in Winston, P. H. (ed.), *The Psychology of Computer Vision*, McGraw-Hill, New York, 1975, pp. 211–277.

Minsky, M., 'Plain Talk About Neurodevelopmental Epistemology', *Intern. Joint Conference on A.I.* 1977, 1083–1092.

Minsky, M., 'Computer Science and Representation of Knowledge', in Dertouzos, M. L. and Moses J. (eds.), *The Computer Age: A Twenty-Year View*, MIT Press, Cambridge, Mass., 1979, pp. 392–421.

Minsky, M., 'K-lines: A Theory of Memory', *Cognitive Science* 4 (1980), 117–133.

Minsky, M. and Papert, S., *Artificial Intelligence. Progress Report*, M.I.T. AI Memo No. 252, Cambridge, Mass., 1972.

Mises, von R., *Positivism* (1951), Dover, New York, 1968.

Mitroff, I., 'On the Norms of Science: A Report of a Study of the Apollo Moon Scientists', *Communication and Cognition* 7 (1974), 125–151.

Mitroff, I. I. and Kilmann, R. H., 'Systemic Knowledge: Toward an Integrated Theory of Science', *Theory and Society* 4 (1977), 103–129.

Moray, N., 'Attention in Dichotic Listening: Affective Cues and the Influence of Instructions', *Quarterly Journal of Experimental Psychology* 9 (1959), 56–60.

Moray, N., *Attention: Selective Processes in Vision and Hearing*, Academic, New York, 1970.

Moray, N. and Fitter, M., 'A Theory and the Measurement of Attention', in Kornblum, S. (ed.), *Attention and Performance, IV*, Academic Press, New York, 1973, pp. 3–19.

Moreno, J. L., *Who Shall Survive?*, Nervous and Mental Disease Monograph, Washington, D.C., 1934.

Morin, E., *Le paradigme perdu: la nature humaine*, du Seuil, Paris, 1973.

Moruzzi, G. and Magoun, H. W., 'Brain Stem Reticular Formation and Activation of the EEG', *EEG Clinical Neurophysiology* 1 (1949), 455–473.

Mulkay, M., *The Social Process of Innovation*, Macmillan, London, 1972.

Mulkay, M., 'Conceptual Displacement and Migration in Science: A Prefatory Paper', *Science Studies* 4 (1974), 205–234.

Mulkay, M., *Science and the Sociology of Knowledge*, Allen & Unwin, London, 1979.

Mulkay, M. J., Gilbert, G. N., and Woolgar, S., 'Problem Areas and Research Networks in Science', *Sociology* 9 (1975), 187–203.

Muller-Freienfels, R., 'Zur Psychologie der Psychologie', *Acta Psychologica* 1 (1934), 157–174.

Mullins, N. C., *Social Communication Networks among Biological Scientists*, Doctoral dissertation, Harvard University, 1966.

Mullins, N. C., 'The Development of a Scientific Specialty: The Phage Group and the Origins of Molecular Biology', *Minerva* 10 (1972), 51–82.

Mullins, N. C., *Theories and Theory Groups in Contemporary American Sociology*, Harper and Row, New York, 1973.

Mullins, N., Hargens, L. *et al.*, 'The Group Structure of Cocitation Clusters, A Comparative Study', *American Sociological Review* 42 (1977), 552–562.

Musgrave, A., 'Kuhn's Second Thought', *British Journal for Philosophy of Science* 22 (1971), 287–306.

Myrdal, G., 'How Scientific Are the Social Sciences', *Journal of Social Issues* 28 (1972), 151–170.

Naess, A., *The Pluralist and Possibilist Aspect of the Scientific Enterprise*, Universitetsforlaget, Oslo, 1972.

Nagel, E., *The Structure of Science*, Harcourt & Brace, New York, 1961.

Neisser, U., *Cognitive Psychology*, Appleton, New York, 1967.

Neisser, U., *Cognition and Reality*, Freeman, San Francisco, 1976.

Newell, A., 'Artificial Intelligence and the Concept of Mind', in Schank, R. C. and Colby, K. M. (eds.), *Computer Models of Thought and Language*, Freeman, San Francisco, 1973, pp. 1–60.

Newell, A., 'Physical Symbol Systems', *Cognitive Science* 4 (1980), 135–183.

Nilsson, N. J., *Problem-solving Methods in Artificial Intelligence*, McGraw-Hill, New York, 1971.

Norman, D. A., 'Discussion', in Solso, R. L. (ed.), *Contemporary Issues in Cognitive Psychology. The Loyola Symposium*, Wiley, New York, 1973.

Norman, D. A., *Memory and Attention. An Introduction to Human Information Processing*, second edition, Wiley, New York, 1976.

Norman, D. A., 'Analysis and Design of Intelligent Systems', in Klix (ed.), *Human and Artificial Intelligence*, North Holland, Amsterdam, 1979, pp. 37–43.

Norman, D. A. (ed.), *Perspectives on Cognitive Science*, Ablex, Norwood, N.J., 1980.

Norman, D. A. and Rumelhart, D. E. (eds.), *Explorations in Cognition*, Freeman, San Francisco, 1975.

Norton, B. J., 'Karl Pearson and Statistics', *Social Studies of Science* 8 (1978), 3–34.

Nowotny, H., 'On the Feasibility of a Cognitive Approach to the Study of Science', *Zeitschrift für Soziologie* 2 (1973), 282–296.

Nowotny, H. and Schmutzer, M., *Gesellschaftliches Lernen*, Wissenserzeugung und die Dynamik von Kommunikationsstrukturen, Herder & Herder, Frankfurt, 1974.

Nowakowska, M., 'Epidemical Spread of Scientific Objects', *Theory and Decision* 3 (1973) (Additional note in *Theory and Decision* 7 (1976), 141–142.)

Oettinger, A. G., 'The Uses of Computers in Science', in McCarthy, J., Evans, D. C. et al., *Information*, Freeman, San Francisco, 1966, pp. 113–130.

Orwell, G., *Animal Farm* (1945), Penguin, Harmondsworth, 1951.

Ossowska, M. and Ossowski, S., 'The Science of Science', *Organon* 1 (1936), 1–12.

Osterrieth, P., Piaget J. (eds.), *Le problème des stades en psychologie de l'enfant*, Presses Universitaires de France, Paris, 1956.

Pagel, W., *William Harvey's Biological Ideas*, Karger, Basel, 1967.

Pagel, W., *New Light on William Harvey*, Karger, Basel, 1976.

Palmer, S. E., 'Visual Perception and Knowledge: Notes on a Model of Sensory Cognitive Interaction', in Norman and Rumelhart (eds.), *Explorations in Cognition*, Freeman, San Francisco, 1975, pp. 279–307.

Palmer, S. E., 'Hierarchical Structure in Perceptual Representation', *Cognitive Psychology* 9 (1977), 441–474.

Panofsky, E., *Early Netherlandish Painting* (1953), Harper & Row, New York, 1971.

Papert, S., *Mindstorms. Children, Computers and Powerful Ideas*, Harvester, Brighton, 1980.

Pearson, K., *The Grammar of Science* (1892), Meridian, New York, 1957.

Petrie, H. G., 'The Strategy Sense of "Methodology" ', *Philosophy of Science* 35 (1968), 248–257.

Petrie, H. G., 'Comments on Chapter 1 by Broudy', in Anderson, R. C., Spiro, R. J., and Montague, W. E. (eds.), *Schooling and the Acquisition of Knowledge*, Erlbaum, Hillsdale, N.J., 1977, pp. 19–26.

Piaget, J., 'Psychologie et critique de la connaissance', *Archives de Psychologie* 19 (1925), 193–210.

Piaget, J., *La causalité physique chez l'enfant*, Alcan, Paris, 1927.

Piaget, J., *Introduction à l'épistémologie génétique, I–III*, Presses Universitaires de France, Paris, 1950.

Piaget, J., 'Autobiography', In Murchison (ed.), *A History of Psychology in Autobiography*, Vol. IV, Clark University Press, Worcester, 1952, pp. 237–256.

Piaget, J., 'Le développement des perceptions en fonction de l'âge', in Fraisse, P. and Piaget, J. (eds.), *Traité de psychologie expérimentale, VI*, Presses Universitaires de France, Paris, 1963, pp. 66–107.

Piaget, J., *Sagesse et illusions de la philosophie*, Presses Universitaires de France, Paris, 1965.

Piaget, J., *La prise de conscience*, Presses Universitaires de France, Paris, 1974 (a).

Piaget, J., *Réussir et comprendre*, Presses Universitaires de France, Paris, 1974 (b).

Piaget, J., *Le comportement, moteur de l'évolution*, Gallimard, Paris, 1976.

Piaget, J., 'Autobiographie', *Revue Européenne des sciences sociales* 14 (1976), 1–43.

Piaget, J., 'Relations Between Psychology and Other Sciences', *Annual Review of Psychology* 30 (1979), 1–8 (a).

Piaget, J., 'La psychogenèse des connaissances et sa signification épistémologique', in Massimo Piattelli-Palmarini (ed.), *Théories du langage – Théories de l'apprentissage*, du Seuil, Paris, 1979, pp. 53–64 (b).

Piaget, J. and Inhelder, B., *Le développement des quantités physiques chez l'enfant*, Delachaux et Niestlé, Neuchâtel, 1941.

Piaget, J. and Szeminska, A., *La genèse du nombre chez l'enfant*, Delachaux et Niestlé, Neuchâtel, 1941.

Polanyi, M., *The Tacit Dimension*, Doubleday, New York, 1965.

Popper, K., 'Normal Science and Its Dangers', in Lakatos, I. and Musgrave, A. (eds.), *Criticism and the Growth of Knowledge*, Cambridge Univ. Press, London, 1970, pp. 51–58.

Portugal, F. H. and Cohen, J. S., *A Century of DNA. A History of the Discovery of the Structure and Function of the Genetic Substance*, M.I.T. Press, Cambridge, Mass., 1977.

Price, D. de Solla, *Science since Babylon*, Yale University Press, New Haven, 1961.

Price, D. de Solla, *Little Science, Big Science*, Columbia University Press, New York, 1963.

Price, D. de Solla, 'The Science of Science', in Goldsmith and Mackay, 1964, pp. 195–208.

Price, D. de Solla, 'Citation Measures of Hard Science, Soft Science, Technology and Nonscience', in Nelson, C. and Pollock, D. (eds.), *Communication among Scientists and Engineers*, Heath & Co., Lexington, 1970, pp. 3–22.

Price, D. de Solla, 'Invisible College Research: State of the Art', in Crawford, S. (ed.), *Informal Communication among Scientists*, Proceedings of a Conference on Current Research, American Medical Association, 1971, pp. 4–14.

Price, D. de Solla, 'Editorial Statement', *Scientometrics* 1 (1978), 7–8.

Price, D. de Solla and Beaver, D. B., 'Collaboration in an Invisible College', *American Psychologist* 21 (1966), 1011–1018.

Prigogine, Y., 'Entretien avec J. C. Bringuier', in Bringuier, J. C. (ed.), *Conversations libres avec Jean Piaget*, Lafont, Paris, 1977, pp. 156–161.

Prigogine, Y. and Stengers, I., *La nouvelle alliance. Métamorphose de la science*, Gallimard, Paris, 1979.

Pylyshyn, Z., 'Computation and Cognition: Issues in the Foundation of Cognitive Science', *The Behavioral and Brain Sciences* 3 (1980), 111–169.

Radnitzky, G., *Contemporary Schools of Metascience*, Regnery, Chicago, 1972.

Radnitzky, G., 'Towards a Theory of Traditions in Science', *Communication & Cognition* 6 (1973), 15–46.

Radnor, M., Feller, I., and Rogers, E. (eds.), *The Diffusion of Innovations: An Assessment*, Northwestern University, Center for the Interdisciplinary Study of Science and Technology, Evanston, Ill., 1978.

Ravetz, J. R., *Scientific Knowledge and Its Social Problems*, Clarendon Press, Oxford, 1971.

Reichenbach, H., *Experience and Prediction*, Univ. of Chicago Press, Chicago, 1938.

Restivo, S., 'Notes and Queries on Science, Technology and Human Values', *Science, Technology and Human Values*, No. 31 (1981), 20–22.

Restivo, S., 'Commentary: Some Perspectives in Contemporary Sociology of Science', *Science, Technology and Human Values*, No. 35 (1981), 22–30.

Rip, A., 'The Social Context of "Science, Technology and Society" Courses', *Studies in Higher Education* 4 (1979), 15–26.

Roe, A., 'Patterns in Productivity of Scientists', *Science* 176 (1972), 940–941.

Roessner, J. D., 'Technological Diffusion Research and National Policy Issues', *Knowledge* 2 (1980), 179–201.

Rogers, E. M., *Diffusion of Innovations*, Free Press, New York, 1962.

Rogers, E. M., *Bibliography of Research on the Diffusion of Innovations*, Michigan State Univ., Diffusion of Innovations Report 1, East Lansing, 1964.

Rogers, E. M., *Bibliography of the Diffusion of Innovations*, Council of Planning Librarians, Monticello, Ill., 1977.

Rogers, E. M. and Shoemaker, F. F., *Communications of Innovations*, Free Press, New York, 1971.

Rogers, E. C. and Agarwala-Rogers, R., *Communication in Organizations*, Free Press, New York, 1976.

Rosenblueth, A., Wiener, N. and Bigelow, J., 'Behavior, Purpose and Teleology', *Philosophy of Science* 10 (1943), 18–24.

Rosenfeld, A., 'Image Processing and Recognition', in Yovits (ed.), *Advances in Computers*, Vol. 18, Academic 264, New York, 1979, pp. 1–57.

Rosenthal, R., *Experimenter Effects in Behavioral Research*, enlarged edition, Irvington, New York, 1976.

Rosenthal, R. and Rubin, D. B., 'Interpersonal Expectancy Effects: The First 345 Studies', *The Behavioral and Brain Sciences* 1 (1978), 377–415.

Rothman, B., *Jean Piaget: Psychologist of the Real*, Harvester Press, Hassocks, 1977.

Royce, J. R., 'Psychology Is Multi-: Methodological, Variate, Epistemic, World View, Systemic, Paradigmatic, Theoretic, and Disciplinary', in Arnold, W. J. (ed.), *1975 Nebraska Symposium on Motivation*, Univ. of Nebraska Press, Lincoln, 1976, pp. 1–63.

Rubin, E., *Visuell wahrgenommene Figuren* (1915), Gyldendalske, Kopenhagen, 1921.

Russell, B., *Mysticism and Logic* (1917), Allen & Unwin, London, 1963.

Russell, B., *The Scientific Outlook* (1931), Norton, New York, 1962.

Ryle, G., *The Concept of Mind* (1949), Barnes & Noble, New York, 1964.

Sarton, G., *The Renaissance: Six Essays*, New York, 1962.

Saunders, J. B. de C. M. and O'Malley, C. D., *The Illustrations from the Works of Andreas Vesalius of Brussels*, Dover, New York, 1973.

Scandura, J. M. and Scandura, A. B., *Structural Learning and Concrete Operation: An Approach to Piagetian Conservation*, Praeger, New York, 1981.

Schank, R., *Conceptual Information Processing*, North-Holland, Amsterdam, 1975.

Schank, R., 'Using Knowledge to Understand', in Schank, R. and Nash-Webber, B. (eds.), *Theoretical Issues in Natural Language Processing*, ACL, Cambridge, Mass., 1975, pp. 117–121.

Scheffler, I., *Science and Subjectivity*, Bobbs-Merrill, Indianapolis, 1967.

Scheffler, I., 'The Wonderful Worlds of Goodman', *Synthese* 45 (1980), 201–209.

Schlick, M., *Gesammelte Aufsätze* (1938), G. Olms, Hildesheim, 1969.

Searle, J. R., 'Minds, Brains, and Programs', *The Behavioral and Brain Sciences* 3 (1980), 417–457.

Senders, J., 'The Human Operator As a Monitor and Controller of Multi-degree-of-freedom Systems', *IEEE Transactions on Human Factors in Electronics* (1964), H-F-E-5, 2–5.

Sexton, Misiak (eds.), *Historical Perspectives in Psychology: Readings*, Brooks Coole, Belmont, 1971.

Shannon, C. and Weaver, W., *The Mathematical Theory of Communication*, Univ. of Illinois Press, Urbana, 1949.

Shapere, D., 'The Structure of Scientific Revolutions', *The Philosophical Review* 73 (1964), 383–374.

Shapere, D. (ed.), *Philosophical Problems of Natural Science*, Macmillan, New York, 1965.

Shapere, D., 'Meaning and Scientific Change', in Colodny (ed.), *Mind and Cosmos. Essays in Contemporary Science and Philosophy*, Univ. of Pittsburgh Press, Pittsburgh, 1966, pp. 41–85.

Shaver, K. G., *An Introduction to Attribution Processes*, Winthrop, Cambridge, Mass., 1975.

Shaw, R. and Bransford, J. (eds.), *Perceiving, Acting and Knowing*, Erlbaum, Hillsdale, N.J., 1977.

Shepard, R., 'Review of U. Neisser, Cognitive Psychology', *American Journal of Psychology* 81 (1968), 285–289.

Shera, J., *The Foundations of Education for Librarianship*, Becker and Hayes, New York, 1972.

Shimony, A., 'Is Observation Theory-laden? A Problem in Naturalistic Epistemology', in Colodny (ed.), *Logic, Laws and Life*, Univ. of Pittsburgh Press, Pittsburgh, 1977, pp. 185–208.

Shirai, Y., 'A Context-Sensitive Line Finder for Recognition of Polyhedra', *Artificial Intelligence* 4 (1973), 95–120.

Simon, H. A., *The Sciences of the Artificial*, MIT Press, Cambridge, Mass., 1969.

Singer, B. F., 'Toward a Psychology of Science', *American Psychologist* 26 (1971), 1010–1015.

Sloman, A., *The Computer Revolution in Philosophy: Philosophy, Science, and Models of Mind*, Harvester, Hassocks, 1978.

Small, H., 'Co-citation in the Scientific Literature: A New Measure of the Relationship Between Two Documents', *Journal American Society for Information Science* 24 (1973), 265–269.

Small, H., 'A Citation Model for Scientific Specialties', *Proceedings American Society for Information Science* 12 (1975), 34–35.

Small, H., 'Structural Dynamics of Scientific Literature', *International Classification* 2 (1976), 67–73.

Small, H., 'A Co-citation Model of a Scientific Specialty: A Longitudinal Study of Collagen Research', *Social Studies of Science* 7 (1977), 139–166.

Small, H., 'Cited Documents As Concept Symbols', *Social Studies of Science* 8 (1978), 327–340.

Small, H. and Crane, D., 'Specialties and Disciplines in Science and Social Science: An Examination of Their Structure Using Citation Indexes', *Scientometrics* 1 (1979), 445–461.

Small, H. and Greenlee, E., 'Citation Context Analysis of a Co-citation Cluster: Recombinant-DNA', *Scientometrics* 2 (1980), 277–301.

Sneed, J. D., 'Philosophical Problems in the Empirical Science of Science: A Formal Approach', *Erkenntnis* 10 (1976), 115–146.

Snow, C. P., *The Two Cultures* (1959), Cambridge University Press, Cambridge, 1969.

Solso, R. (ed.), *Contemporary Issues in Cognitive Psychology: The Loyola Symposium*, Wiley, New York, 1973.

Spence, K. W., 'The Differential Response in Animals to Stimuli Varying Within a Single Dimension', *Psychological Review* 44 (1937), 430–444.

Sridharan, N. S. (ed.), 'Special Issue on Applications to the Sciences and Medicine', *Artificial Intelligence* 11 (1978), 1–193.

Stanford University, *Heuristic Programming Project 1980*, Computer Science Department, Stanford University, Stanford, 1980.

Stehr, N. and König, R. (eds.), 'Wissenschaftssoziologie', *Kölner Zeitschrift für Soziologie und Sozialpsychologie*, Sonderheft 18, 1975.

Stern, N., 'Age and Achievement in Mathematics: A Case-study in the Sociology of Science', *Social Studies of Science* 6 (1978), 127–140.

Stevens, S. S., 'The Operational Definition of Psychological Concepts', *Psychological Review* 42 (1935), 517–527.

Stevens, S. S., 'Psychology: The Propaedeutic Science', *Philosophy of Science* 3 (1936), 90–104.

Stevens, S. S., 'Psychology and the Science of Science', *Psychological Bulletin* 36 (1939), 221–263.

Sullivan, D., White, H., and Barboni, E., 'Co-citation Analyses of Science: An Evaluation', *Social Studies of Science* 7 (1977), 223–240.

Suppe, F. (ed.), *The Structure of Scientific Theories*, Univ. of Illinois Press, Urbana, 1974.

Sussman, G. J., *A Computer Model of Skill Acquisition*, American Elsevier, New York, 1975.

Sutherland, N. S., 'Object Recognition', in Carterette and Friedman (eds.), *Handbook of Perception*, Vol. III, Academic, New York, 1973, pp. 157–185.

Sutherland, N. S. and Mackintosh, N. J. (eds.), *Mechanisms of Animal Discrimination Learning*, Academic, New York, 1971.

Swanson, D. R., 'On Improving Communication Among Scientists', *Bulletin of the Atomic Scientists*, February (1966), 8–12.

Thompson, D. A. W., *Growth and Form* (1917), Cambridge University Press, Cambridge, 1942.

Toulmin, S., *Foresight and Understanding*, Indiana Univ. Press, Bloomington, 1961.

Transgard, H., 'A Cognitive System Approach to Methodology: An Outline', *Quality and Quantity* 6 (1972), 137–151.

Treisman, A., 'Contextual Cues in Selective Listening', *Quarterly Journal of Experimental Psychology* 12 (1960), 242–248.

Turing, A., 'On Computable Numbers, With An Application to the Entscheidungsproblem', *Proceedings of the London Mathematical Society* 42 (1936), 230–265.

Uhr, L., *Pattern Recognition, Learning and Thought. Computer Models of Mental Processes*, Prentice Hall, Englewood Cliffs, 1973.

Ullman, S., 'Against Direct Perception', *The Behavioral and Brain Sciences* 3 (1980), 373–415.

Van den Daele, W., Krohn, W., and Weingart, P. (eds.), *Geplante Forschung*, Suhrkamp, Frankfurt am Main, 1979.

Van Norren, B., *Original and Derived Creativity in Scientific Thinking*, Afdelingen voor Sociale Wetenschappen aan de Landbouwhogeschool, Wageningen (Netherlands), 1976.

Van Velthoven, G. M., *Trends in Economische Psychologie*, Lecture presented at staff meeting of Department of Psychology, University of Tilburg (Netherlands), 1976.

Vlachy, J., 'Mobility in Science. A Bibliography of Scientific Career Migration, Field Mobility, International Academic Circulation and Brain Drain', *Scientometrics* 1 (1979), 201–228.

Waddington, C. H., *The Nature of Life* (1961), Unwin, London, 1963.
Waddington, C. H., *Behind Appearance*, M.I.T. Press, Cambridge, Mass., 1970.
Walentynowicz, B., *Social Studies of Science* 5 (1975), 213–222.
Waltz, D. L. (ed.), *TINLAP 2. Theoretical Issues in Natural Language Processing*, Association for Computing Machinery, New York, 1978.
Wason, P. C., 'In Real Life Negatives Are False', *Logique et analyse* 15 (1972), 17–38.
Wasson, C. R., *Dynamic Competitive Strategy and Product Life Cycles* (3rd ed.), Austin Press, Austin, Texas, 1978.
Watson, R. and Campbell, D. T., 'Editor's Foreword', in Boring, E. G., *History, Psychology and Science: Selected Papers*, Wiley, New York, 1963, pp. V–VIII.
Wax, M. L., 'Myth and Interrelationship in Social Science: Illustrated Through Anthropology and Sociology', in Sherif, M. and Sherif, C. W. (eds.), *Interdisciplinary Relationships in the Social Sciences*, Aldine, Chicago, 1969, pp. 77–99.
Webb, E. J., Campbell, D. T., Schwartz, R. D., and Sechrest, L., *Unobtrusive Measures: Nonreactive Research in the Social Sciences*, Rand MacNally, Chicago, 1966.
Weber, Max, *Die protestantische Ethik und der Geist des Kapitalismus* (1905), transl. by T. Parsons as *The Protestant Ethic and the Spirit of Capitalism*, Scribner's Sons, New York, 1958.
Wechsler, J. (ed.), *On Aesthetics in Science*, MIT, Cambridge, Mass., 1978.
Weimer, W., 'The Psychology of Inference and Expectation: Some Preliminary Remarks', in G. Maxwell and R. Anderson (eds.), *Induction, Probability and Confirmation*, University of Minnesota Press, Minneapolis, 1975.
Weimer, W. B. and Palermo, D. S. (eds.), *Cognition and the Symbolic Processes*, N. J., Erlbaum, Hillsdale, 1974.
Weinberg, B. H., 'Bibliographic Coupling: A Review', *Information Storage and Retrieval* 10 (1974) 189–196.
Weingart, P., 'On a Sociological Theory of Scientific Change', in Whitley (ed.), *Social Processes of Scientific Development*, Routledge & Kegan Paul, London, 1974.
Weingart, P., *Wissensproduktion und Soziale Struktur*, Suhrkamp, Frankfurt am Main, 1976.
Weingart, P., 'Comment on the "Fury Conference"', *4S Newsletter* 5 (1980), 32–33.
Weiss, C., 'Knowledge Creep and Decision Accretion', *Knowledge: Creation, Diffusion, Utilization* 1 (1980), 381–404.
Weiss, P., 'Knowledge: A Growth Process' (1960), reprinted in Kochen, M. (ed.), *The Growth of Knowledge*, Wiley, New York, 1967, pp. 209–215.
Weizenbaum, J., *Computer Power and Human Reason*, Freeman, San Francisco, 1976.
Wersig, G., *Informationssoziologie. Hinweise zu einem informationswissenschaftlichen Teilbereich*, Fischer Athenäum, Frankfurt am Main, 1973.
Whitley, R., 'Communication Nets in Science: Status and Citation Patterns in Animal Physiology', *Sociological Review* 17 (1969), 219–233.
Whitley, R., 'Cognitive and Social Institutionalization of Scientific Specialties and Research Areas', in Whitley (ed.), *Social Processes of Scientific Development*, Routledge & Kegan Paul, London, 1974.
Whitley, R., 'Components of Scientific Activities, Their Characteristics and Institutionalisation in Specialties and Research Areas: A Framework for the Comparative Analysis of Scientific Developments', in Knorr, Strasser and Zilian (eds.), *Determinants and Controls of Scientific Development*, Reidel, Dordrecht, 1975, pp. 37–73.

Whitley, R., 'The Sociology of Scientific Work and the History of Scientific Developments', in Blume, S. (ed.), *Perspectives in the Sociology of Science*, Wiley, Chichester, 1977, pp. 21–50.

Whitley, R., 'The Context of Scientific Investigation', in Knorr, Krohn, and Whitley (eds.), *The Social Process of Scientific Investigation. Sociology of the Sciences*, Vol. IV, 1980, pp. 297–321.

Whitley, R., 'The Organisation and Establishment of the Professionalised Sciences', in *Scientific Establishments and Hierarchies. Sociology of the Sciences*, Vol. V, 1981 (in press).

Whitteridge, G., *William Harvey and the Circulation of the Blood*, Macdonald, London, 1971.

Whitteridge, G., 'Introduction and Notes to William Harvey', *The Movement of the Heart and Blood*, Blackwell, Oxford, 1976.

Wilks, Y., 'Philosophy of Language', in Charniak and Wilks (eds.), *Computational Semantics*, North Holland, Amsterdam, 1976, pp. 205–233.

Winograd, T., *Understanding Natural Language*, Academic, New York, 1972.

Winograd, T., 'A Procedural Model of Language Understanding', in Schank, R. C. and Colby, K. M. (eds.), *Computer Models of Thought and Language*, Freeman, San Francisco, 1973, pp. 152–186.

Winograd, T. 'Frame Representations and the Declarative-Procedural Controversy', in Bobrow, D. G. and Collins, A. (eds.), *Representation and Understanding. Studies in Cognitive Science*, Academic Press, New York, 1975, pp. 185–210.

Winograd, T., 'Artificial Intelligence and Language Comprehension', in National Institute of Education, *Artificial Intelligence and Language Comprehension*, N.I.E., Washington, D.C., 1976, pp. 3–25.

Winograd, T., 'Towards a Procedural Understanding of Semantics', *Revue Internationale de Philosophie* 30 (1976), 260–303.

Winograd, T., 'On Some Contested Suppositions of Generative Linguistics About the Scientific Study of Language', *Cognition* 5 (1977), 151–179.

Winston, P., *Artificial Intelligence*, Addison-Wesley, Reading, Mass., 1977.

Wisdom, J., *Philosophy and Psychoanalysis*, Blackwell, Oxford, 1957.

Wittgenstein, L., *Tractatus Logico-Philosophicus* (1921), transl. by Pears, D. E. and McGuinness, B. F., Humanities, New Jersey, 1974.

Wittgenstein, L., *Philosophical Investigations*, transl. by Anscombe, G. E. M., Macmillan, New York, 1953.

Wolf, P., 'La théorie sensori-motrice de l'intelligence et la psychologie du développement général', in Bresson, F. and M. de Montmollin (eds.), *Psychologie et épistemologie génétiques. Thèmes Piagetiens*, Dumont, Paris, 1966, pp. 235–250.

Woolgar, S. W., *The Emergence and Growth of Research Areas in Science With Special Reference to Research on Pulsars*, Doctoral dissertation, Emmanuel College, Cambridge, 1978.

Yellin, J. A., 'Model for Research Problem Allocation Among Members of a Scientific Community', *Journal of Mathematical Sociology* 2 (1972), 1–36.

Zaltman, G., Duncan, R., and Holbek, J., *Innovations and Organizations*, Wiley, New York, 1973.

Ziman, J. M., *Public Knowledge. An Essay Concerning the Social Dimension of Science*, Cambridge Univ. Press, Cambridge, 1968.

Zuckerman, H., *Scientific Elite. Nobel Laureates in the United States*, Free Press, New York, 1977.

Zuckerman, H. and Merton, R. K., 'Patterns of Evaluation in Science: Institutionalization, Structure and Functions of the Reference System', *Minerva* 9 (1971), 66–100.

Zuckerman, H. and Miller, R. B., 'Indicators of Science: Notes and Queries', *Scientometrics* 2 (1980), 347–353.

INDEX

AAAI (American Association for Artificial
    Intelligence) 145
accommodation 229
accuracy 98
acoustics 114
action and understanding (Piaget) 232
adaptation 230
    – and science 236
AI (Artificial Intelligence) xiii, xvi, 4–17,
    31, 71, 89, 106, 126, 128, 144,
    145, 151, 176, 187, 192, 201,
    224, 229, 256
Alampay, D. A. 191
Alberti, L. B. 225, 280
alchemy 154
Altman, I. 277
ambiguity (and disambiguation) 7, 8, 10,
    11, 53
    ambiguous figure 90, 181, 191
Amosov, N. M. 24
analogy 96, 200
analysis by synthesis 9, 16
anatomy 143, 220, 221
    Padua school in – 220
anomaly 87, 192, 196, 199
anthropology xiii, xv
APA (American Psychological Association)
    132, 275
Apostel, L. xiii, xix, 262
appearance 47, 239
Aristotle 76, 197, 200, 220, 230, 247,
    248, 280
Arnheim, R. 263, 278
art and science 257
ascetism 65
Asch, S. E. 137, 276
assimilation 229
Astbury, W. T. 141
astrology 154
astronomy 68, 82

optical – 144
radio – 144
attention 90, 91, 153, 176, 202, 237
    – and cognitive orientation 177–181
attribution theory 29
Austin, J. L. 32
Avenarius, R. 26

Barber, B. 138
Bar-Hillel, Y. 15, 262
Barnes, B. 268
Barr, A. 279
Barron, F. 72
Bartlett, F. C. 173, 175, 200, 204
Bateson, G. 16
Beaver, D. B. 134
Beck, A. T. xiv
Beckman, L. 73, 268, 269
behaviorism xiii, 22, 154, 176, 179, 216
Ben-David, J. 35, 142, 145, 267, 271
Berkeley, G. 40, 265
Berlin, I. 256
Bernal, J. 78, 141
Bertelson, P. 179
Beveridge, W. I. B. 62
Bhaskar, R. 280
bibliographic coupling 125
bibliography as secondary literature 116,
    118, 119
bibliometrics 112ff, 130, 131
    bibliometric indicator 122
    overview of bibliometric indicators 123
Bitterman, M. E. 99
Block, R. A. 263
Bloor, D. xiii, 260, 282
Blume, S. S. 274
Boas, F. xx
Bobrow, D. xiv
Boden, M. A. xix, 20, 261, 262, 264,
    281

Bohm, D. 30, 99, 100
Böhme, G. 159, 160
Bohr, N. 142, 146
Boneau, C. A. 104, 273
Boole, G. 241
Booth, A. D. 263
Boring, E. G. 80, 105, 176
bottom up 37, 189
 — combination with top down 37, 190, 191, 198
Bourbaki 281
Boutroux, E. 229
Boyd, R. 276
Boyle, R. 134, 195, 197, 276
Brainerd, C. J. 281
Breusegem, E. xix
Broadbent, D. E. 178
Brookes, B. C. 112
Brown, R. 106, 107, 277
Brunelleschi, F. 225, 280
Bruner, J. S. xiii, 91, 180, 210, 264, 282
Brunschvicg, M. 78, 229
bubble chamber photograph 5
Bugelski, B. R. 191
Bunge, M. 149, 266
Buridan, J. 281
Bush, M. 73
Bylebyl, J. J. 196, 219–221, 279

Campbell, D. T. xvii, 80, 236, 260, 268, 270, 271, 281
Cantrill, H. 276
capitalism 64
Carmichael, L. 10
Carnap, R. 35, 56, 57, 59, 60, 61, 104, 266, 267, 277
Carroll, J. B. 273
Cartwright, D. 166
case grammar 206
Castañeda, H. 263
Cattell, J. M. 183
cause 47
centration 250–251
Chandrasekaran, B. 145
Chaplin, C. 29
Charniak, E. xiv, 20, 21
Cherry, E. C. 176, 178

Chomsky, N. xiv, 60, 61, 152, 267, 277
Chubin, D. 276
Cicourel, A. V. xiii, 136
circulation, discovery of – 192–201, 219
citation index 126
citation network 124–130
 — studies of DISISS, Univ. of Bath 125
Clagett, M. 241
Clark, T. N. 149
clinical method (Piaget) 228, 230
closure of conceptual systems 86, 160, 161, 238, 240, 241, *see also* conservation
Clowes, M. 7, 262
Cochran, T. C. 75
co-citation 125
 — cluster 126, 155
 — map 126–127
cocktail-party-phenomenon 176, 202
cognitive science xiv,xvi, 3, 22, 25, 30, 31, 32, 36, 37, 144–145, 151, 253, 261
Cognitive Science Society xiv, 260, 261
cognitive structure xviii, 15, 30, 36, 89, 98, 105, 106, 129, 138, 140, 146, 158, 169, 202, 223, 224, 240, 241, 243, 247, 253
cognitive view (stage) xiv, xvi, 4, 5, 9, 15, 16, 31, 36, 41, 180, 214, 229, 247
Cohen, D. K. 259
Cohen, I. B. 75, 241, 271, 281
Cohen, J. S. 279
Colby, B. N. xiii
Cole, F. J. 276
Cole, J. R. 130, 137, 157
Cole, S. 137, 157
Collins, A. xiv
Collins, R. 142, 143
Columbus, R. 196
communication xvii, 3, 117
 informal – 136
 informal – network 133
 mass – 138, 147
 formal *vs* informal – 138, 139
computer revolution xiv, 144, 252
Comte, A. 34, 39, 40, 78, 79, 264, 265, 269
Conant, J. 84, 93, 103, 228, 272

concept driven 190, 191, 198, 233
conservation 238–244, 245, 255
conservative focussing 91
consistency 98
context 8, 9, 14, 15, 16, 71, 76, 80, 81,
    83, 213, 253, 254
    – boundary problem 8, 14, 36, 75
    – of discovery 60
    – of justification 60
    anatomical – 219, 220
    historical – 78
    medical – 219, 220
    philosophical – 219
contextual analysis 8
contextual view (stage) 5, 14, 16, 31, 35,
    36, 179, 180
continuity theory 77, 84
coordinating definition 58
Copernicus, N. and Copernican Revolution
    74, 81–83, 91, 98, 192, 221, 269,
    271, 276
correlation 48
Coward, R. M. 130, 155, 166
Craig, J. 75
Craik, K. W. xv, 4, 23
Crane, D. 111, 123, 135, 148, 149
Crawford, S. 134, 139
Crick, F. H. C. 160

Darwin, C. 256
data 39, 51, 91, 208
    – driven 190, 191, 198
    – structure 203, 204
debugging 211, 212
default 207, 208
Delbrück, M. 136, 142, 146, 150
Dember, W. N. 254
Democritus 43
Dertouzos, M. L. 262
Descartes, R. 45, 76
De Sloovere, M. xx
Deutch, D. 178
Deutch, J. A. 178
De Vlieger, E. xx
dichotic listening 178
diffusion of innovations 118, 119, 147,
    162–164, 169

digital representation 6
discipleship 145–146
disciplinary matrix 96–100
discipline, scientific – 79, 100, 114, 124,
    142, 145
discontinuity theory 77, 84, 85
Dornic, S. 278
Downey, K. J. 140
Drott, M. C. 275
dualism 44, 46, 257
    Cartesian – 26
    psycho-physical – 70
duck-rabbit 11, 90, 175, 177, 180, 189,
    192, 193
Duhem, P. M. M. 32, 34
Duncker, K. 189
Durkheim, E. 140

Edge, D. O. 131, 141, 144, 165, 275
Edgerton, S. Y. 257, 280
education 249
    scientific – 77, 84, 100–102
    – and mastery teaching 146
effect 47
Eiduson, B. T. 35, 72, 268, 269
Einstein, A. 50, 63, 91, 146, 247, 248
Eisenstein, E. L. 174, 252, 277
Elkana, Y. 275
empirical system (of a theory) 58
empiricism xiv, 16, 38, 41, 65, 183, 187,
    245, 247
    logical – see logical positivism
epidemiological model of spread of ideas
    158
epistemology, genetic, see Piaget, J.
essence 47
ethnomethodology 29, 136
ethology 99
ethos of science (Merton's values) 66–67,
    73, 137, 138
exemplar xvii, 96, 97, 208, 209, 216, 218
    – and exercises 100, 101, 216, 217
experts, mind as a community of – 201,
    203, 215, 222
explanation, scientific – 47–48
exploitation model (Holton) 155–157
extern, see externalism

externalism   25, 35, 36, 67–71, 81, 83,
   84, 88, 130, 201, 253, 258
extraordinary science, *see* revolutionary
   science
extrinsic factor, *see* externalism

Fabricius, H.   195
fact   34, 49
Fairbank, J. K.   75, 269
Farr, R. M.   269
feature analysis   7, 16, 31
Feibel, W.   xix
Feigenbaum, E. A.   279
Feigl, H.   57, 59
Feldman, J.   11
Feyerabend, P.   92, 103, 272, 273
figure-ground   244, 253, 254
Fillmore, C. J.   206, 262
filter theory (attention)   178
finalization   159–161, 169, 237, 238, 251
Fisch, R.   72, 74, 268
Fitter, M.   153, 180
Flavell, J.   281
Fleck, J.   261
Fleck, L.   103–104, 273
Fodor, J. A.   xiv
Fores, M.   80
formal system (of a theory)   58
Forman, P.   161, 279
frame   204–208, 209, 210, 218, 229
Frazer, J. G.   77, 265
Freud, S.   27

Galen   192–195, 196, 200, 201, 220, 223,
   241, 242
Galileo   3, 82, 155, 228, 244, 276, 281
Gallant, J. A.   165
Galle, J. G.   59
Galperin, P. J.   278
Garcia, R.   227
Garfield, E.   129, 275
Garfinkel, H.   29, 136
Garvey, W. D.   123, 274
Gaston, J.   274
generative grammar   60
Gestalt psychology   90, 106, 176, 183,
   191, 244

gestalt effect   254
gestalt switch   90, 175, 176, 192, 195
Gibson, E.   181, 278
Gibson, J.   185, 224, 278
Gilbert, G. N.   160
Gilbert, R. M.   177
Goffman, W.   148, 158
Goldmann Eisler, F.   12
Goldsmith, M.   80
Goldstein, I.   206
Gombrich, E.   38, 188, 228, 278, 280
Goodman, C. C.   180
Goodman, N.   224, 263, 278, 282
Goodwin, B.   xiii
Graham, L.   xix
Greenlee, E.   129, 130
Greiner, L. E.   158, 159
Griffith, B.   127, 135, 273, 274, 275
Grosz, B. J.   262
group, informal –   136
   small –   136, 147
   reference –   137, 139
growth, concepts of –   113, 114
   exponential –   113
   linear –   113
   logistic –   114
growth in science, metabolism of   119
   regulative mechanisms of   154
growth points in science   115
Gruber, H. E.   231, 247, 281, 282
Guthrie, E. R.   203

Hagstrom, W. O.   111, 133, 137, 138, 276
half life, of citations   120
   of specialties   157
Halle, M.   9
hallucination   202
Hanson, N. R.   33, 34, 90, 103, 173, 227
Harmon, G.   158
Harvey, W.   192–201, 215, 218, 220, 221,
   223, 241–243, 278, 279, 281
Haugeland, J.   xiv
Hawthorne studies   136
Hayes, P. J.   xiii
Hayes-Roth, B.   144, 145
Heidbreder, E.   105
Heimdahl, C.   273

Heisenberg, W.　161, 279
Helmholtz, H.　41, 265
Hempel, C. G.　35, 59, 282
heterarchy　212–215, 218
Hewitt, C.　215
Higham, J.　68, 69
Hilbert, D.　58
Hilpinen, R.　263
Hintikka, J.　263
Hippocrates　220
history of science　xviii, xx, 35, 59, 70, 75–78, 79, 82, 84, 100, 101, 111, 198, 230, 248
Holton, G.　103, 147, 155–157
Homans, G. C.　141
Huey, E. B.　183
Hume, D.　40, 41, 182, 265

idea 41
idealism　45, 46, 59
illusion　9, 48
imagination　9, 190
impression　41
indexical expression　14, 15, 16
Infeld, L.　63
information　38
　– explosion　112
　– flow　138–140
　– processing　xv, 4, (stages) 16
　– processing system　24, 25, 36
　– theory　4, 176
Inhelder, B.　239
injection model (mass communication) 139
innovation, scientific –　135, 136
　innovator　100, 145, 146
intelligence　203, 211
interdisciplinarity　140
intern, see internalism
internalism　35, 67–71, 81, 83, 84, 85, 88, 130, 201, 253, 258
intrinsic factor, see internalism
introspection　22, 26
invisible college　104, 133–135
ISI (Institute for Scientific Information) 126
Isis　xx, 274
Ivins, W.　225

James, W.　xiv, 21, 93, 274
Jastrow, J.　175
Johnson-Laird, P. N.　176, 278
Johnston, R.　268, 271
Josselson, H. H.　262

Kahneman, D.　279
Kant, I.　25, 38, 231, 276
Kasher, A.　263
Katz, E.　276
Keele, S. W.　237
Kemble, E. C.　76
Kepler, J.　74, 269
Kessen, W.　53
Kessler, M. M.　125
Kilmann, R. H.　80, 81
Kimerly, J. R.　277
King, M. D.　103
Klima, R.　260
Knight, D.　269
Knorr, K.　138, 147, 200, 260, 274, 282
knowing how vs knowing that　218
　knowing how　230, 235, 237, see also skill
　knowing that　235, 237
knowledge　3
　prerequisite –　15, 36
　tacit –　130, 205, 222, 231
Koefoed, P.　xix
Koestler, A.　219, 221
Köhler, W.　176
Koks, E.　xx
König, R.　274
Kosslyn, S.　278
Kotarbinsky, T.　269
Krantz, D. L.　154
Krohn, R.　276, 282
Krohn, W.　160
Kuhn, T. S.　xvi, 36, 81, 82–107, 111, 112, 121, 122, 134, 146, 147, 149, 154, 160, 162, 168, 175, 192, 199, 203, 204, 209, 214, 216, 217, 218, 224, 228, 229, 237, 241, 248, 261, 271, 272, 273, 280, 282

Lachman, J.　62, 129
Lachman, R.　62, 129

Lakatos, I. 273
Lakoff, G. xiii
Langley, P. 280
language 50, 90, 91, 117
    formal – 54, 55
    meta – 53
    natural – 51, 53
    – processing 11, 102, 181
Larkin, J. 280
Latour, B. 138, 147, 248, 282
Laudan, L. 260
Lavoisier, A. L. 160
law, scientific 35, 58, 81
Law, J. 140, 141
Lazarsfeld, P. 276
Leeper, R. W. 181
Lehman, H. C. 268
Lehrer, K. 263
Lemaine, C. 274
Lenat, D. B. 271
Levelt, W. J. M. 152
levels of analysis, multiple 31, *see also*
    perception, multiple level model
Leverrier, U. L. L. 59
Liberman, A. M. 9
library science, *see* bibliometrics
life cycle, analysis 155
    – of specialties 115, 148–168
    paradigm – 148
Lin, N. 123, 139, 274
Lindblom, C. E. 259
Lindsay, P. H. 261
linear approximation 211, 212
Lipkin, B. S. 278
literature, scientific 105, 131
    abstract 116
    archival – 121
    current – 100, 101, 122
    index 116
    primary – 116–118, 122
    research front – 121
    review 116, 117
    secondary – 115–119
Locke, J. 40, 265
logical atomism 54
logical positivism 34, 35, 39, 50, 59–
    62

logical reconstruction 54, 56, 57
Lorenz, K. 99

Mach, E. 26, 45–49, 58, 64, 241, 265
Machlup, F. 277
Mackay, A. 80
Mackintosh, N. J. 177
MacLeod, R. 268
Magoun, H. W. 179
Mahoney, M. J. xiii, 268
Mandelbaum, M. 264
Mandler, G. 53
map(s) of science 126–130
Marshakova, T. V. 126
Martins, H. 103
Maruyama, M. 106
marxism 64, 78
Maslow, A. 72
Masterman, M. 102
materialism 46, 64, 69
Matthew-principle 133
McCarthy, J. 263
McClelland, D. 72, 74
McCulloch, W. xv, 212
McGrath, J. E. 277
McGuire, W. J. 269
Meadows, A. J. 3
Medawar, P. 263, 282
meeting(s), scientific – 133, 135
memory 90, 202
Menard, H. 114, 116, 121, 123
Mendeleev, D. I. 86
Mendelsohn, E. 103, 260, 274
Menzel, H. 138
Merton, R. K. 35, 63–71, 75, 123, 133,
    137, 138, 260, 267, 268, 269, 274,
    275
metaphor 36, 104, 200, 218
metaphysics, *and* metaphysical beliefs 39,
    43, 48, 53, 77, 88, 94, 96, 257
method, scientific – *and* methodology xvi,
    3, 49, 62, 63, 67, 98, 99, 149–154,
    162
    methodological values 96, 98
Meyerson, E. 78
Michie, D. 4, 261
microscope xiv, 3, 4

migration, intellectual *and* social – 141–145
Miles, R. H. 277
Mill J. *and* J. S. 40
Miller, G. A. 4, 176, 202, 203, 264, 278
Miller, R. B. 275
Mills, C. W. 266
Milton, S. 276
Minkowski, H. 146
Minsky, M. 99, 106, 151, 204, 205, 207–210, 213, 256–257, 262, 264, 280
Mises, R. von 266
Mitroff, I. I. 80, 81, 138
M.I.T.-school in AI 203, 229, 248, 256
model, internal – *or* world – xv, 4, 89, 180, *see also* world view
monadic view (stage) 4, 6, 12, 16, 31, 48, 179
monism 46
    neutral – 44, 49, 64
monotheism 77
Moore, G. E. 51
Moray, N. 153, 178, 180, 181, 278
Moreno, J. L. 274
Morgan, L. 265
Morin, E. 104, 273
Moruzzi, G. 179
Moses, J. 262
Mulkay, M. 141, 144, 148, 160, 165, 261, 276
Muller-Freienfels, R. 111
Muijsenberg, L. xx
Müller, J. 42
Mullins, N. C. 111, 126, 135, 136, 142, 149, 150
Musgrave, A. 273
Myrdall, G. xvii

Naess, A. 63
Nagel, E. 266
navigation 68, 71
Necker cube 11, 90
Neisser, U. 9, 17, 181, 182, 203, 261
Nelson, C. E. 123
neopositivism, *see* logical positivism
Newell, A. 126, 128, 145, 208, 210, 223, 263

New Look (in perception) 180
Newton, I. 70, 75, 227, 247, 248, 281
Nietsche, F. W. 72
Nilsson, N. J. 261
Nobel prize 146, 247
normal science 72, 84, 85, 88, 101, 223
Norman, D. A. xiv, 154, 158, 186, 260, 261, 279
norms, Mertonian – 137, 138
    Piaget 250
Norton, B. J. 266
Nowakowska, M. 158
Nowotny, H. xiii, xix, 19, 260

object, notion of – 185–189, 244, 253, 255
observation 3, 32, 33, 38, 193, 197, 198, 201, 233, 234
Occam, W. of 45, 47
    – 's razor, 265
oceanography 116
O.E.C.D. 131
Oettinger, A. G. 13
O'Malley, C. D. 193
operationism 22, 57
opinion leader 139, 140
optics 114
Ortega hypothesis, *see* populistic view
orthodox view 57, 255, *see also* received view
Orwell, G. 131
Ossowska, M. 79
Ossowski, S. 79

Pagel, W. 63, 197, 198, 221, 279
Palmer, S. E. 185, 186, 187, 278
Panofsky, E. 132
Papert, S. 151, 204, 213, 218, 262, 280
paradigm xiv, xvi, 34, 36, 85–107, 111, 114, 115, 129, 146, 153, 154, 175, 199, 204, 223, 224, 226, 247, 248
    paradigmatology 106
    paradigm-detection 104–105, 130
    paradigm-dissection 105–106, 130
    paradigm-hunting 104
paradox, cognitive – 17, 18, 181, 182, 201, 247

liar − 52
parsimony 99, 151, 208, 209, *see also*
    Occam, W. of
particle-physics 5, 114
Pasteur, L. 3
pattern recognition 5−11
Pauling, L. 202
Pavlov, I. P. 177
Pearson, K. 34, 45−49, 58, 265, 266
Peferoen, L. xix
pelican-antelope 11, 90
perception xvii, 9, 90, 181, 182, 202,
    221, 233, 244, 255
  direct − 26
  multi-level model of − 32, 182−192,
    214
  gestalt − 173, 175, 191, 192
personality, scientific − 71−74
perspective 224−226, 244−246, 258
Petrie, H. G. 260, 263
Pfungst, O 268
Phage group 135, 136, 142
philosophy of science xv, xvi
  logical positivist − 59−62, 63, 79
physiology 43, 143
  animal − 134
Piaget, J. xiii, xvii, 3, 77, 78, 82, 106,
    145, 207, 225, 227−251, 255, 258,
    260, 264, 277, 280, 282, 282
picture processing 5, 71, 181, 185
Pitts, W. xv
Place, T. xix
Plato 228
Polanyi, M. 222
polytheism 77
Ponzo illusion 254
Popper, K. 149, 273
populistic view 157
Portugal, F. H. 279
positivism 34, 40, 48, 49, 50, 84, *see also*
    logical positivism
  systematic − 39
  critical − 40, 48, 49
prediction 48
pre-paradigmatic 88
preprint 4, 135
preprocessor 6

presupposition 14
Price, de Solla D. 80, 105, 111, 112, 113,
    116, 120, 121, 123, 133, 134, 168,
    273
Price's index 120
Prigogine, Y. 251, 252, 259
printing press 252
protestantism 64, 65, 67, 68, 70, 72
Prothero, J. W. 165
psycholinguistics 61, 117, 152, 153, 154
psychology xiii, xvi, 4, 22, 26, 79, 95,
    117, 132, 142−145, 151, 176, 180,
    211, 216, 229, 241
  animal − 99
  cognitive − xiii, xvi, 25, 60, 105, 144,
    145, 190, 203, 228
  theoretical − 145
  − of science xvi, xviii, 62, 71−75
Ptolemy 83, 221
puzzle-solving xviii, 85, 88, 97
  jigsaw − 214−215, 218
Pylyshyn, Z. xiv

Quillian, M. R. 204

Radnitzky, G. 62, 149, 150, 151, 152,
    154
Radnor, M. 277
Raphael, B. 261, 264
rate of decay (of references), *see* rate of
    obsolescence
rate of obsolescence 119−121
rationalism xiv, 38, 65
rational reconstruction, *see* logical recons-
    truction
rat-man 190, 191
Ravetz, J. R. 146, 216−218
reading, psychology of − 7−11, 183−
    184, 212−213
  proofreading errors 8, 183
realism 224
  naive − 43
  scientistic − 43
received view 57, *see also* orthodox view
receptive fields (Hubel and Wiesel) 7
reflexive abstraction 232, 236
Reichenbach, H. 267

relativism 256, 257
religion 80
  sociology of – 64
representation xv, 4, 5, 24–27, 29, 94,
  181, 205, 209, 215, 224, 235, 236,
  245, 246
  pictorial – 9–10, 187, 224
  procedural – of knowledge 203, 215–
  218
representational model, see world view
Restivo, S. 260
revolutionary science 72, 84, 86, 87, 88,
  101, 223
Reymond, A. 78
Rich, R. F. 277
Rip, A. xix
risky shift 127, 166
Roe, A. 35, 123
Roessner, J. D. 277
Rogers, E. M. 118, 119, 136, 162–164,
  274, 277
Rohrschach test 180
role xv, 28
  scientific – 142
  – hybridization 143
Rosenblueth, A. xv
Rosenfeld, A. 278
Rosenthal, R. 268, 269, 275
Rothman, B. 281, 282
Royal Society 68, 134
Royce, J. R. 106
Rubin, D. B. 268
Rubin, E. 176
rule of correspondence, see coordinating
  definition
Rumelhart, D. E. xiv, 186, 261
Russell, B. 48, 49, 51–55, 225, 266
Ryle, G. 198, 280

Sacks, H. 29
Sarton, G. xx, 148, 252, 277
Saunders, J. B. de C. M. 193
Scandura, A. B. 281
Scandura, J. M. 281
Schank, R. xiv, 20, 21
Scheffler, I. 103, 282
scheme, Bartlett 204

conceptual – 85, 93, 95, 97
Piaget 231, 232
Schmutzer, M. xiii, 19
Schrödinger, F. 276
science of science xv, xvi, 34, 39, 78–81,
  248, 252
science indicator 131
science policy 79, 80, 158
scientific community 87, 98, 102, 111,
  129
scientific method, see method
scientism 40
scientometrics 130–131
Searle, J. R. 263
self-concept xv, 25, 45, 219, 258, 259
  self-world segmentation 221, 222,
  224
Selfridge, O. G. 8
Senders, J. 180, 181
sensation, or sense datum, or sense impression 34, 45, 47, 50, 51
  sign character of – 41, 42
sentence, isolated – 12, 13, 14, 71, 152,
  153
Shannon, C. 4, 176
Shapere, D. 50, 92, 272
Shaver, K. G. 264
Shelley, J. 263, 282
Shepard, R. 203, 261
Shera, J. 111–112
Shirai, Y. 20
Shoemaker, F. F. 119, 162–164, 277
similarity network 209, 210
Simon, H. A. 20, 280
simplicity 98
Singer, B. F. 79
skill (as know how) 215–217, 235–237,
  238
Sloan, P. R. 274
Sloman, A. 262, 278
Small, H. 123, 126, 127, 129, 130, 275
Smeets, M. xx
Sneed, J. D. 271
Snow, C. P. 274, 277
social control (in scientific groups) 137
social epistemology, library science as –
  111, 112

sociology of science 62, 71, 79, 111, 132, 147, 161, 170
solipsism 31, 45, 255
Solso, R. 25
specialty, scientific – 114, 115, 124, 130, 133, 134, 135, 137, 139, 140–145, 148–169, 241, 248
Spence, K. W. 177
Spencer, H. 39, 264
Spiegel-Rösing, I. 273
Sridharan, N. S. 204, 279
Starnberg model 159–161
Stehr, N. 274
Stengers, I. 252, 259
Stern, N. 268
Stevens, K. 9
Stevens, S. S. 22, 79
Storer, N. W. 267
structural view (stage) 5, 7, 12, 16, 31, 56–61, 179
Sullivan, D. 126, 267
Suppe, F. 57, 62, 266
Sussman, G. J. 211, 212, 215, 217
Sutherland, N. S. 177, 278
Swanson, D. R. 132
symbolic generalization 96, 97
syntax 12
  syntactic analysis 16, 35, 60
Szeminska, A. 238

Tarski, A. 52
telescope xiv, 3, 4
testability 98
template matching 6, 16, 31
textbook 100
theology 39, 65, 77
theory, scientific – 35
theory-laden perception 92, 197, 198
Thompson, H. xiii
Thompson, D'Arcy W. 155
Titchener, E. B. 176
top down 37, 189, 233
  combination with bottom up 37, 190, 191, 198
Toulmin, S. 93, 103
transducer 24
Transgard, H. 260

Treisman, A. 178
Turing, A. M. xv
two-step flow (mass communication) 139, 162

Uexküll, J. von 25, 263
Uhr, L. 95, 278
utilitarism 65

Vandamme, F. xix
Van den Daele, W. 160
Van Norren, B. 260
Van Velthoven, G. M. xiv
verifiability-criterion 57
Vesalius, A. 193, 220
Vlachy, J. 276
Vonèche, J. J. 231, 247, 281, 282

Waddington, C. H. 231, 282
Walentynowicz, B. 79, 269
Waltz, D. L. 262
Warren, K. S. 158
Wason, P. 153
Wasson, C. R. 166, 167
Watson, J. B. 26, 263
Watson, J. D. 160
Watson, R. I. 80, 268, 270, 271
Wax, M. L. 281
Webb, E. J. 275
Weber, M. 64, 65, 74, 154
Wechsler, J. 263, 282
Weimer, W. B. 260
Weinberg, B. H. 125
Weingart, P. xix, 130, 260
Weiss, C. 259
Weiss, P. 274
Weisstein, N. 25
Weizenbaum, J. 262
Welles, O. 276
Wersig, G. 275
White, M. xix
Whitehead, A. N. 55
Whitley, R. xix, 130, 131, 134, 165, 260, 274
Whitteridge, G. 196-198, 279
Whorf, B. 106, 273
Wiener, N. xv

Wilks, Y. 267
Winograd, T.   14, 15, 20, 22, 60, 203,
      214, 261, 280
Winston, P. 210, 222, 279, 280
Wisdom, J. 272
Wittgenstein, L. 17, 51, 54, 55, 272
Wolf, P. 248
Woolgar, S. 138, 147, 160, 165, 248
word-to-word translation 12
world view, *or* world model   xv, 4, 5, 9,
      16, 29, 27–30, 203, 219, 255
      multiple -s 20–22, 30–32, 219–226

Wundt, W. 26, 142, 143

X-ray protein crystallography 140, 141

Yellin, J. A. 148, 157, 158, 160
young lady/mother-in-law 11, 176, 182,
      189

Zaltman, G. 277
Ziman, J. M. 81
Zuckermann, H. 123, 130, 146, 268, 275